LIFE OF THE PAST   James O. Farlow, editor

THE WHITE RIVER BADLANDS

GEOLOGY and PALEONTOLOGY

RACHEL C. BENTON

DENNIS O. TERRY JR.

EMMETT EVANOFF

H. GREGORY McDONALD

INDIANA UNIVERSITY PRESS   Bloomington & Indianapolis

This book is a publication of

Indiana University Press
Office of Scholarly Publishing
Herman B Wells Library 350
1320 East 10th Street
Bloomington, Indiana 47405 USA

iupress.indiana.edu

Manufactured in the United States of America

Library of Congress Cataloging-in-Publication Data

The White River Badlands : geology and
paleontology / Rachel C. Benton, Badlands
National Park, Dennis O. Terry Jr., Temple
University, Emmett Evanoff, University of Northern
Colorado, H. Gregory McDonald, Park Museum
Management Program, National Park Service.
    pages cm.–(Life of the past)
    Includes bibliographical references and index.
    ISBN 978-0-253-01606-5 (cl : alk. paper)–ISBN
978-0-253-01608-9 (eb) 1. Fossils–Collection
and preservation–South Dakota–White River
Region. 2. Paleontology–South Dakota–White
River Region. I. Benton, Rachel. II. Terry,
Dennis O., [date] III. Evanoff, Emmett. IV.
McDonald, H. Gregory (Hugh Gregory), [date]
    QE718.W54 2015
    560.9783'9–dc23
                                              2014044309

2 3 4 5   20 19 18

We wish to dedicate this book to the Jones Family of Quinn, South Dakota. For over 26 years, Kelly, Mary, and Doug provided a home away from home for the authors and many of their students. Be it providing a place to sleep while conducting fieldwork, hosting a group of researchers for a barbecue, or simply providing a welcoming respite from the heat of the day, the Jones family and the logistical support that they provided over the years helped to make this book possible. Thank you.

Viewed at a distance, these lands exhibit the appearance
of extensive villages and ancient castles, but under forms
so extraordinary, and so capricious a style of architecture,
that we might consider them as appertaining to some new
world, or ages far remote.

Fray Pierre Jean De Smet, 1848

But it is only to the geologist that this place can have any
permanent attractions. He can wind his way through the
wonderful canons among some of the grandest ruins in the
world. Indeed, it resembles a gigantic city fallen to decay.
Domes, towers, minarets, and spires may be seen on every
side, which assume a great variety of shapes when viewed
in the distance. Not unfrequently, the rising or the setting
sun will light up these grand old ruins with a wild, strange
beauty, reminding one of a city illuminated in the night
when seen from some high point. It is at the foot of these
apparent architectural ruins that the curious fossil treasures
are found.

F. V. Hayden, 1880

Is it of interest to you that the White River Badlands are
the most famous deposits of the kind in the world? Do you
know that aside from their picturesque topography they tell
a marvelous nature story; a story of strange climate, strange
geography, and strange animals; of jungles, and marshes,
and tranquil rivers, of fierce contests for food, and life, and
supremacy; of varied series of events, through ages and ages
of time.

C. C. O'Harra, 1920

Contents

C

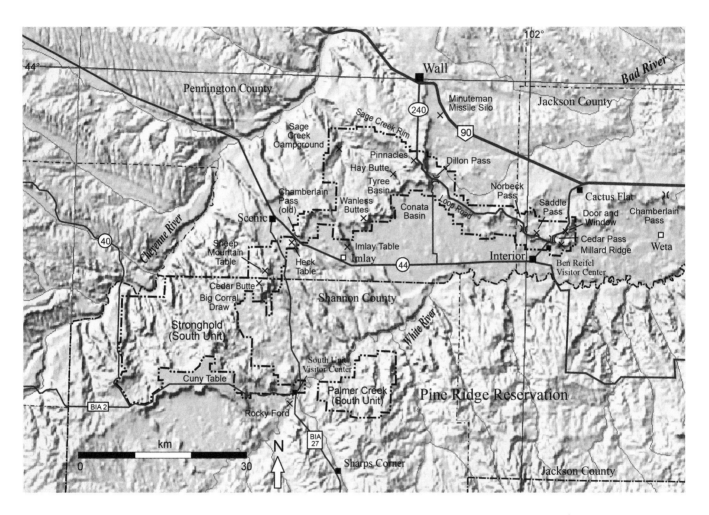

P.1. Map of the Big Badlands of South Dakota showing locations of specific places and features discussed in the text. The boundary of Badlands National Park is shown by the heavy dash–dot–dot line. The northern area of Badlands National Park in Pennington and Jackson counties is the North Unit. The base map is from the U.S. National Atlas Web site (http://nationalatlas.gov/mapmaker).

Preface

MAKOSICA (MAH-KOH SHEE-JAH) IS THE LAKOTA WORD for "badlands," or the barren and rough country of buttes and cliffs that are cut by multitudes of deep canyons and ravines. The term *badlands* does not refer to anything evil about the lands but rather to the difficulty of crossing the country on foot or horse. Modern travelers crossing the Badlands Wall of South Dakota in cars on paved highways do not appreciate the difficulty these landforms posed to early travelers. The French name for this country, *mauvaises terres á traverser*, "the bad lands to traverse," was an even more explicit description. In places in Badlands National Park, one can still walk for over 10 km at the base of the Badlands Wall and not find even a game trail that crosses the wall. Nevertheless, the Big Badlands of South Dakota is one of the most spectacular landforms in the United States and is cut in rocks containing some of the most abundant vertebrate fossils of any rocks of the Age of Mammals (Cenozoic Era) in North America. Fossils from the White River Badlands can be found in every major natural history museum in the world. Badlands National Monument (later Badlands National Park) was established to protect the unique landforms of the White River Badlands and the "vast storehouse of the biological past" (Badlands National Park, Statement for Management, 1992).

The Badlands, with a capital "B," represents the Badlands of Western South Dakota; it is a place-name and the original basis for the geomorphic term. The word *badlands* has entered the geological vocabulary (when written in lowercase) as a geomorphic term describing a highly eroded landscape with little vegetative cover in arid to semiarid climates. Within the context of this book, *badlands* in this sense is used as a generic descriptive term as any topographic area that meets these criteria. The terms *White River Badlands, Big Badlands*, or just *the Badlands* will be used interchangeably throughout the text to refer to these exposures throughout southwestern South Dakota. The Big Badlands of western South Dakota is unquestionably the most famous of all the areas around the globe referred to as badlands, and it is certainly the most prolific in terms of fossils that have been collected and placed in museums. The White River Badlands represents all the badlands within the White River drainage basin of western South Dakota and Nebraska. This book will focus mostly on the White River Badlands of South Dakota. Badlands National Park is a 244,000-acre National Park Unit established to protect a portion of the White River Badlands, and it is the central focus of this book (Fig. P.1).

Since 1846, with the first scientific report of a partial fossil jaw from the White River Badlands, these deposits have been an important focus of paleontological research. The diversity of fossils recovered by researchers over the past 167 years from strata that span 9 million years of Earth history has provided valuable data on the evolution of North American mammals during the late Eocene and Oligocene epochs. The rocks and fossils from the White River Badlands have also provided valuable information on climate change during one of the greatest global drops in temperature during the Cenozoic. This climatic change contributed to the evolutionary changes of the fauna and flora and produced major changes in both local communities and the global Eocene/Oligocene biosphere.

In 1920 Cleophas C. O'Harra published *The White River Badlands*. At the time he wrote the book, O'Harra was president of the South Dakota School of Mines in Rapid City, but it was as professor of mineralogy and geology at the School of Mines that O'Harra gathered the information upon which his book was based. When *White River Badlands* was published, it was considered cutting-edge research, and it has been reprinted many times since its initial publication. O'Harra included data collected from the field expeditions of the late nineteenth and early twentieth centuries, including many led by him. As he mentions in his preface, the book was written with the layperson in mind, and since its publication, it has been the definitive work on the geology and paleontology of the Big Badlands in southwestern South Dakota.

The goal of our book is to build on the foundation laid by O'Harra and, like O'Harra, we summarize the research conducted by many geologists and paleontologists (including the authors) that took place over the decades after his contribution. We continue in the spirit of *White River Badlands* by directing our text to the many enthusiasts, both amateur and professional, with an interest in the geology and paleontology of the Big Badlands. Recognizing that this diverse audience also reflects a diversity in the amount of formal training in geology and paleontology, we have tried to provide general summaries of the subject matter, specific information, and detailed lists of references and glossaries with the sincere hope that this book will serve as a gateway for those who wish to investigate further. This book provides a broad overview of the geology and paleontology of the Badlands, and we urge all of those with a strong interest to pursue the primary literature upon which this book is based.

This book is not primarily intended as a textbook, although it could certainly serve as a supplemental text for a class on local geology or paleontology. It is a synthesis that provides the reader with a solid introduction to a classic geological area based on the research that has been completed by multiple researchers over the last 167 years. We assume that the reader has a basic background in geology and paleontology and an avid interest in the White River Badlands. As our understanding of the diversity and taxonomy of the fossils has evolved, our concepts of the geology have also evolved. Some new stratigraphic concepts have been introduced without a previous published record. These are based on many years of fieldwork and research in the Badlands by some of us. Those familiar with the geology of the Big Badlands may encounter differences in the geologic interpretations.

As a result of the enormous amount of information relating to the geology and paleontology of the White River Badlands that has been published since O'Harra's original 1920 volume, it became obvious that certain limits had to be set. The area-level scope of the current book encompasses most of the published record of paleontological localities within Badlands National Park and extends in a 100-mile radius, with Cedar Pass as the center (Fig. P.1). The only exception is chapter 7, "The Big Badlands in Space and Time," which compares the central features of this book with areas similar in age in the western United States and around the world. The temporal scope of this project is limited to the late Eocene and earlier Oligocene epochs, with only minor discussion of pre-Eocene geology and regional geologic history in order to establish a framework for discussion.

Chapter 1, "History of Paleontologic and Geologic Studies in the Big Badlands," explores the history of science as it relates to the original discoveries and surveys of the White River Badlands and the individuals who have contributed to our understanding of the geology and paleontology of this area. It also discusses many of the early interpretations of how the late Eocene and Oligocene rocks in this area were deposited and how our understanding of this region has changed as the science of geology has matured.

Because a working knowledge of the regional geology is critical to understanding the fossil record and provides the primary context within which fossils are preserved, it is covered in three different but complementary chapters.

Chapter 2, "Sedimentary Geology of the Big Badlands," outlines the depositional environments and sediment sources which produced the rocks included today within the White River Badlands. Each formal rock unit within the White River Group will be described in great detail. Within the science of geology, it is crucial to be able to recognize individual rock units and correlate them across broad expanses. A preliminary discussion of the Sharps Formation within the Arikaree Group will also be discussed.

Chapter 3, "Paleoenvironmental and Paleoclimatic Interpretations from Paleosols," explores the process in which paleosols (ancient soils) were formed and preserved in the Badlands and what role they play in interpreting ancient environments and climate. This chapter also summarizes much of the paleosol research that has been completed since 1983.

Chapter 4, "Postdepositional Processes and Erosion of the White River Badlands," examines the post-Oligocene geologic features of the White River Badlands. Many of the features now exposed in the White River rocks were formed after burial of the sediments and while they were turning into sedimentary rocks (diagenesis). About 5 million years ago, the major geologic processes in this area switched from depositional to erosional, eventually creating many of the famous landforms in the Big Badlands of today. Faulting associated with post-Oligocene extensional tectonics in the Great Plains has had a profound impact on the preservation and distribution of Cenozoic rocks the White River Badlands. Finally, although wind played a large role in the origin of the White River rocks millions of years ago, it still has a role in forming sand dunes across the region and redistributing the ancient dust into the agricultural fields of eastern South Dakota and Iowa.

By far the most significant scientific features of the Big Badlands are its fossils, primarily mammal fossils. The next two chapters introduce the fauna and discuss how the fossils accumulated across the ancient landscapes of the White River Badlands.

Chapter 5, "Bones That Turned to Stone: Systematics," focuses on the fossil plants, animals, and trace fossils of the White River Badlands. These discussions are based on the published record of body and trace fossils found in and around Badlands National Park, with our discussions organized as genera, including seven invertebrates, one fish, one amphibian, 14 reptiles, seven birds, and 88 mammals. This chapter is written in the style of a field guide so that the reader has a summary of important features to identify a particular fossil at the genus level. This chapter also includes photo plates of many of the fossils from the White River Badlands and the diagnostic features that help with identifications. Important aspects of the evolution and paleoecology of individual taxa are also discussed.

Chapter 6, "Death on the Landscape: Taphonomy and Paleoenvironments," explores the interrelated nature of fossil preservation and paleoenvironments, as well as how scientists can extract data from the rocks and fossils in order to interpret the paleoforensics of fossil bones. Two important fossil localities in Badlands National Park are used as examples to highlight the interdisciplinary nature of this research, and general discussions are provided of fossil distribution and controls on the fossilization process.

Chapter 7, "The Big Badlands in Space and Time," places the White River Badlands into a larger context. We explore global events that occurred during the late Eocene and Oligocene epochs, and how ancient records from across the globe can be combined in order to develop an overall picture of paleoclimatic change during this critical interval of Earth's history.

Chapter 8, "National Park Service Policy and the Management of Fossil Resources," focuses on the management of paleontological resources at Badlands National Park. This chapter explores ongoing park projects and how we protect fossil resources, and the interface between the visitor and the abundant fossil resources—something unique to Badlands National Park.

Acknowledgments

THIS BOOK WOULD NOT HAVE BEEN POSSIBLE WITHOUT the support of many professional colleagues, museum personnel, artists, former students, National Park Service employees, members of the Oglala Sioux Tribe, and friends and family. This endeavor goes beyond the four authors to a vast network of professionals who care deeply about the White River Badlands and have invested many years in their study and protection.

It is with great appreciation that we thank the following colleagues for their careful study and detailed observations that have provided valuable information on which this book is based. We would like to thank Marty Becker, Phil Bjork, Clint Boyd, John Chamberlain, Dan Chure, Joe DiBenedetto, Jim Evans, John Flynn, Ted Fremd, Matt Garb, Jacques Gauthier, Lance Grande, David Grandstaff, Bob Hunt, Howard Hutchison, Matthew Kohn, Bill Korth, Hannan LaGarry, Leigh Anne LaGarry, Alvis Lisenbee, James Martin, Al Mead, Jason Moore, Darrin Pagnac, Dennis Parmley, David Parris, Don Prothero, Greg Retallack, Foster Sawyer, Bill Simpson, Ellen Starck, Phil Stoffer, Richard Stucky, Bill Wall, Xiaoming Wang, Ed Welsh, and Alessandro Zanazzi.

Many of the observations and interpretations included herein are the result of interactions with numerous graduate and undergraduate students over the years. We would especially like to thank Katie Card, Anthony Cerruti, Jim and Jeff Childers, Amanda Drewicz, Leslee Everett, Lew Factor, Neil Griffis, Reko Hargrave, Patricia Jannett, Raymond Kennedy, Paul Kosmidis (posthumous), Eve Lalor, Justin Little, Bill Lukens, Matt McCoy, Christine Metzger, Jason Mintz, Doreena Patrick, Justin Spence, Gary Stinchcomb, and Brandt Wells. The field help for many years (and saving Evanoff after a bad fall) by Terry Hiester (posthumous) is especially appreciated.

Without the continued support of the National Park Service family, this book would not have been written. We would like to thank the following National Park Service personnel for their support and encouragement. From Badlands National Park, we thank Eric Brunnemann, Larry Johnson, Steve Thede, Brian Kenner, Eddie Childers, Milt Haar, Mark Slovek, Mike Carlbom, Josh Delger, Laniece Sawvell, Mindy Householder, Levi Moxness, Wayne Thompson,

Adam Behlke, Christine Gardner, Lainie Fike, Phil Varela, Amanda Dopheide, Delda Findeisen, Lee Vaughan, Paul Roghair, Jenny Albrinck, Megan Cherry, Julie Johndreau, Aaron Kaye, Tyler Teuscher, Ian Knoerl, Connie Wolf, Chris Case, Steve Howard, Wolf Schwarz, Robert McGee-Ballinger, Ken Thompson, Casey Osback, Vince Littlewhiteman, Ryan Frum, Eric Yount, Pam Griswold, Tyson Nehring, Danny Baker, Linda Livermont, Jill Riggins, Valerie Reeves, Heather Tucker, and Pam Livermont. A very special thank-you goes to David Tarailo for his work on plate layouts and design. Vince Santucci, senior geologist and Washington liaison, has also provided invaluable support for this project. We would also like to thank Jerrilyn Thompson and Julie Stumpf from the National Park Service Midwest Regional Office for helping us manage some large computer image files. The Badlands Natural History Association board members and executive director, Katie Johnston, and from the Albright-Wirth Program, Katherine Callaway, also provided financial support for this project. Thanks are also due to Barbara Beasley of the U.S. Forest Service and Brent Breithaupt of the Bureau of Land Management for permits and access to other exposures of the Badlands across Nebraska and Wyoming.

Thanks also go to the many curators and collections managers that facilitated our work in the various museums housing vertebrate material from the White River Group. We would like to thank all of them, particularly the South Dakota School of Mines and Technology, Laurie Anderson, Sally Shelton, Samantha Hustoft, Carrie Herbel, Amy Wright, Bill Schurmann (posthumous), and Mike Ryan; the American Museum of Natural History, Ruth O'Leary, Judy Galkin, Carl Mehling, and Alex Ebrahimi-Navissi; the Yale Peabody Museum, Christopher Norris, Daniel Brinkman, Ethan France, and Annette Van Aken; the Smithsonian National Museum of Natural History, David Bohaska, Charyl Ito, Mike Brett-Surman, Thomas Jorstad, Michelle Pinsdorf, and Matthew Miller; the Denver Museum of Nature and Science, Richard Stucky, Logan Ivy, and Rene Payne; the University of Colorado Museum of Natural History at Boulder, Jaelyn Eberle, Toni Culver, and Katie McComas; the Field Museum of Natural History; and the University of Texas at

Austin, Tim Rowe, Jessica Maisano, and Matthew Colbert, for granting us permission to use our photographs and some of their photographs of their specimens for this book.

We would like to extend a special thank-you to the Oglala Sioux Tribe Tribal Historic Preservation Advisory Council for their careful review of the tribal specimen photos considered for this publication. Members include Mike Catches Enemy (ex officio advisory member/director/tribal historic preservation officer), Jhon Goes In Center (advisory member/ chair), Garvard Good Plume Jr. (advisory member), Wilmer Mesteth (advisory member), Dennis Yellow Thunder (ex officio advisory member/cultural resource specialist), and Hannan LaGarry (ex officio advisory member/paleontologist).

Several artists have contributed magnificent images to both the cover and colored plates found within the book. We extend our gratitude to Jim Carney, Laura Cunningham, Robert Hynes, and Diane Hargreaves for their creative works.

We are very grateful to Bob Sloan, Jim Farlow, Jenna Lynn Whittaker, Daniel Pyle, and Nancy Lila Lightfoot at Indiana University Press, as well as copy editor Karen Hellekson, for their help on this project. We also thank Paula Douglass for developing the index.

Finally, we extend thanks to our family and friends for their support during the project.

Institutional Acronyms

AMNH   American Museum of Natural History, New York, New York

BADL   Badlands National Park, Interior, South Dakota

DMNS   Denver Museum of Nature and Science, Denver, Colorado

FMNH   Field Museum of Natural History, Chicago, Illinois

SDSM   South Dakota School of Mines and Technology, Rapid City, South Dakota

UCM   University Colorado Museum of Natural History Boulder, Boulder, Colorado

USNM   National Museum of Natural History, Washington, D.C.

YPM   Yale Peabody Museum, New Haven, Connecticut

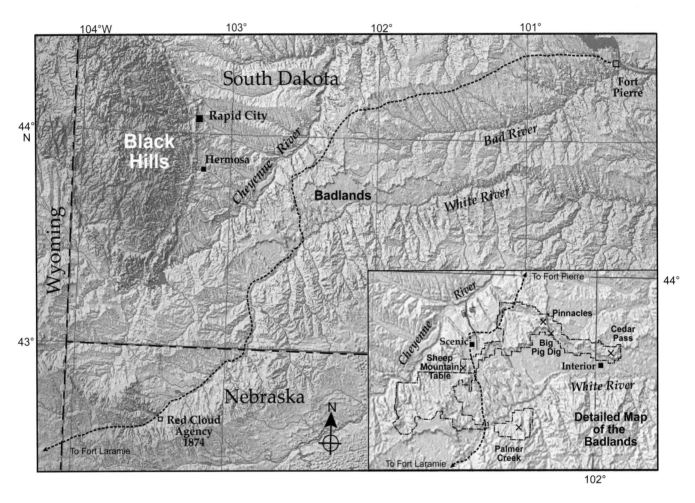

1.1. Regional map of southwest South Dakota and adjacent states showing the features related to the history of the geologic and paleontologic studies of the Big Badlands. The dashed line indicates the route of the Fort Pierre–Fort Laramie road, plotted after the map of Warren (1856). The inset map of the details of the Badlands area includes the boundaries of the three units of present Badlands National Park (dash–dot lines). The base map is from the U.S. National Atlas Web site (http://nationalatlas.gov/mapmaker).

THE FIRST FOSSILS FROM THE BADLANDS OF SOUTH Dakota were collected by employees of the American Fur Company and sent to scientists in the eastern United States. The fur company had opened up a wagon road between Fort Pierre on the Missouri River and Fort John, later known as Fort Laramie, on the Platte River (Fig. 1.1). This was a much shorter route from the Missouri than the long Platte River road, and it crossed the Big Badlands in the headwaters of Bear Creek near the modern town of Scenic, then went south on the east side of Sheep Mountain Table to the White River. Various fur company employees may have collected fossils from this area in the 1840s, but it was the chief agent of the upper Missouri posts for the fur company, Alexander Culbertson, who sent fossils to St. Louis and to his father and uncle in Pennsylvania. Dr. Hiram Prout of St. Louis had been sent a lower jaw fragment of a huge mammal that he identified as *Palaeotherium* because of its similarity to figured specimens of this European fossil mammal. He sent a cast of this specimen and a letter to Yale University in 1846. The letter and a crude drawing of the specimen's teeth were published in 1846. Prout described the specimen in greater detail the following year, 1847, and this became the first White River fossil mammal to be described in the scientific literature (Fig. 1.2A). The other fossils sent to Culbertson's father and uncle eventually made their way to Dr. Joseph Leidy of the Philadelphia Academy of Sciences (Fig. 1.3). One of Leidy's many academic talents was vertebrate paleontology, and beginning with the description of the first fossil camel skull found in the United States, which he named *Poebrotherium* in 1847 (Fig. 1.2B), he started a long career as the preeminent vertebrate paleontologist of the United States.

These first publications on the fossils from the Badlands piqued the interest of geologists and naturalists, some of whom eventually visited the Badlands. Among these was Dr. David Dale Owen, who was making a geologic survey of Wisconsin, Minnesota, and Iowa. In 1849 Owen sent one of his assistant geologists, Dr. John Evans, to the Badlands to collect fossils and to determine the age relations of the fossil-bearing rocks. Evans and his field party spent about a week in the Badlands, and all the fossils were sent to Leidy. The results of Evan's expedition plus descriptions of the vertebrate

fossils by Leidy were published in 1852. This report included the first map of the region (Fig. 1.4) and the first diagram of the Badlands (Fig. 1.5). Evans described the Badlands as follows:

> To the surrounding country . . . the Mauvaises Terres present the most striking contrast. From the uniform, monotonous, open prairie, the traveler suddenly descends, one or two hundred feet, into a valley that looks as if it had sunk away from the surrounding world; leaving standing, all over it, thousands of abrupt, irregular, prismatic, and columnar masses, frequently capped with irregular pyramids, and stretching up to a height of from one to two hundred feet, or more.
>
> So thickly are these natural towers studded over the surface of this extraordinary region, that the traveler threads his way through deep, confined, labyrinthine passages, not unlike the narrow, irregular streets and lanes of some quaint old town of the European Continent. Viewed in the distance, indeed, these rocky piles, in their endless succession, assume the appearance of massive artificial structures, decked out with all the accessories of buttress and turret, arched doorway and clustered shaft, pinnacle, and finial, and tapering spire. (Evans, 1852:197)

Thaddeus A. Culbertson was the younger half-brother of Alexander Culbertson and was educated at what is now Princeton University. He decided to travel in the summer of 1850 to the upper Missouri country to study the Native Americans along the river and to collect natural history specimens. He discussed his trip with Spencer F. Baird, who was soon to become a curator at the Smithsonian Institution. Baird urged the younger Culbertson to make a visit to the Badlands to collect fossil vertebrates. Culbertson traveled with two guides to the upper Bear Creek drainage and spent about a day collecting. He returned to Fort Pierre with a small but good collection of fossils. Culbertson returned to Washington in August 1850 but died 3 weeks after his return from complications related to tuberculosis. Baird had some of Culbertson's journal of the trip published in 1851, but Culbertson's entire journal was not completely published until 1952 by McDermott. All of the fossils that Evans and the Culbertson brothers collected were sent to Leidy, who published his first monograph on the Mauvaises Terres fauna in 1853.

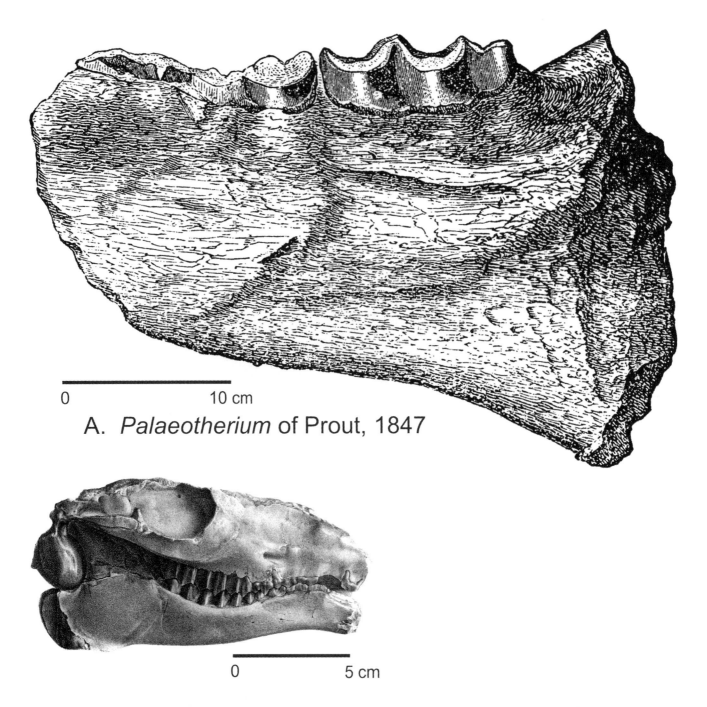

0   10 cm

A. *Palaeotherium* of Prout, 1847

0   5 cm

B. *Poebrotherium wilsoni* Leidy, 1847, type specimen

1.2. Illustrations of the first two fossils described from the Big Badlands. (A) Diagram of "gigantic *Palaeotherium*" jaw published in Prout (1847). (B) Type specimen of *Poebrotherium wilsoni* published by Leidy (1847). This diagram is from Leidy (1853:plate 1, fig. 1). All early diagrams of White River fossils were made from specimens, not from reconstructed complete skulls and jaws.

Eighteen fifty-three was the year not only of Leidy's first monograph but also of the second trip of Evans to the Badlands, and the first trip to the Badlands by Fielding B. Meek and Ferdinand V. Hayden (Fig. 1.3). Meek became the preeminent invertebrate paleontologist in the United States, specializing in the invertebrate fossils of the west, and Hayden would later become the director of one of the five

great geologic surveys of the American West after the Civil War. In 1853 both were assistants of James Hall of the New York Geological Survey. Hall wanted collections from the upper Missouri basin, including fossils from the Badlands. Sent by Hall to St. Louis, Meek and Hayden initially met opposition to their proposed collecting trip to the Badlands by Evans, who considered the two to be interlopers in the fossil

Joseph Leidy

Fielding Bradford Meek

Ferdinand Vandeveer Hayden

John Bell Hatcher

1.3. Some of the important early paleontologists who worked in the Badlands. Joseph Leidy worked at the Philadelphia Academy of Sciences and was the first to publish a monograph of the fossils from the Big Badlands. Fielding B. Meek and Ferdinand V. Hayden collected in the Badlands in 1853 and named the White River Group in 1859. John Bell Hatcher worked as a collector for O. C. Marsh at Yale and later taught at Princeton University.

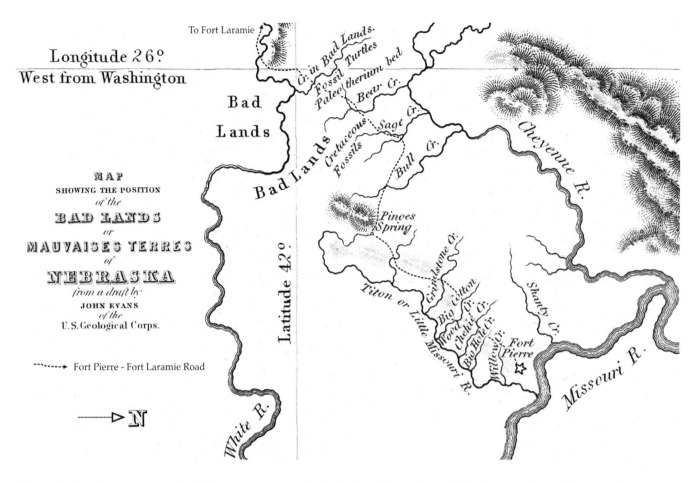

1.4. Details of the Evans map compiled in 1849 showing the route to the Badlands, published in Evans (1852). The perspective on this map is to the west, as if one were traveling to the Badlands from Fort Pierre. The road from Fort Pierre to Fort Laramie is plotted with the thin dashed line (cf. Fig. 1.1).

beds. However, the two groups finally cooperated and spent about a month collecting along the Fort Pierre–Fort Laramie road (Fig. 1.6). Hayden would return to the Badlands in May 1855, traveling along buffalo trails along the south side of the White River, and collecting at such areas as the Palmer Creek area of the modern South Unit of Badlands National Park (Hayden, 1856). Hayden served for the Union army as a surgeon during the Civil War, and after the war he would make one last trip to the Badlands. In May 1866 Hayden traveled from Fort Randall along the Missouri River up the Niobrara River, across the Pine Ridge, to his old fossil-collecting areas at Palmer Creek, the south end of Sheep Mountain Table, and the upper Bear Creek drainage. Hayden (1869) was disappointed in the relatively small numbers of fossils that he found in the areas that he had collected in 1853 and 1855. Apparently there had not been enough erosion to uncover fossils in the numbers that he had found in his earlier surveys. In 1869 Hayden wrote the geology discussion to Leidy's great monograph, *The Extinct Mammalian Fauna of Dakota and Nebraska.* This monograph summarized all of the fossil mammals from the White River Group (named as the White River Series by Meek and Hayden in 1858) that had

been collected over the previous two decades (Fig. 1.7). This work would be the best description of White River fossils for the next 70 years.

Collecting parties from East Coast universities and museums dominated the studies of the Badlands in the late nineteenth century and first half of the twentieth century. Othniel C. Marsh of the Yale Peabody Museum collected in the White River Badlands in 1874 (Schuchert and LeVene, 1940). Marsh collected primarily in northwest Nebraska, though he may have made excursions as far north as the South Dakota Badlands. While at the Red Cloud Agency, Marsh learned of the following Lakota tale from a friend, Captain James H. Cook. Cook had been shown a huge molar from a brontothere by the Lakota, and Cook's friend, American Horse, told the following legend about the beast:

American Horse explained that the tooth had belonged to a "Thunder Horse" that had lived "away back" and that then this creature would sometimes come down to earth in thunderstorms and chase and kill buffalo. His old people told stories of how on one occasion many, many years back, this big Thunder Horse had driven a herd of buffalo right into a camp of Lacota [*sic*] people during a bad thunderstorm, when these

1.5. First image of the topography of the Badlands, published in the Evans report (1852:196). The image was made from a sketch by Eugene de Girardin, the artist on the Evans expedition. De Girardin's sketches reproduced the topographic features of the Badlands quite well, but somehow the badland slopes and cliffs became translated by the engraver into a series of vertical pillars not seen in the Badlands.

people were about to starve, and that they had killed many of these buffalo with their lances and arrows. The "Great Spirit" had sent the Thunder Horse to help them get food when it was needed most badly. This story was handed down from the time when the Indians had no horses. (Osborn, 1929:xxi)

Not long after, Marsh named one of the genera of these huge relatives of the rhinoceros *Brontotherium*, "thunder beast." Though this genus name is not widely used today, the group is still referred to as brontotheres.

After 1874 Marsh hired collectors to send him fossils from the West. The most capable and renowned of these collectors was John Bell Hatcher (Fig. 1.3). In 1886 Marsh sent Hatcher out to the Great Plains to collect skulls and skeletons of brontotheres that occur in the lower deposits of the White River Group. Hatcher started his work in northwest Nebraska and adjacent Wyoming, but in 1887 he traveled to the Badlands east of Hermosa, South Dakota, where he collected 13 skulls, including three skulls in a single day. In the 15 months that he collected brontothere fossils during the three field

seasons of 1886, 1887, and 1888, Hatcher collected 105 skulls and numerous skeletons and isolated bones of brontotheres (Hatcher, 1893:214) that totaled about 24.5 tons of fossil materials (estimated from the figures given in Schuchert and LeVene, 1940). No other collector of White River fossils has matched the volume of materials collected by Hatcher.

Between 1890 and 1910 many major museums and universities sent collecting parties to the Badlands. These included the American Museum of Natural History in New York; the Carnegie Museum in Pittsburgh, Pennsylvania; the Field Museum of Natural History in Chicago; Princeton University; Amherst College in Massachusetts; the University of Nebraska in Lincoln; the University of Kansas in Lawrence; the University of South Dakota in Vermillion; and the South Dakota School of Mines (O'Harra, 1910). Meek and Hayden (1858) had given the name White River Series to the rocks of the South Dakota Badlands, and they subdivided the rocks by their fauna into the lower titanothere beds and the overlying turtle–*Oreodon* beds. Jacob L. Wortman, while

1.6. Fielding B. Meek's sketch of the Badlands made in 1853 and published in Hayden's geology report in Leidy's 1869 monograph. The area shown is of the large buttes near the modern access road to Sheep Mountain Table.

working in the South Dakota Badlands in 1892 as a collector for the American Museum of Natural History, recognized an additional faunal subdivision for the White River (Wortman, 1893). He subdivided Hayden's turtle–*Oreodon* beds into the lower *Oreodon* beds (dominated by the oreodont *Merycoidodon*) with the *Metamynodon* channels and the upper *Leptauchenia* beds (another kind of oreodont) with the *Protoceras* channels. *Metamynodon* is a large, primitive, odd-toed ungulate (perissodactyl) related to the rhinoceros, and *Protoceras* is a medium-size, even-toed ungulate (artiodactyl) that is a member of an extinct group related to the camels and deer. This three-part faunal division of the White River sequence was later formalized as the Chadronian, Orellan, and Whitneyan land mammal ages (Wood et al., 1941). The first subdivisions of the White River rocks on the basis of rock types (lithology) was made by N. H. Darton (1899) of the U.S. Geological Survey, who recognized the lower Chadron Formation as well as the upper Brule Formation in western

Nebraska and South Dakota. The Chadron Formation included the basal red beds and the overlying greenish-gray claystone beds of the lower White River, while the Brule Formation included the tan mudstone and siltstone beds of the upper White River. Together, the Chadron and Brule formations make up the White River Group of South Dakota and Nebraska.

Paleontologists could position their fossil localities to within these broad faunal subdivisions, but the lack of detailed maps in the 1800s prevented the detailed recording of geographic locations of fossil sites, and only rudimentary sedimentology concepts were understood. Evans (1852), Hayden (1869), and most nineteenth-century geologists thought the White River sediments had been deposited in a huge lake. Fine-grained, fairly well-bedded rocks were thought to have been deposited in quiet water, and the presence of freshwater snails and clams were used as evidence of the existence of a lake that covered a huge area of the Great Plains and

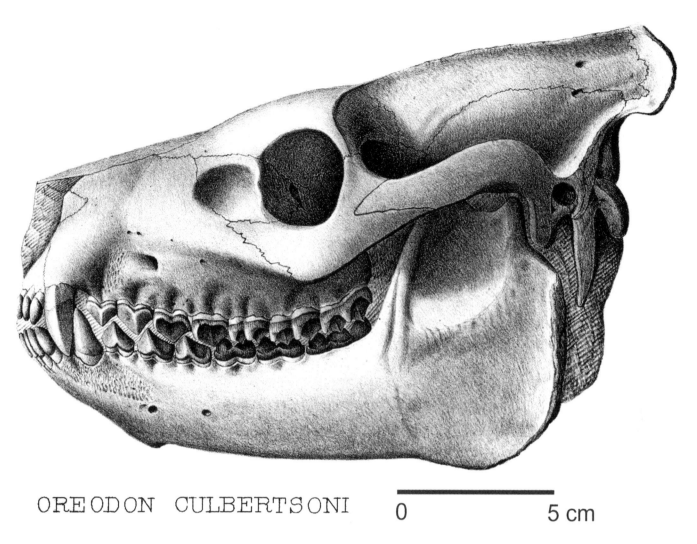

OREODON CULBERTSONI

0   5 cm

1.7. A diagram of a complete skull of *Oreodon culbertsoni,* published by Leidy (1869:plate 6, fig. 1). By 1869 enough complete skulls were available for complete reconstructions of some White River mammals.

butted up against the flanks of the Black Hills and Rocky Mountains. The bones of mammals and other land-dwelling organisms were thought to have washed into the lake from rivers during floods. This so-called lacustrine theory for the origin of the White River and other Tertiary rocks of the Great Plains was questioned as early as 1869 by Leidy, who found few aquatic vertebrates in the White River fossil record to support the existence of the lake. The lacustrine theory was finally debunked by Hatcher in 1902. Hatcher made his argument that the White River rocks were deposited by rivers because of the presence of ancient river channels represented by long, thin, sinuous gravel and coarse sand deposits scattered throughout the White River fine-grained mudrocks. The White River fauna included almost all land-dwelling organisms, along with extremely few aquatic vertebrates and invertebrates except in channel deposits or the thin limestone deposits. The few plant fossils (hackberry seeds, fossil roots, and rare tree stumps) also were widely distributed in the White River mudrocks. Because of his arguments and

professional stature as a well-respected vertebrate paleontologist, Hatcher put the lacustrine theory to rest.

For six decades, paleontologists and geologists from Princeton University made extensive studies of the rocks and fossils from the South Dakota Badlands. William Berryman Scott led the first Princeton students into the Badlands in 1882. While publishing extensively on the White River mammals, he returned with students to the area in 1890 and 1893. John Bell Hatcher had been hired as a curator of vertebrate paleontology by Princeton in 1893, and he joined Scott and the students in the Badlands that summer. Scott turned the student field camp duties over to Hatcher, who made many more extensive collections for Princeton. Hatcher left Princeton in 1900, and in 1905 Dr. William J. Sinclair was hired as vertebrate paleontologist. In 1920 he started a major study of the fossils and geology of the lowest beds of the Brule Formation in the Badlands, then called the red layer. Sinclair was not only an excellent vertebrate paleontologist but also an excellent geologist. Sinclair (1923) was one of the first to consider the detailed origins of the White River bone beds on the basis of the lithologic context and postmortem (taphonomic) features of the fossil bones. He carefully recorded the vertical positions of the fossils within the lower Brule Formation and documented vertical changes in the faunas. Sinclair's first student, Harold R. Wanless, made one of the most extensive studies of the White River rocks in the South Dakota Badlands. Wanless (1921) studied the lithologic features of the White River Group and the distribution and origin of the rocks over a large area, mainly west of Sheep Mountain Table (Wanless, 1923). Wanless carefully recorded his observations and interpretations and his papers are still essential reading for anyone who studies the geology of the Badlands. Sinclair was to work with William Berryman Scott on a monographic study of the White River fauna, but Sinclair died in 1935. Sinclair's student Glenn L. Jepsen took over as Scott's colleague in the monumental monograph *The Mammalian Fauna of the White River Oligocene*, published between 1936 and 1941. The well-illustrated five-part set (Fig. 1.8) was divided into the following taxonomic groups; insectivores and carnivores (Scott and Jepsen, 1936), rodents (Wood, 1937), lagomorphs (Wood, 1940), artiodactyls (Scott and Jepsen, 1940), and perissodactyls, edentates and marsupials (Scott, 1941). This was the first extensive White River monograph since Leidy (1869) and has never been duplicated.

Jepsen took over as the vertebrate paleontologist at Princeton but his interest drifted into older Paleogene faunas, ending White River studies at Princeton. His student John Clark continued the work on the White River Group from the 1930s well into the 1970s. Clark's doctoral dissertation, published in 1937, was on the geology and paleontology of the Chadron Formation in the South Dakota Badlands. Similar in methods to those of Sinclair and Wanless, Clark also made analyses of the vertebrate fauna of the Chadron formation. Working primarily for the Carnegie Museum of Natural History in Pittsburgh, Pennsylvania, and later at the Field Museum of Natural History in Chicago, Clark continued his studies of the Chadron Formation and expanded into studying the lower Brule Formation (Scenic Member). Clark (1954) gave names to the three parts of the Chadron Formation west of Sheep Mountain Table: the Ahearn, Crazy Johnson, and Peanut Peak members. After 30 years of study, Clark and two contributors, James R. Beerbower and Kenneth K. Kietzke, published a memoir in 1967 about their work in White River rocks and faunas of the Badlands titled *Oligocene Sedimentation, Stratigraphy, Paleoecology, and Paleoclimatology in the Big Badlands of South Dakota*. This memoir discusses such topics as paleoclimatology as interpreted from the rocks and fauna, paleoecology from the distribution of the fauna in the rocks and the rocks' depositional environments, and details of fluvial sedimentology. These discussions predated

modern detailed sedimentologic and faunal analyses by decades.

The school with the longest continual record of study of the White River rocks and fossils in the Badlands is the South Dakota School of Mines and Technology in Rapid City. Cleophas C. O'Harra was the first School of Mines geology professor to take students into the Badlands, primarily for geologic studies and secondarily to collect fossils. His first trip was in 1899 to the Sheep Mountain Table area, and O'Harra continued to take students on yearly trips to the Badlands for the next two decades. In 1920 O'Harra wrote a popular guide to the geology and paleontology of the Badlands and surrounding areas that is still available in print and has served as a model for this volume. In 1924 Glenn Jepsen, then an instructor at the School of Mines organized the first trip to solely collect vertebrate fossils in the Badlands for the School of Mines Museum. When Jepsen left for his studies at Princeton, James D. Bump took over the role of paleontologist, becoming the director of the museum in 1930. In 1940 Bump and other faculty at the School of Mines were awarded a grant from the National Geographic Society to collect exhibit-quality White River fossils for the Museum. Bump and his crew spent 3 months collecting in the upper Brule rocks in the Palmer Creek area, which is now in the South Unit of Badlands National Park. These fossils became the basis of exhibits in the new museum hall in the O'Harra Building that was completed in 1944, though the final exhibits were constructed through the 1960s. The exhibits of White River fossil vertebrates are still on display in this building and include some of the finest White River mammal skeleton reconstructions in the nation. In 1956 Bump formally named the two members of the Brule Formation, the Scenic Member of the lower Brule Formation and the Poleslide Member of the upper Brule Formation, and designated type sections near the town of Scenic and on the south side of Sheep Mountain Table, respectively. The rocks that overlie the Brule Formation were named and described as the Sharps Formation in 1961 by J. C. Harksen, J. R. Macdonald, and W. D. Sevon, all from the School of Mines. Macdonald had been hired as the first vertebrate paleontology curator for the school in 1949. The volcanic tuff on the top of the Brule Formation, the Rockyford "Ash," was named and described by a J. M. Nicknish and J. R. Macdonald in 1962. Later, paleontologists from the School of Mines Museum, Robert W. Wilson and Philip R. Bjork, worked extensively in the Badlands during the 1960s through the 1980s. Dr. Bjork was especially active in collecting White River fossils from the upper Brule Formation (the Poleslide Member) in the Cedar Pass and Palmer Creek areas of Badlands National Park. To date, there have been eight completed Master's theses on geology and 18 theses on paleontology of the Badlands at the School of

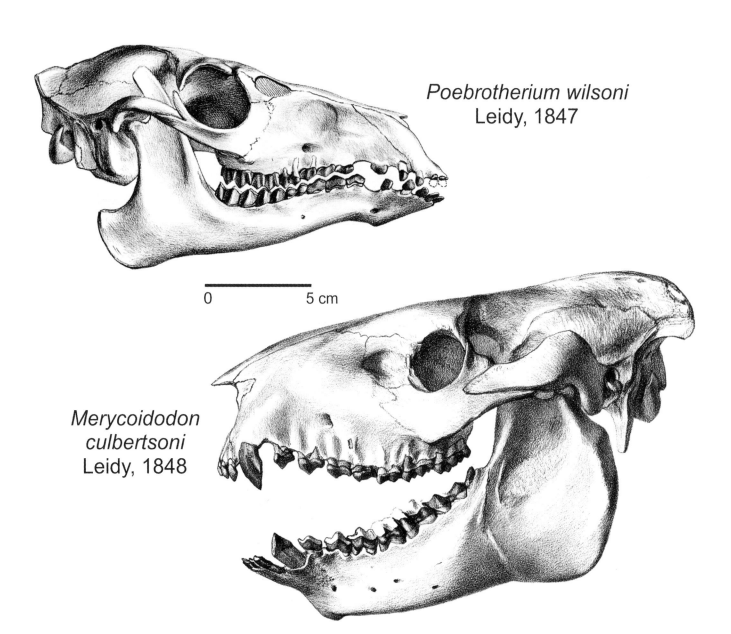

Poebrotherium wilsoni
Leidy, 1847

0   5 cm

Merycoidodon culbertsoni
Leidy, 1848

1.8. Diagrams of the skulls of *Poebrotherium wilsoni* and *Merycoidodon culbertsoni* by the artist R. Bruce Hornsfall. These were made for part 4 of the *Artiodactyla* of the *Mammalian Fauna of the White River Oligocene* monograph by Scott and Jepsen (1940:plate 64, fig. 1, and plate 69, fig. 1). Almost all the diagrams in the Scott and Jepsen monographs are reconstructions based on complete skeletons, skulls, and bones from the vast collections of White River vertebrates amassed by the mid-twentieth century. Compare these with the original images of the type specimen of *Poebrotherium* (Fig. 1.2B) and the first reconstruction of the skull of *Oreodon* (Fig. 1.7).

Mines. As a result of this research, the South Dakota School of Mines Museum has perhaps the largest collection of fossil vertebrates from the Badlands in the United States. As a partner repository with Badlands National Park it is the primary repository for the fossils collected in the park and surrounding areas.

During the 1980s, new methods were developed to analyze the rocks of the White River Group. One such technique is the analysis of ancient soils (or paleosols). Sedimentary rocks are typically described by lithology in depositional packages. Terrestrial sedimentary rocks deposited by rivers or as dust deposits (eolian sediments) are greatly modified by weathering and the action of soil organisms (plants and animals) to form soils. The remains of these soil processes are preserved in the rock, and require detailed analysis. In 1983 Dr. Greg J. Retallack of the University of Oregon described in great detail the ancient soil features of White River rocks in the upper Conata basin and Pinnacles area of Badlands National Park. From these data Retallack related the ancient soils to modern soil types that form under specific climatic and depositional environments (Fig. 1.9). He documented changes through time in the vegetation and climate as recorded in the White River Group in a single section located south of the Pinnacles Overlook. His report, published in 1983, was a landmark study showing the potential of paleosol studies for detailed paleoenvironmental interpretations.

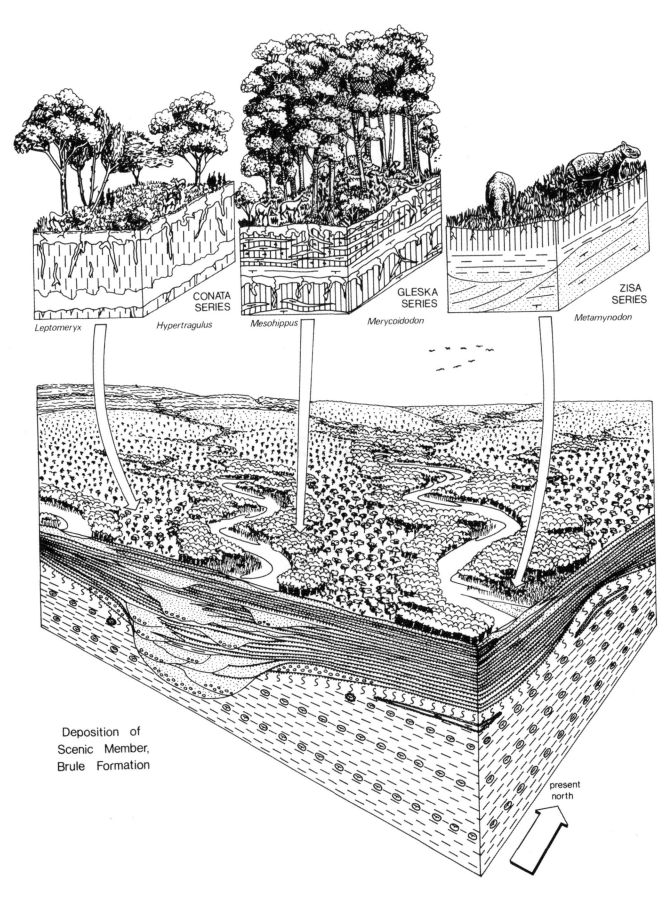

CONATA
SERIES

Leptomeryx   *Hypertragulus*

GLESKA
SERIES

Mesohippus   *Merycoidodon*

ZISA
SERIES

Metamynodon

Deposition of
Scenic Member,
Brule Formation

present
north

1.9. Reconstruction of ancient environments during the deposition of the Scenic Member made by Retallack (1983a:fig. 7). The Zisa, Gleska, and Conata series are types of ancient buried soils that were formed in active channels, heavily forested river riparian (near channels), and open distal overbank (far from channels) environments.

Retallack's study was the predecessor to the work of Dr. Dennis O. Terry Jr. and his students from Temple University, Philadelphia, who have studied many additional paleosol sequences in the Badlands throughout the North Unit of Badlands National Park (see the discussion in chapter 3). Terry, along with Dr. James E. Evans, proposed in 1994 a new formation at the base of the White River Group, the Chamberlain Pass Formation.

Another advanced technique of studying the White River rocks is the study of the ancient magnetic signals in the rocks, called paleomagnetism analysis. This technique measures in rocks the magnetic orientation of tiny magnetic minerals. These magnetic minerals record the properties of the Earth's magnetic field at the time the rock was formed. The dynamics of the Earth's magnetic field will sometimes allow the magnetic poles to switch position. This switching occurs relatively quickly relative to geologic time and its effects are global. The times when the Earth's magnetic field has switched have been calibrated by radiometric dates, so the changes in the paleomagnetic signals in rocks give us a proxy for time. Donald R. Prothero, emeritus of Occidental College in Los Angeles, is a pioneer in the study of paleomagnetism of Cenozoic sedimentary rocks in the western United States. He has analyzed the paleomagnetism of the White River rocks in the Badlands and has published a series of publications on this paleomagnetic record (Prothero, 1985; Tedford et al., 1996; Prothero and Whittlesey, 1998).

The Big Badlands were first protected in 1939 by the establishment of Badlands National Monument. The Monument included what is now known as the North Unit, but in 1978 Badlands became a national park with the inclusion of 130,000 acres of lands to the south that was to be jointly administered by the National Park Service and the Pine Ridge Indian Reservation. This southern addition is now the South Unit of Badlands National Park. In 1994 the National Park Service hired Rachel C. Benton as the first park paleontologist for Badlands National Park. Dr. Benton has developed an extensive paleontological resource management program that has included locating, surveying, and monitoring fossil localities; evaluating and permitting proposed geological and paleontological research projects in Badlands National Park; evaluating and recording new fossil sites discovered by researchers and visitors to the park; and overseeing monitoring of fossil resources that are uncovered during construction projects in the park. To do all of this work, Dr. Benton oversees as many as 12 seasonal employees and interns each summer. She is also involved with public education, teaching seasonal interpreters about the fossils of Badlands; overseeing new additions to the exhibits in the Ben Reifel Visitor Center; opening and maintaining quarries so visitors can see fossil excavations in progress; and establishing and overseeing a fossil preparation laboratory in the Ben Reifel Visitor Center that has been open for public viewing.

One of the most popular visitor attraction and a major scientific study organized by Dr. Benton was the development and excavation of the Big Pig Dig. This is a bone bed situated near the Conata Picnic Ground that contained a large number of fossil bones and skulls, primarily of the large piglike entelodont *Archaeotherium*, and the rhinoceros, *Subhyracodon*. The site was discovered in 1993 and was excavated for 15 field seasons. Over 19,000 bones, teeth, and skulls were excavated from the site and are now stored at the Museum of Geology at the South Dakota School of Mines and Technology. The site was open to the public as it was excavated by SDSM seasonal paleontologists, students, and interns, and it was visited by between 5000 and 10,000 visitors per year. The fossils of the bone bed were eventually all collected, and the site was closed in 2008.

Two research projects organized by Dr. Benton that were extremely important for the science and the management of fossil resources in Badlands National Park were two multiyear fossil and geologic surveys within the Brule Formation in the North Unit of the park. The first, informally known as the Scenic Bone Bed Project, was a survey of the detailed paleontology and geology of the lower and middle Scenic Member in the western half of the North Unit. The project was active during the summers of 2000 to 2002, with the final report compiled by Dr. Benton in 2007. The second project, the Poleslide Bone Bed Project, was a survey of the paleontology and geology of the lower Poleslide Member in the eastern third of the North Unit. The Poleslide Project research was active in the summers of 2003 to 2005, and the final report was compiled in 2009 (Benton et al., 2009). The research for both projects included the location and description of fossils sites by two paleontology crews, one overseen by Dr. Benton and the other by Carrie L. Herbel, then the collections manager at the South Dakota School of Mines Museum. The paleosols associated with many of the individual bone beds were described by Dr. Dennis Terry and his students, contributing to our understanding of the origins of the bone beds. Dr. Emmett Evanoff of the University of Northern Colorado worked out the detailed stratigraphy (distribution and sequence) of rock layers in the Scenic and Poleslide members of the North Unit of Badlands National Park. This work allows the widespread and scattered bone beds to be placed in the proper order in time and space. Much of the following discussion of the geology, taphonomy, paleoenvironments, and paleontological resource management of this book is derived from these studies and subsequent work in the Badlands National Park.

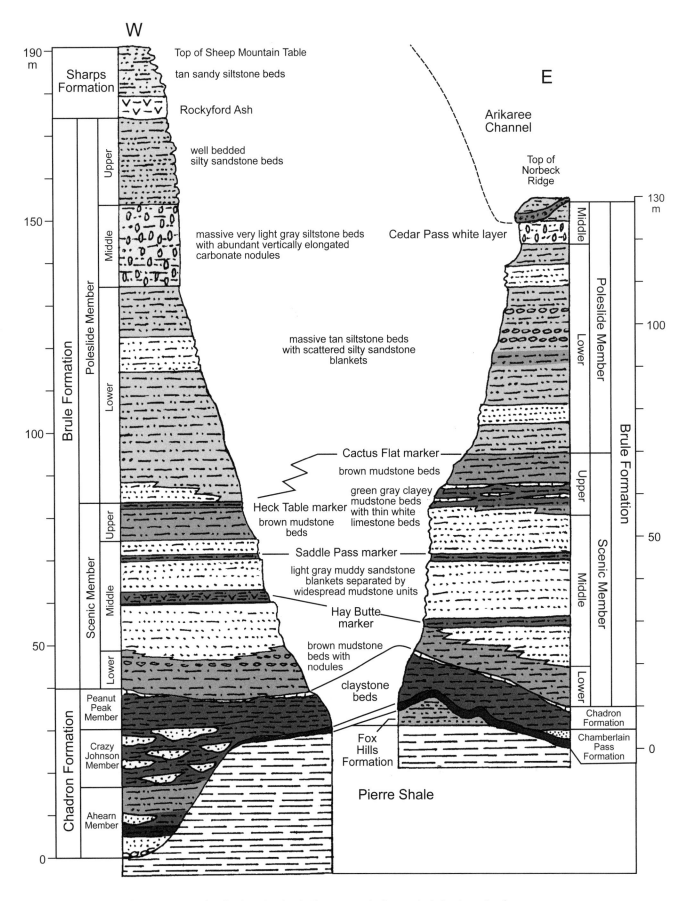

2.1. Stratigraphic units of the North Unit of Badlands National Park. The sequence in the west includes the rocks of the Red River paleovalley of Clark, Beerbower, and Kietzke (1967) and the rocks exposed on Sheep Mountain Table. The sequence in the east is a composite of features from the Dillon Pass area east to Norbeck Ridge.

Sedimentary Geology of the Big Badlands

THE ROCKS OF BADLANDS NATIONAL PARK RECORD THE end of the great Western Interior Seaway near the end of the age of dinosaurs (Mesozoic Era) and, after a 30-million-year gap, record some of the features on land during the greatest volcanic eruptive intervals and one of the greatest climatic changes in the age of mammals (Cenozoic Era). To understand the geologic history of the sedimentary rocks exposed in the South Dakota Badlands, we need to see how the rocks are distributed in space and time (stratigraphy) and how they were deposited (sedimentology). We start our story by building the framework of the rocks in space and time by discussing the stratigraphy of the Badlands. The rocks in the Badlands were deposited in the sea, in river deltas, by rivers, in lakes, and by the winds, and we will discuss the evidence for these depositional environments. To understand the ancient environments and how they changed through time, we need to start by understanding the origins of the sediments that now make up the 190 m thick sequence of rocks exposed in Badlands National Park (Fig. 2.1).

STRATIGRAPHY: GEOLOGIC FRAMEWORK OF THE WHITE RIVER BADLANDS

Sedimentary geologists recognize formal stratigraphic units, or distinct packages of rocks that are carefully described and named for permanent geographic features and their associated lithologies. These formal stratigraphic units include formations, members, and groups. The fundamental stratigraphic unit is a formation, which is a package of sedimentary rock with unique lithologic (rock) features that separate it from other formations above and below it. Formations must be widespread and thick enough to be easily plotted on a map. There are six formations recognized in the Badlands, including, from the oldest to the youngest, the Pierre Shale, the Fox Hills Formation, the Chamberlain Pass Formation, the Chadron Formation, the Brule Formation, and the Sharps Formation. Formations of similar origins and closely related through time can be combined to create groups. The Chamberlain Pass, Chadron, and Brule formations are part of the White River Group, the subject of most of this discussion. The 365 m thick Pierre Shale that lies below the Fox

Hills Formation has traditionally been considered to be a formation, but Martin and Parris (2007) have subdivided the Pierre Shale into seven formations and elevated the unit to the Pierre Shale Group. (We will continue to call this unit simply the Pierre Shale in the following discussion.) Formations can be subdivided into members, which are thinner, less distinctive, but widespread units within a formation. The history and nomenclature of the White River Group and its formal subdivisions is presented in Table 2.1. These names all refer to distinct packages of rocks that have unique features and contain fossils that record changes in environments through time. Our discussion will focus on the features of the White River Group, the rocks, and the main source of fossils in the intricately carved Badlands.

Sedimentary geologists also recognize a variety of informal stratigraphic units. These tend to be widespread distinctive individual layers, too thin to be named as members but nonetheless important markers for stratigraphic positioning. Such marker beds can include volcanic tuff beds, widespread limestone beds, and, in the case of the White River Group, thin but widespread mudstone layers that can be traced for many tens of kilometers through the Badlands. Volcanic tuffs, or just tuffs, are by far the most distinctive and stratigraphically significant marker beds. They are formed from the lithification of a volcanic ash, or a deposit of fine-grained volcanic material blown out of large volcanic eruptions, carried downwind as ash clouds, and deposited on the land surface, into lakes, or on the seafloor. Volcanic ashes fall within a day to a few months, which is essentially instantaneous in geologic timescales measured in millions of years. Tuffs are considered lithologic timelines, or layers that represent a single geologically brief event. The tuffs contain volcanic glass, an amorphous material similar to window glass formed from the quenching of tiny bubbles of lava during an eruption. Volcanic ash clouds also carry tiny crystals of minerals, such as zircon, biotite, hornblende, sphene, apatite, pyroxene, and the potassium feldspar sanidine, all found in tuffs in the White River Group in the Badlands. Each individual tuff contains a unique set of these minerals, and the minerals often have unique geochemistries that allow the tuffs to be identified from isolated outcrops scattered over broad

Table 2.1. Origin of stratigraphic names used in the Big Badlands

Stratigraphic unit	Original source	Type area and comments
White River Group	Meek and Hayden (1858)	Originally named for the exposures along the White River, now in South Dakota. It was later divided faunally into the *Titanotherium,* turtle–*Oreodon, Protoceras,* and *Leptauchenia* beds.
Pierre Shale	Meek and Hayden (1862)	Named for exposures around Fort Pierre, now in central South Dakota.
Fox Hills Formation	Meek and Hayden (1862)	Named for exposures in the Fox Hills on the divide between the Cheyenne and Moreau rivers, now in north-central South Dakota.
Chadron Formation	Darton (1899)	Named for exposures at the base of the Pine Ridge in the vicinity of Chadron, Nebraska. Originally represented the *Titanotherium* bed of the lower White River Group.
Brule Formation	Darton (1899)	Named for the Brule Lakota tribe for exposures in the Pine Ridge of northwest Nebraska. Originally represented the turtle–*Oreodon* and *Protoceras–Leptauchenia* beds of the upper White River Group.
Ahearn Member of the Chadron Formation	Clark (1954)	Named for the Ahearn ranch near the mouth of Indian Creek, the type area, in Pennington County, South Dakota.
Crazy Johnson Member, Chadron Formation	Clark (1954)	Named for Crazy Johnson Butte in Pennington County, South Dakota.
Peanut Peak Member of the Chadron Formation	Clark (1954)	Named for Peanut Peak Butte on the south fork of Indian Creek, Pennington County, South Dakota.
Scenic Member of the Brule Formation	Bump (1956)	Named for the town of Scenic, Pennington County, South Dakota.
Poleslide Member of the Brule Formation	Bump (1956)	Named for Poleslide Canyon on the south side of Sheep Mountain Table, Shannon County, South Dakota.
Sharps Formation	Harksen, Macdonald, and Sevon (1961)	Named for Sharps Corners in Shannon County, South Dakota. It was originally placed in the Arikaree Group because of its Arikareean fauna. However, its dominant lithology of sandy siltstone beds suggest that it is part of the White River Group. The brown siltstone beds of the uppermost White River Group in Nebraska is thought to be equivalent to the Sharps Formation (J. Swinehart, pers. comm., 2010).
Rockyford Ash Member of the Sharps Formation	Nicknish and Macdonald (1962)	Basal marker unit of the Sharps Formation. Named for Rocky Ford in Shannon County, South Dakota. Type locality is on the south end of Sheep Mountain Table.
Chamberlain Pass Formation	Evans and Terry (1994)	Named for Chamberlain Pass east of Badlands National Park in Jackson County, South Dakota. Includes the lowest beds of the White River Group.

areas (Larson and Evanoff, 1998). In the field, tuffs are fine-grained rock layers of mudstone or claystone with noticeable amounts of floating sand-size crystals with well-preserved crystal forms (called euhedral crystals). The most obvious of these is the six-sided crystals of biotite, a black mica that is so flat that large crystals of biotite are floated in air far beyond the other, more compact crystals. Volcanic ashes fall on to a land surface, so tuffs typically have sharp or distinct bottom contacts but gradually mix with background sediments to develop diffuse top contacts. Finally, many of the minerals in a tuff, especially zircon and sanidine, carry radioactive elements in their crystalline structure that can be dated by precise radioisotopic dating techniques.

Numeric geologic dates, typically given as millions of years ago (Ma), are determined by radiometric dating, which analyzes the ratios of original radioactive materials (the parent element) to concentrations of new materials (the daughter element) generated by the radioactive decay process (such as the conversion of the radioisotope potassium-40 to stable argon-40). Radiometric dating of White River rocks is best accomplished on volcanic tuffs that contain feldspar and zircon crystals that formed in the magma chamber just before the eruption. These crystals act as miniature time capsules and preserve the parent/daughter element ratios. Modern radiometric dating techniques can analyze single crystals, providing precise determinations of ages, often to within a few tens of thousands of years. So far no tuffs in the South Dakota Badlands have been radiometrically dated, but White River tuffs have been dated in Wyoming and Nebraska, thus providing numeric date calibration for the transitions of the land mammal ages (Fig. 2.2).

The fossil mammal assemblages of the White River Group changed over time and can be used to determine the relative ages of rocks in three biostratigraphic zones. The three fossil mammal assemblages include the lower mammal fauna characterized by the huge brontotheres (*Megacerops*). A second fauna lacks brontotheres but is dominated by the oreodont *Merycoidodon* and is associated with the rhino *Metamynodon* found in river-channel deposits. The highest and youngest fauna is dominated by the oreodont *Leptauchenia* with the odd-looking ungulate *Protoceras* in river channels. The early paleontologists working in the Badlands recognized this succession of fossil faunas as the *Titanotherium* bed, the turtle–*Oreodon* bed with the *Metamynodon* channels, and the *Leptauchenia* beds with the *Protoceras* channels (Hayden, 1858, 1869; Wortman, 1893). In 1941 Wood et al. formalized these faunal zones into the Chadronian, Orellan, and Whitneyan land mammal ages. The land mammal ages are essentially time zones defined by their unique fossil mammal assemblages (Woodburne, 2004; Fig. 2.2). The most recent work on land mammal ages in the White River Group has documented the detailed stratigraphic ranges of fossil taxa across the boundaries of these biostratigraphic units (Zanazzi, Kohn, and Terry, 2009; Mathis and MacFadden, 2010).

Another technique for determining the age of the White River rocks is the study of the ancient magnetic signals in the rocks, called paleomagnetic analysis. Tiny magnetic minerals, such as magnetite, will orient like bar magnets in saturated sediment to the Earth's magnetic field at the time of deposition. As the sediments are lithified into sedimentary rocks, the orientation of the magnetic minerals is locked into the rock. What is curious about the Earth's magnetic field is that the dynamics in the Earth's liquid nickel–iron core will sometimes allow the north and south magnetic poles to switch, even though the Earth's orientation and rotation remain the same. The modern orientation of magnetic poles is called normal magnetic polarity, and when the north magnetic pole switches to the south geographic pole, it is called reversed magnetic polarity. The tiny magnetic grains in sediments will orient either to the normal or reversed magnetic orientations, and these polarity zones will be preserved in the rocks. Because these polarity zones affect the entire globe, they can be used to correlate rocks or to match rocks of the same age around the world. The boundaries between polarity zones on the global magnetopolarity timescale have been

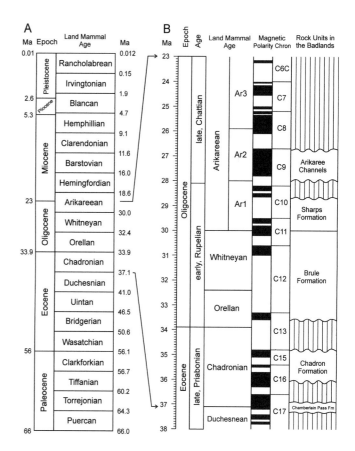

2.2. North American Land and Mammal Ages and the age relations of the rocks in the Big Badlands. (A) List of all the land mammal ages throughout the Cenozoic recognized in North America. The vertical scale does not reflect the actual duration of the individual land mammal ages. (B) Age chart of the ages and magnetostratigraphic features of 15 million years during the late Eocene through Oligocene. The vertical scale is in 100,000-year increments. Ar1 to Ar3 indicate subages of the Arikareean land mammal age. Estimated ages of the late Eocene and Oligocene rocks exposed in the Big Badlands are provided at right. The age dates for the epochs and ages are from the International Stratigraphic Chart for 2013 (http://www.stratigraphy.org/ICSchart/ChronostratChart2013-01.pdf). The numeric ages for the land mammal ages are from Woodburne and Swisher (1995), adjusted for current ages of the radiometric standard samples (Fish Canyon Tuff, 28.02 Ma). The ages of the Arikareean subages are from Albright et al. (2008). The ages of the paleomagnetic polarity zones is from Luterbacher et al. (2004) and Albright et al. (2008). Black intervals in the Magnetic Polarity column represent normal polarity intervals; white intervals represent reversed polarity intervals.

calibrated with radiometric dates (Cande and Kent, 1992; Berggren et al., 1995; Luterbacher et al., 2004). The trick is to determine which polarity zone – normal or reverse – in a rock sequence belongs to which polarity zone on the global timescale. The magnetic signal in White River rocks is weak, but sensitive magnetometers can determine the polarity of oriented rock samples. Prothero (1985), Tedford et al. (1996), and Prothero and Whittlesey (1998) have analyzed the rocks of the White River Group in the Badlands and recorded four normal polarity zones and three reversed polarity zones. By matching these zones with the global magnetopolarity timescale (Fig. 2.2), Prothero estimated a duration of about

5 million years, from about 35 Ma to 30 Ma, for the deposition of the middle Chadron Formation through the end of the Brule Formation in the Badlands.

The oldest strata exposed in the Badlands is the Pierre Shale, an extremely thick sequence of black, organic-rich shale with scattered thin bentonites (volcanic ashes altered to clays at the bottom of the sea) and limestone nodules that often formed from groundwater precipitation of calcite around fossil marine shells. The fossils of the Pierre Shale include ammonites, large marine clams, and various types of marine reptiles, such as mosasaurs and giant sea turtles (Fig. 2.3). The fossils occur in zones that have been calibrated by radiometric dates to approximately 70 million years. The Pierre Shale represents the deposits of the shallow Western Interior Seaway, which covered the central part of North America and stretched from the Gulf of Mexico north to the Arctic Ocean, cutting North America into two separate landmasses. The Pierre Shale was named for exposures of this rock unit around Fort Pierre along the Missouri River by Meek and Hayden in 1862. The Pierre Shale is exposed in and around the Sage Creek Campground (Plate 1; see Fig. P.1 for the location of this site and other locations mentioned in the following discussion).

Overlying the Pierre Shale is the Fox Hills Formation. The Fox Hills Formation is a sequence of alternating gray to orange silty shale beds and brown to orange, fine-grained, thin to thick sandstone beds (Plate 1). Fossils are numerous in the Fox Hills Formation, including impressions of plant material, petrified wood, clams, fish scales, ammonites, crabs, shark teeth, and belemnites, a relative of the squid (Stoffer et al., 2001; Jannett and Terry, 2008). The mixture of terrestrial and marine fossils reflects the origin of the Fox Hills Formation as an advancing river delta into the Western Interior Seaway. There was continuous deposition between the Pierre Shale and the Fox Hills Formation with sandstone sheets increasing in number and thickness upward from the top of the Pierre Shale into the Fox Hills Formation. Throughout the entire package of strata that comprise the delta sequence, sediments will coarsen upward overall as areas that were once the toe of the delta were covered by additional sediment that built out into the ocean basin. The sand sheets show evidence of water flow as ripples from tide, storm, and river currents. Measurements of the alignment of these ripples show that this delta complex advanced from the north into the Badlands region. On the basis of the age of these fossils, the Fox Hills Formation in the Badlands persisted to about 67 Ma, just before the extinction of the dinosaurs. The Fox Hills Formation was named for outcrops in north-central South Dakota by Meek and Hayden (1862). It is best exposed in Badlands National Park at the upper Conata basin near Dillon Pass and below the Sage Creek Rim Road (Plate 1).

After the retreat of the Western Interior Seaway, the Badlands region experienced a prolonged episode that included both erosion and nondeposition, forming a surface that geologists call an unconformity, or contacts that represent major gaps in the geologic record. In the Badlands, this gap persisted for 30 million years. Both the older and younger rocks are essentially horizontal in the Badlands, but toward the Black Hills the Pierre Shale was tilted and eroded before the deposition of White River sediments. White River rocks filled valleys that cross older and older rocks to the west, eventually resting on Precambrian crystalline rocks in the center of the Black Hills (DeWitt et al., 1989). The tilting and uplift of the Black Hills was during an episode of mountain building across the eastern part of western North America, referred to as the Laramide Orogeny. The Laramide Orogeny was different from typical mountain-building events. Long mountain chains, such as the Cascades along the west coast of North America, are built by the plate tectonic process of subduction, during which dense oceanic crust slides beneath lighter continental crust and eventually dives down steeply into the mantle, melts, and provides magma to create volcanoes. The Laramide Orogeny was also caused by subduction of the east Pacific plate (called the Farallon Plate) during the latest Cretaceous and early Paleogene along what was the west coast of North America. However, instead of diving steeply into the mantle and melting, the Farallon Plate subducted at a low, shallow angle, like a spatula sliding under a pancake. This low angle of subduction did not produce volcanoes, but it compressed and shortened the overlying continental crust. This shortening forced large blocks of crust to pop up far to the east. These include the Rocky Mountains of south-central Montana, Wyoming, Colorado, and northern New Mexico, but further to the east, smaller blocks were also elevated by this compression to form localized mountains, such as the Black Hills. As the Black Hills pushed upward, their sedimentary rock cover was stripped to eventually expose the Precambrian core. The Black Hills rose during the Paleocene and early Eocene, as shown by sediments derived from Paleozoic and Precambrian rocks in the Black Hills that were shed to the northwest into the Powder River Basin of Wyoming (Curry, 1971; Flores and Ethridge, 1985).

During this extended period of time the Badlands region underwent an extensive episode of soil formation under humid and tropical conditions, similar to those in a modern-day rain forest. This altered the normally tan, beige, and brownish strata of the Fox Hills Formation and the underlying black Pierre Shale into a thick zone of bright yellow, red, and purple strata that can be seen at the base of the Badlands in Dillon Pass (Plate 2). This period of intense soil formation has been documented from North Dakota to Colorado and is generally referred to as the Interior Zone in

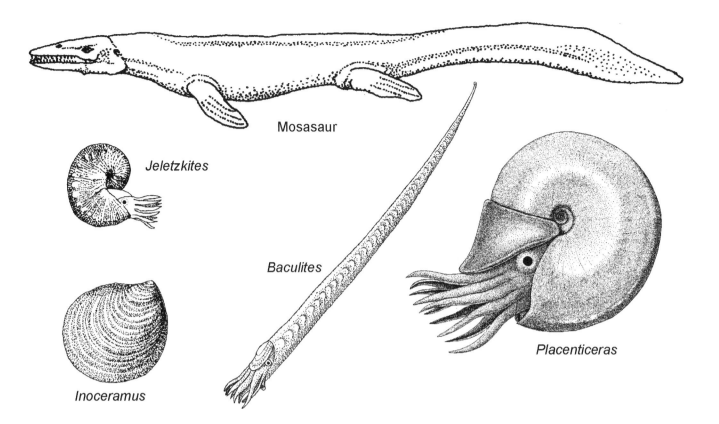

2.3. Reconstructions of some Cretaceous animals found in the vicinity of the Big Badlands. They include the 6 m long marine reptile mosasaur, the straight ammonite *Baculites* (shell length 1.2 m), and the coiled forms *Placenticeras* (80 cm shell diameter) and *Jeletzkytes* (9 cm shell height), which are related to modern squid and octopi. The clam shown is *Inoceramus sagensis* (maximum length 10 cm), named by F. B. Meek for Sage Creek. The figures are not to scale. *Inoceramus* was the most distinct and numerous bivalve of the Cretaceous Western Interior Seaway.

the Badlands (Pettyjohn, 1966; Terry, 1998). At the base of the Interior Zone, Retallack (1983b) also a described a thick paleosol that he called the Yellow Mounds paleosol after the Yellow Mounds Overlook near Dillon Pass. This paleosol is more of a deep weathering zone in the Fox Hills Formation extending down from the red Interior Zone of stacked paleosols. Bedding features and original materials that would be removed within a soil are retained in the Fox Hills rocks that are in the Yellow Mounds weathering zone. An unusual gravel, composed almost entirely of quartz or other silica-rich rocks, was formed by the intense weathering that formed the Interior Zone soils. These gravels are scattered across the unconformity surface. It was not until the late Eocene (approximately 37 Ma; Fig. 2.2) that deposition resumed in the Badlands region. From this point forward, the sedimentary deposits that accumulated to create the various layers in the Badlands were totally terrestrial in origin, forming in rivers and lakes and as windblown accumulations of dust.

The Chamberlain Pass Formation is the basal unit of the White River Group and represents the return of deposition in this region 30 million years after the retreat of the Western Interior Seaway. The Chamberlain Pass Formation is recognized by its red mudstone and brilliant white lenticular bodies of sandstone directly overlying the bright yellow and red Interior Zone (Plate 2). The Chamberlain Pass Formation was deposited on the relatively flat surface of the unconformity. Although this unit is thin in the Badlands, it becomes progressively thicker, up to tens of meters, to the southwest in Nebraska. The age of the Chamberlain Pass Formation is constrained to the late middle Eocene on the basis of the rare occurrence of vertebrate fossils of the latest Duchesnean or earliest Chadronian land mammal age in South Dakota and Nebraska (Clark, Beerbower, and Kietzke, 1967; Terry, 1998). The formation is named for its type section at Chamberlain Pass, a gap in the Badland Wall near Kadoka, South Dakota (Evans and Terry, 1994; Fig. P.1). The Chamberlain Pass Formation is best seen in the Dillon Pass area, along the Sage Creek Rim Road and just west of the town of Interior along Highway 44 (Fig. P.1).

The Chadron Formation was named for the town of Chadron, Nebraska, by Darton (1899). Darton never named a type section for the Chadron Formation, but most workers consider the outcrops in northwest Nebraska to be the type area. Darton (1899) did recognize the formation's presence in the South Dakota Badlands, replacing the old term, *Titanotherium* bed, with a lithologic unit. The mammalian fauna is a typical Chadronian fauna characterized in part by the presence of fossils of the huge brontotheres. The Chadron

Formation is late Eocene in age, as based on a combination of vertebrate fossil data, paleomagnetic correlations, and radiometric dating of tuffs outside the South Dakota Badlands (Prothero and Emry, 2004).

The Chadron Formation in the Badlands is dominated by greenish-gray massive claystone beds and minor amounts of lenticular and sheet sandstones, conglomerates, and thin limestone sheets and stringers. The Chadron Formation includes the Ahearn, Crazy Johnson, and Peanut Peak members, all named by Clark (1954). The two older members are constrained to limited exposures west–southwest of Sheep Mountain Table. The Ahearn and Crazy Johnson members (Plate 3) filled a west-to-east paleovalley, called the Red River valley, which was cut 27.4 m below the surface of the Interior Zone (Clark, 1937; Clark, Beerbower, and Kietzke, 1967). The basal Ahearn Member rests directly on black Cretaceous Pierre Shale that was unaltered by the Interior Zone episode of soil formation. The Ahearn Member is composed of coarse basal gravel beds, multicolored sandstone beds, and upper tan to orange mudstones locally cut by lenticular sandstones (Clark, Beerbower, and Kietzke, 1967). The overlying Crazy Johnson Member includes a combination of numerous sandstone channels and sheets separated by olive mudstone and claystone beds (Clark, Beerbower, and Kietzke, 1967). The Peanut Peak Member is the most widespread unit seen throughout the Badlands. It directly overlies the Chamberlain Pass Formation and includes massive beds of olive-gray claystone that display haystack-mound morphology and a thick popcorn-textured weathered surface (Plate 4, bottom). Only the Peanut Peak Member of the Chadron Formation is recognized regionally (Terry, 1998). The three-part subdivision of the Chadron Formation is restricted to the Red River paleovalley and has limitations for regional stratigraphy (Harksen and Macdonald, 1969). For example, an unnamed fluvial sequence of channel sandstones, overbank sheet sands, and thin claystone sheets occur in the upper Chadron Formation in the vicinity of the town of Scenic and the Cain Creek drainage.

The Brule Formation directly overlies the Chadron Formation and is composed of alternating mudstone, muddy sandstone, and siltstone intervals with rare thick to thin limestone sheets. It is easily recognized as the brownish-tan, cliff-and-spire-supporting strata with prominent reddish-brown striping in its lower part and steep near-vertical cliffs in its upper part (Plate 4). The formation is divided into the lower Scenic Member and higher Poleslide Member (Fig. 2.1). The Scenic Member is composed of two thick intervals of massive, typically nodular mudstones separated by thick and widespread blankets of light gray well-bedded muddy sandstones. These sandstone blankets contain interbedded thin brown to red mudstone and claystone sheets exposed as the middle striped zone of the Badlands Wall. The overlying Poleslide Member is a more uniform tan to light gray and is composed mainly of thick siltstone and secondarily of rare mudstone beds and scattered, thick, well-bedded sandstone blankets. Sandstone blankets are widespread sequences of stacked sandstone sheets that can extend for 80 km along the Badlands Wall. Vertebrate fossils and paleomagnetic correlations (Prothero, 1996; Prothero and Emry, 2004) indicate that the Brule Formation is early Oligocene in age. The Brule Formation is capped by the Rockyford Ash of the Sharps Formation on Sheep Mountain Table (Nicknish and Macdonald, 1962).

The Brule Formation was named by Darton in 1899 with no designated type section, though most geologists recognize the Toadstool Geologic Park area of northwest Nebraska to be the type area. As with the Chadron Formation, Darton recognized the presence of the Brule Formation in the South Dakota Badlands in his 1899 paper. The Scenic and Poleslide members were named by Bump (1956) for the town of Scenic and for Poleslide Canyon on the south side of Sheep Mountain Table. The Poleslide refers to a canyon where poles, cut from the cedars on Sheep Mountain Table, were slid down to the lower flats. Poleslide Canyon and outcrops near the town of Scenic were designated by Bump (1956) as type sections for the members. The detailed stratigraphy of the Brule Formation in the North Unit of Badlands National Park has been a major focus of study for over a decade and a half by Emmett Evanoff (Benton et al., 2007, 2009). Previous studies on the detailed stratigraphy of the Brule Formation in the Badlands were done by Wortman (1893), Sinclair (1921b), Wanless (1923), Bump (1956), and Clark, Beerbower, and Kietzke (1967).

The basal contact of the Scenic Member is where green-gray claystone beds of the upper Chadron Formation abruptly change into brown mudstone beds of the basal Scenic. Claystones tend to have abundant clay with little or no silt grains and have a smooth, nongritty texture. Mudstones contain roughly equal amounts of silt and clay and thus have a slightly gritty texture. The contact also has scattered thin white limestone or silica stringers at the top of the Chadron rocks. The Chadron/Scenic contact is not flat but was an erosional contact with high relief, as seen by thickness variations in Scenic Member. The Scenic Member ranges from a minimum total thickness of 39.5 m to a maximum of 76.6 m in the North Unit of Badlands National Park. This variation in thickness is largely caused by ancient topographic relief on the basal contact of the Scenic Member (Fig. 2.4).

If the uppermost Chadron and the lowest Scenic rocks are within normal polarity zones, as indicated by Tedford et al. (1996) and Prothero and Whittlesey (1998), then an estimate of the amount of time missing in the Chadron/Brule erosional contact can be made. The Chadron rocks are latest Eocene in age and the Brule rocks are early Oligocene (Tedford et al., 1996; Prothero and Emry, 2004). The youngest late Eocene normal polarity zone is zone C15n (Fig. 2.2). The oldest early Oligocene normal polarity zone is C13n, which had a duration of 1.05 million years (between 34.8 Ma and 33.75 Ma; see Luterbacher et al., 2004). Thus, at least 1.05 million years is missing at this contact. Because the Eocene–Oligocene boundary occurs within C13r (Luterbacher et al., 2004), the complete rock and fossil record across the Eocene–Oligocene boundary is not preserved in Badlands National Park.

The Scenic Member has three subdivisions: the brown mudstone beds of the lower Scenic, the muddy sandstone beds middle of the middle Scenic, and the gray to brown mudstone beds of the upper Scenic (Plate 4). The lower Scenic sequence is characterized by thick beds of brown mudstone. The mudstones are typically calcareous and can contain globular carbonate nodules, especially in the upper part of the unit, where they are typically arranged into nodule layers. The lower Scenic brown mudstone beds outcrop as low, rounded hills below the steep cliffs of the middle Scenic muddy sandstone beds. Channel sandstones are rare in the lower brown mudstone beds and tend to be isolated ribbons. Thin but locally widespread limestone sheets containing fossil freshwater snail shells, ostracod carapaces, and algal fossils are scattered in the lower Scenic brown mudstones. The lower Scenic brown mudstone unit is one of the most widespread units in the North Unit of Badlands National Park, blanketing the basal Brule contact. Its thickness ranges from as little as 12.6 m on Chadron highs to as much as 65.3 m in the base of the large paleovalleys cut into the Chadron Formation (Fig. 2.4). Fossils are scattered throughout this unit as discrete bone beds, such as the Big Pig Dig quarry. The lower brown mudstone beds were well known to the early vertebrate paleontologists, who called it the red layer (Wortman, 1893; Sinclair, 1921b) for the red hematite staining of the fossils in this unit, or the Lower Nodules or Lower Nodular Zone for the carbonate nodules (Wanless, 1921, 1923; Bump, 1956).

The middle Scenic is characterized by multiple stacks of very thick, widespread, light gray muddy sandstone blankets that are separated by thin to thick, brown or red mudstone and claystone beds (Plate 4). These sandstone blankets are well bedded and are composed of thick to thin sheets of alternating very light gray, very-fine-grained sandstone interlayered with light brownish-gray muddy sandstone. The individual thin sandstone sheets locally surround scattered

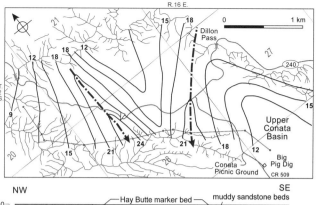

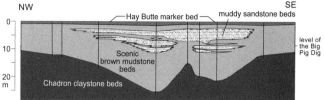

2.4. Topography cut into the Chadron Formation in the Upper Conata Basin, Badlands National Park. *Top,* Map showing thicknesses between the base of the Scenic Member of the Brule Formation and the Hay Butte marker. The contours are lines of equal thicknesses (in meters) called isopachs. Isopachs that are low indicate Chadron hills and those that are high values indicate valleys, with dot–dash arrows showing the drainages at the start of Scenic deposition. The thin black lines show ridge crests along the Badlands Wall. The arrow along State Highway 240 shows the viewpoint location and direction of Plate 4. *Bottom,* Cross section showing the thicknesses of various rock types in the lower Scenic below the Hay Butte marker along the badland ridge extending northeast from the Big Pig Dig. The location of the cross section is shown on the map with the dashed line. The dots along the cross section line on the map show the location of the sections shown by the vertical lines on the cross section. The cross section has a vertical exaggeration of 32.5×.

thick lenticular sandstone ribbons. The individual blankets are separated by thin to thick brown to red mudstone and claystone sheets that rarely extend for more than a kilometer along the outcrop. Bedding is not obvious in the mudstone and claystone sheets, but Clark, Beerbower, and Kietzke (1967:78) report that the claystone beds show thin horizontal bedding on fresh-cut surfaces when coated with kerosene. Together, the white, well-bedded sandstone blankets that are separated by brown to red mudstone beds give the middle Scenic outcrops a distinct banded pattern on steep, step-like slopes. Fossils are rare in the middle Scenic sandstone blankets, but they do occur in bone beds at a few localities at the bottoms of a sandstone ribbons and in carbonate-rich mudstones below the middle Scenic mudstone marker beds.

Two widespread mudstone marker units occur in the middle Scenic, the Hay Butte and Saddle Pass markers. The lower and most distinct of the widespread mudstone marker units is the Hay Butte marker. This informal stratigraphic unit is a thick complex of brown to gray mudstone beds that combine to form a unit averaging 1.9 m thick. The Hay Butte marker extends all along the Badlands Wall from

2.5. Hay Butte and Saddle Pass marker beds within the middle Scenic Member sandstone beds at the Conata Picnic Ground, Badlands National Park. Photo by the authors.

Sheep Mountain Table to the northeast edge of Badlands National Park. The Hay Butte marker is one of the thickest and most pronounced brown bands on the Badlands Wall (Fig. 2.5). In the middle of the Hay Butte marker is an individual mudstone bed that contains abundant euhedral, six-sided biotite crystals as large as the size of medium sand. Other euhedral crystals of apatite, monazite, and rare sphene occur in this mudstone bed. This crystal-rich mudstone bed is most obvious on the west side of the Badlands National Park around Sheep Mountain Table but becomes indistinct on the eastern margin of the park. This bed is a volcanic tuff and represents a timeline within the Hay Butte marker. The older topography that was cut into the Chadron Formation was buried by the time of the deposition of the Hay Butte marker. The marker rests on sandstone blankets of the lower middle Scenic in the Brule paleovalleys and on the lower brown mudstones of lower Scenic on the Chadron highs (Fig. 2.4). Because the beds of the Hay Butte marker were deposited essentially on a level plain, the thickness of the Scenic rocks between the Chadron/Scenic contact and the Hay Butte marker records the topographic relief cut into the Chadron Formation. The Scenic rocks below the Hay Butte marker range in thickness from a minimum of 10.7 m to a maximum of 38.1 m. Therefore, the erosional contact cut into the Chadron Formation had a local maximum relief of 27.4 m. The Hay Butte marker is named for the long butte

just west of the southeast branch of Sage Creek and west of the Pinnacles. The type section of the Hay Butte is in the type section of the Scenic Member, just south of the town of Scenic (Fig. 2.6).

The second marker bed in the middle Scenic is the Saddle Pass marker. This informal stratigraphic unit is composed of a thick, lower, brown mudstone bed and a thinner upper red to brown mudstone or claystone bed. Together they average 1.5 m thick, though the marker thins to a minimum of 0.6 m to the west and has a maximum thickness of 2.2 m in the east. East of the Wanless Buttes (Fig. P.1), the Saddle Pass marker always overlies a widespread carbonate nodule bed. West of the Wanless Buttes, the Saddle Pass marker thins and the underlying carbonate nodules disappear. However, like the Hay Butte marker, the Saddle Pass marker occurs along the entire Badlands Wall in the North Unit of Badlands National Park, typically appearing as a persistent brown band occurring on average 11 m above the Hay Butte marker (Fig. 2.5). The Saddle Pass marker is named for Saddle Pass, a notch through the Badlands Wall with a steep trail (the Saddle Pass Trail) north of the Loop Road about 2.5 km northwest of the Ben Reifel Visitor Center. Its type section is located adjacent to the Saddle Pass trail (Fig. 2.7).

The sequence of rocks in the upper Scenic is characterized by a series of gray to brown mudstone beds that weather into smooth buff-colored slopes above the steplike cliffs of

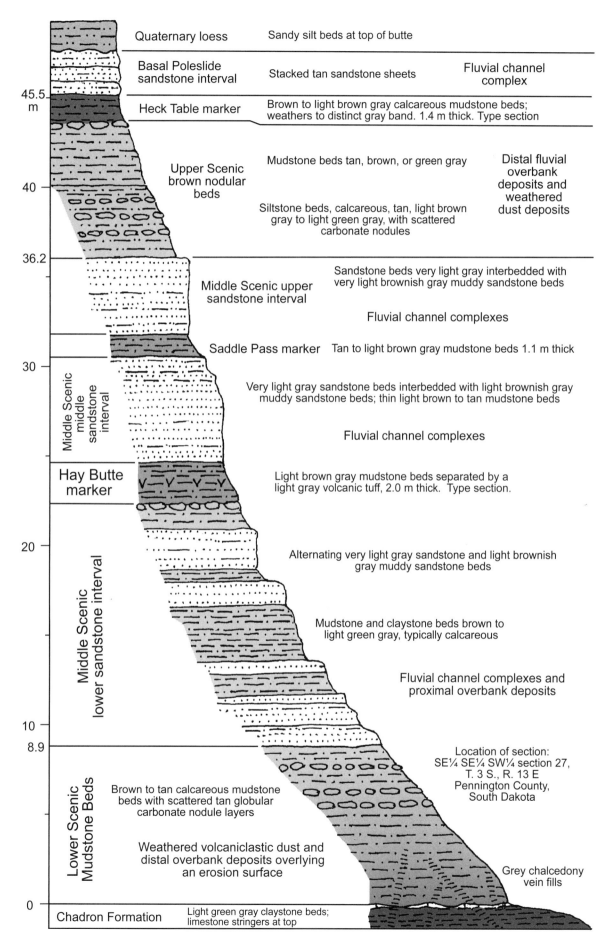

2.6. Type section of the Scenic Member of the Brule Formation at a location just south of Scenic, South Dakota, as given by Bump (1956). The section includes the type sections for the Hay Butte and Heck Table markers of the Scenic Member.

Quaternary loess — Sandy silt beds at top of butte

Basal Poleslide sandstone interval — Stacked tan sandstone sheets — Fluvial channel complex

Heck Table marker — Brown to light brown gray calcareous mudstone beds; weathers to distinct gray band. 1.4 m thick. Type section

Upper Scenic brown nodular beds — Mudstone beds tan, brown, or green gray

Siltstone beds, calcareous, tan, light brown gray to light green gray, with scattered carbonate nodules

Distal fluvial overbank deposits and weathered dust deposits

Middle Scenic upper sandstone interval — Sandstone beds very light gray interbedded with very light brownish gray muddy sandstone beds

Fluvial channel complexes

Saddle Pass marker — Tan to light brown gray mudstone beds 1.1 m thick

Middle Scenic middle sandstone interval — Very light gray sandstone beds interbedded with light brownish gray muddy sandstone beds; thin light brown to tan mudstone beds

Fluvial channel complexes

Hay Butte marker — Light brown gray mudstone beds separated by a light gray volcanic tuff, 2.0 m thick. Type section.

Middle Scenic lower sandstone interval — Alternating very light gray sandstone and light brownish gray muddy sandstone beds

Mudstone and claystone beds brown to light green gray, typically calcareous

Fluvial channel complexes and proximal overbank deposits

Location of section: SE¼ SE¼ SW¼ section 27, T. 3 S., R. 13 E Pennington County, South Dakota

Lower Scenic Mudstone Beds — Brown to tan calcareous mudstone beds with scattered tan globular carbonate nodule layers

Weathered volcaniclastic dust and distal overbank deposits overlying an erosion surface

Grey chalcedony vein fills

Chadron Formation — Light green gray claystone beds; limestone stringers at top the middle Scenic rocks (Plate 4). West of the long ridge extending north of the old town of Imlay (the Imlay Table, Fig. P.1), the upper mudstone beds are brown to tan and contain abundant globular carbonate nodules. Fossils are fairly abundant in these nodular mudstones and were called the Upper Nodular Zone by Bump (1956). These western mudstone beds are capped by a third marker unit, the Heck Table marker, which is characterized by a thick basal brown mudstone bed capped by a claystone bed. Together these two beds make a marker bed with an average thickness of 1.4 m and weather into a prominent gray band that occurs at the top of the Scenic Member at the type section (Bump, 1956). The Heck Table marker is named for a butte southeast of Scenic, and its type section is at the top of the Scenic type section (Fig. 2.6). The Heck Table marker disappears to the east at the north end of Imlay Table. East of Imlay Table, the upper Scenic is characterized by a three-part sequence of a lower tan mudstone interval containing abundant globular masses of carbonate, a middle gray clayey mudstone unit riddled with thin stringers and discontinuous thin ledges of limestone, and an upper sequence of medium brown (buff) mudstones below the siltstone beds of the Poleslide Member. Fossils are extremely rare in the middle gray clayey mudstone beds of the eastern upper Scenic sequence. However, fossils can occur in scattered bone beds associated with the lower tan mudstone and uppermost buff mudstone beds of the eastern upper Scenic sequence. The thickness of the upper Scenic sequence averages 9.1 m west of Imlay Table and 14.1 m east of Imlay Table. This abrupt change in thickness is because the contact between the mudstone beds of the upper Scenic to the siltstone beds of the Poleslide rises by 5 m from the Heck Table marker to a new marker bed at the base of the Poleslide Member to the east (Fig. 2.1). The uppermost Scenic buff mudstone beds in the east gradually change into siltstone beds of the lower Poleslide sequence to the west in the vicinity of the north end of Imlay Table. Siltstones are composed of mainly silt grains with less than a third clay, so they are coarser, with a gritty texture relative to mudstones.

The base of the Poleslide Member east of Imlay Table occurs at the base of a marker bed called the "Cactus Flat bentonite bed" by Stinchcomb, Terry, and Mintz (2007). This bed is a light gray structureless siltstone bed averaging 0.7 m thick and is overlain by the first massive tan siltstone beds of the lower Poleslide. Stinchcomb (2007) separated the minerals from this bed and found some euhedral biotite, feldspar, and zircon crystals suggesting its origin as a tuff. However, in the field this bed has no obvious concentration of abundant euhedral crystals, shows no significant concentration of smectite clays relative to the overlying and underlying beds, and its bottom contact is diffuse. Soil development at the Scenic/Poleslide boundary was intense and caused its euhedral minerals to be dispersed and its lower contact to mix with the underlying mudstone beds. In addition, the "Cactus Flat bentonite bed" may be the extreme distal end of a volcanic ash (like the Hay Butte tuff at the east margin of Badlands National Park), but it is more properly called the Cactus Flat marker bed. Its type section is in the Norbeck Pass area.

Like the Scenic Member, the Poleslide Member can be divided into lower, middle, and upper subdivisions, as was recognized by Bump (1956) in his description of the Poleslide type section. The lower Poleslide sequence is characterized by a combination of massive, stacked, thick siltstone beds, widespread blanket sandstones, and a thick interval of mudstone. Locally, the lower Poleslide sequence also has broad limestone sheets containing freshwater snail shells, ostracod carapaces, and algal fossils, but these limestone beds are limited in number. The detailed stratigraphy of the lower Poleslide rocks is remarkably consistent. The lower Poleslide sequence in the eastern half of Badlands National Park is composed of 12 stratigraphic units that always occur in the same order and retain their distinctive features (Fig. 2.8). These units extend from the east side of the North Unit of Badlands National Park to the Wanless Buttes, a straight-line distance of 35 km. The average thickness of the lower Poleslide sequence in these eastern outcrops is 50.9 m. Most of the outcrops of the lower Poleslide rocks in this eastern area are in steep-sided isolated buttes (Fig. 2.9). The most accessible areas to these rocks in the east are in the Cedar Pass and Norbeck Pass areas. Fossils in the lower Poleslide Member are not concentrated as distinct bone beds but are scattered throughout the siltstone beds. In this sequence, Unit 7 (Fig. 2.8) is at the boundary between the *Merycoidodon*-dominated assemblages below and the *Leptauchenia*-dominated assemblages above. This faunal boundary is considered to be the boundary between the Orellan and Whitneyan land mammal ages. To the west at Sheep Mountain Table, the sequence of units is not like those in the east. The base of lower Poleslide sequence in the west is marked by a well-bedded silty sandstone blanket directly above the Heck Table marker. Above this unit the lower Poleslide is characterized by three massive tan siltstone sequences that are separated by a brown mudstone unit and a thick sequence of well-bedded muddy sandstone sheets. The lower Poleslide sequence averages 50 m thick on Sheep Mountain Table, and the rocks are exposed in increasingly steep outcrops at the base of the table. As in the lower Poleslide sequence on the east side of Badlands National Park, the lower thick siltstone sequence contains abundant *Merycoidodon* fossils but no known *Leptauchenia* fossils. The exact stratigraphic position

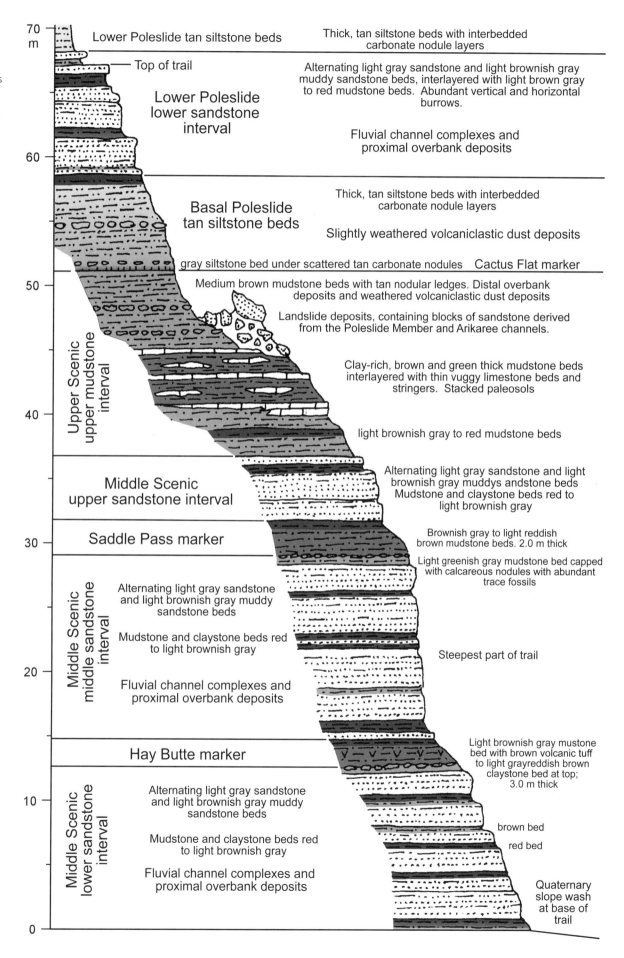

2.7. Section at Saddle Pass along the Saddle Pass trail. This includes the type section of the Saddle Pass marker.

70 m

Lower Poleslide tan siltstone beds — Thick, tan siltstone beds with interbedded carbonate nodule layers

Top of trail

Lower Poleslide lower sandstone interval

Alternating light gray sandstone and light brownish gray muddy sandstone beds, interlayered with light brown gray to red mudstone beds. Abundant vertical and horizontal burrows.

Fluvial channel complexes and proximal overbank deposits

60

Basal Poleslide tan siltstone beds

Thick, tan siltstone beds with interbedded carbonate nodule layers

Slightly weathered volcaniclastic dust deposits

gray siltstone bed under scattered tan carbonate nodules   Cactus Flat marker

50

Medium brown mudstone beds with tan nodular ledges. Distal overbank deposits and weathered volcaniclastic dust deposits

Landslide deposits, containing blocks of sandstone derived from the Poleslide Member and Arikaree channels.

Upper Scenic upper mudstone interval

Clay-rich, brown and green thick mudstone beds interlayered with thin vuggy limestone beds and stringers. Stacked paleosols

40

light brownish gray to red mudstone beds

Middle Scenic upper sandstone interval

Alternating light gray sandstone and light brownish gray muddys andstone beds Mudstone and claystone beds red to light brownish gray

Saddle Pass marker

Brownish gray to light reddish brown mudstone beds. 2.0 m thick

30

Light greenish gray mudstone bed capped with calcareous nodules with abundant trace fossils

Middle Scenic middle sandstone interval

Alternating light gray sandstone and light brownish gray muddy sandstone beds

Mudstone and claystone beds red to light brownish gray

Fluvial channel complexes and proximal overbank deposits

Steepest part of trail

20

Hay Butte marker

Light brownish gray mustone bed with brown volcanic tuff to light grayreddish brown claystone bed at top; 3.0 m thick

10

Middle Scenic lower sandstone interval

Alternating light gray sandstone and light brownish gray muddy sandstone beds

Mudstone and claystone beds red to light brownish gray

Fluvial channel complexes and proximal overbank deposits

brown bed

red bed

Quaternary slope wash at base of trail

0

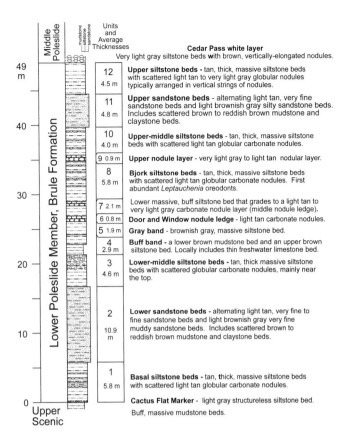

Cedar Pass white layer
Very light gray siltstone beds with brown, vertically-elongated nodules.

Unit	Avg. thickness	Description
12	4.5 m	**Upper siltstone beds** - tan, thick, massive siltstone beds with scattered light tan to very light gray globular nodules typically arranged in vertical strings of nodules.
11	4.8 m	**Upper sandstone beds** - alternating light tan, very fine sandstone beds and light brownish gray silty sandstone beds. Includes scattered brown to reddish brown mudstone and claystone beds.
10	4.0 m	**Upper-middle siltstone beds** - tan, thick, massive siltstone beds with scattered light tan globular carbonate nodules.
9	0.9 m	**Upper nodule layer** - very light gray to light tan nodular layer.
8	5.8 m	**Bjork siltstone beds** - tan, thick, massive siltstone beds with scattered light tan globular carbonate nodules. First abundant *Leptauchenia* oreodonts.
7	2.1 m	Lower massive, buff siltstone bed that grades to a light tan to very light gray carbonate nodule layer (middle nodule ledge).
6	0.8 m	**Door and Window nodule ledge** - light tan carbonate nodules.
5	1.9 m	**Gray band** - brownish gray, massive siltstone bed.
4	2.9 m	**Buff band** - a lower brown mudstone bed and an upper brown siltstone bed. Locally includes thin freshwater limestone bed.
3	4.6 m	**Lower-middle siltstone beds** - tan, thick massive siltstone beds with scattered globular carbonate nodules, mainly near the top.
2	10.9 m	**Lower sandstone beds** - alternating light tan, very fine to fine sandstone beds and light brownish gray very fine muddy sandstone beds. Includes scattered brown to reddish brown mudstone and claystone beds.
1	5.8 m	**Basal siltstone beds** - tan, thick, massive siltstone beds with scattered light tan globular carbonate nodules.
		Cactus Flat Marker - light gray structureless siltstone bed. Buff, massive mudstone beds.

2.8. Units of the lower Poleslide Member of the Brule Formation in the eastern North Unit, Badlands National Park.

of the first abundant *Leptauchenia* fossils is still unknown on the steep flanks of Sheep Mountain Table.

The middle sequence of the Poleslide Member is characterized by very thick, massive, very light gray siltstone beds that contain abundant globular carbonate nodules. These carbonate nodules are typically vertically stacked, suggesting they formed around roots and burrows. The middle Poleslide beds are exposed only on the mostly vertical flanks of Sheep Mountain Table as short sequences at the tops of isolated buttes along the Badlands Wall and in the Cedar Pass area. Like at Sheep Mountain Table, the middle Poleslide rocks are exposed almost exclusively in vertical cliffs in the Cedar Pass area. Thicknesses of the middle Poleslide siltstone beds are 32.0 m at Cedar Pass and 19.4 m on the east flank of Sheep Mountain Table. The variations of thickness of the middle Poleslide sequences between Cedar Pass and Sheep Mountain are caused by the nature of the upper Poleslide sediments. These upper Poleslide sediments are well-bedded fluvial deposits that were deposited in the Sheep Mountain Table area while massive siltstone deposits, representing wind transported dust deposits, were accumulating in the Cedar Pass area.

Overlying the lower Poleslide sequence at Cedar Pass is a prominent white bed that is silty and structureless, contains yellowish-brown vertically stacked carbonate nodules, and is 5.5 m thick (Fig. 2.10). Freeman Ward (1922:29) was the first to describe this bed as a light gray to white very fine "sandstone" that weathers into fine angular fragments that he called "road metal." This bed has since been the focus of a stratigraphic correlation problem for 90 years. Harold Wanless (1923) interpreted this bed to be volcanic ash and correlated it with the white tuff at the top of Sheep Mountain Table. Nicknish and Macdonald named the tuff on top of Sheep Mountain Table the Rockyford Ash in 1962, and Harksen and Macdonald (1969) and Harksen (1974), accepting Wanless's correlation, called the Cedar Pass white bed the Rockyford Ash. This correlation required that the overlying siltstones at Cedar Pass become part of the Sharps Formation, for the Rockyford Ash is the basal unit of the Sharps (Harksen, Macdonald, and Sevon, 1961). These upper rocks at Cedar Pass were mapped as the Sharps Formation by Raymond and King in 1976. However, Wanless in 1923 noted a problem: if these two white beds correlate, then the thickness of the upper White River beds at Sheep Mountain Table (91 m thick; Bump, 1956) was about twice the thickness of the upper White River beds at Cedar Pass (49 m thick). Wanless discussed two geologic possibilities for this discrepancy. The first was greater tectonic subsidence in the Sheep Mountain Table area; the other was a greater accumulation of sediment near the presumed source of the sediment. (Wanless favored the latter hypothesis.) A third possibility is that these two white layers are not the same bed. To test this last hypothesis, the features of the two beds have been compared, and the white bed at Cedar Pass was traced to the west along the high buttes of the Badlands Wall. First, the Rockyford ash on Sheep Mountain Table is a true volcanic tuff with a distinct base, a diffuse top, abundant crystals of euhedral biotite and hornblende, and less abundant zircon, apatite, and green clinopyroxene (E. E. Larson, pers. comm., 1998). The white bed at Cedar Pass has few euhedral biotite grains and no hornblende, and most of the heavy minerals are heavily weathered and abraded, not unlike the heavy minerals in the underlying and overlying siltstone beds. The bottom contact of the Cedar Pass bed looks sharp from a distance but is diffuse when examined up close. As the Cedar Pass white bed is traced to the west, it is still a distinct bed at the Pinnacles Overlook near the north entrance of Badlands National Park. Farther to the west at the Wanless Buttes, the 12 recognizable units of the lower Poleslide are overlain by a thick sequence of massive very light gray siltstone beds, identical to the beds of the middle Poleslide on Sheep Mountain Table (Fig. 2.1). No distinct white bed occurs at the base of these siltstone beds at either the Wanless Buttes or Sheep Mountain Table. The White River tuffs came from sources to the west (Larson

2.9. The Scenic/Poleslide contact as exposed in the butte just east of the Headquarters, Badlands National Park. The contact is between the last mudstone unit of the Scenic Member and the first siltstone unit of the Poleslide Member. The numbers refer to the units within the lower Poleslide Member (Fig. 2.8). Photo by the authors.

and Evanoff, 1998), so the disappearance of a thick volcanic tuff to the west is quite unlikely. This thick white bed is an important and distinctive marker at the base of the middle Poleslide sequence in the east half of the North Unit of Badlands National Park and is an informal stratigraphic unit named the Cedar Pass white layer with its type section at Cedar Pass. In addition, the massive siltstones above the Cedar Pass white layer at Cedar Pass are part of the Poleslide Member, not part of the Sharps Formation.

The upper sequence of the Poleslide Member is characterized by alternating light brown mudstone and light gray siltstone sheets that occur near the top of Sheep Mountain Table. These rocks erode into ledgy vertical cliffs and spires characterized by Wanless (1923:230) as "organ pipe weathering" (Plate 5). Channels with broad lenticular cross sections and filled with coarse sandstone occur in these beds on the south end of Sheep Mountain Table. These channel deposits contain the artiodactyl *Protoceras* (Harksen, 1974). The upper Poleslide sequence is 20.7 m thick on Sheep Mountain Table and is capped by the Rockyford Ash at the base of the Sharps Formation. Upper Poleslide beds occurs 32 m above the Cedar Pass white layer at Millard Ridge on the east side

of the Badlands National Park. Here a widespread fluvial sandstone blanket occurs within typical Poleslide massive siltstone beds. The bottom of this sandstone blanket is the base of the upper Poleslide Member in the Cedar Pass area.

The Sharps Formation conformably overlies the White River Group in the South Dakota Badlands. The Sharps Formation was named by Harksen, Macdonald, and Sevon (1961) for a thick sequence of massive, pinkish-tan, sandy siltstone beds containing globular carbonate nodules. The total thickness of the Sharps Formation at its type section near Sharps Corner, Shannon County, South Dakota, is 120.4 m. The basal bed of the Sharps Formation is a white tuff called the Rockyford Ash, named by Nicknish and Macdonald (1962) for outcrops near Rocky Ford in Shannon County, South Dakota. The type section is on the top of Sheep Mountain Table, where the 9.1 m thick tuff is capped by the basal siltstones of the Sharps. The fauna of the Sharps (Harksen, Macdonald, and Sevon, 1961) is early Arikareean in age.

The high buttes along the Badlands Wall from Millard Ridge to the Wanless Buttes are capped by coarse conglomerates representing basal deposits of paleovalleys filled with rocks containing fossils of middle Arikareean age (Parris and

2.10. Views of the Cedar Pass white layer. (A) Close-up of the Cedar Pass white layer on the east side of Cedar Pass. The thickness of the marker bed in this area is 5.5 m. (B) View from the north-northeast of the butte west of Cedar Pass showing the distinct band of the Cedar Pass white layer. Photos by the authors.

Green, 1969; Harksen, 1974). A typical valley fill has a basal conglomeratic coarse-grained sandstone sheet with pebbles of siltstone and sandstone, overlain by alternating horizontal beds of light gray sandy siltstone and sandstone with rare mudstone and claystone beds. On Millard Ridge (east of Cedar Pass), the cumulative thickness of sandstone beds comprises 45 percent of one of the thick valley-fill sequences. These channel fills cut into the middle and lower Poleslide siltstones to a variety of levels (Fig. 2.11). Most cut down to about the level of the Cedar Pass white layer, but some were incised to 13 m below the base of the Cedar Pass white layer

(Harksen, 1974), while others have a base as high as 32 m above the base of the Cedar Pass white layer. The fauna of these channels is Arikareean in age (Parris and Green, 1969) and includes the late Arikareean giant entelodont *Daeodon* (previously known as *Dinohyus*). Late Arikareean faunas are late Oligocene in age (Tedford et al., 2004) and are younger than the early Arikareean (late early Oligocene) faunas of the Sharps Formation. Though informally called the Sharps Channels, the abundance of sandstone beds and a late Arikareean fauna in these paleovalley fills suggest they are more properly part of the Arikaree Group. These paleovalley fills

2.11. Arikaree sedimentary rocks filling two paleovalley sequences cut into lower Poleslide Member siltstone beds on a butte west of the east end of the Medicine Root Trail, North Unit, Badlands National Park. Here the paleovalleys cut through Unit 11 down to Unit 8 of the lower Poleslide Member. Photo by the authors.

require additional mapping and faunal analysis, but their position in vertical cliffs and on top of high buttes makes them difficult to study.

SEDIMENTOLOGY: DEPOSITIONAL ORIGINS OF THE WHITE RIVER GROUP

Sedimentary rocks form from a series of processes acting at different times. In the case of the rocks of the White River Group, the original sediment was deposited by rivers and streams (fluvial processes), in lakes (lacustrine processes), or by wind (eolian processes). Once the sediment was deposited and exposed at a land surface, weathering occurred from rain and air. Burrowing animals and plant roots churned up and chemically altered the sediment to form a soil that would be buried by the next depositional event. Once the sediment was buried, it was compacted, and the pore space between the grains was filled with mineral cements to become a sedimentary rock, a process known as lithification. The cementation of sediment into rock can occur early or it can take millions of years. In the end, sediments become rocks that are later exposed by erosion for geologists to study.

The sediments of the White River Group came from two sources. One source was the Black Hills, and the other was volcanic eruptions far to the west. The local epiclastic sediment is indicated by such minerals as quartz, hornblende, tourmaline, epidote, and the feldspars microcline, orthoclase, and plagioclase. Unique mineral assemblages in White River sandstone stream deposits reflect the mineralogy of the rocks at the headwaters of the ancient streams in different parts of the Black Hills. Streams from the southern Black Hills transported sand containing a large amount of garnet and black tourmaline, as well as rare orange sphene and brown hornblende (Seefeldt and Glerup, 1958). Streams from the northern Black Hills transported sand containing abundant magnetite and greenish hornblende, less abundant lemon-yellow sphene, and rare gold (Ritter and Wolfe, 1958). These epiclastic minerals were mixed with volcanic materials, primarily volcanic glass and euhedral crystals like those found in the tuff beds.

The White River Group was deposited during one of the greatest episodes of volcanic eruption in the Cenozoic, which geologists call the ignimbrite flare-up. Huge eruptions of volcanic ash occurred from the degassing of magma chambers and the collapse of the overlying rocks, thus forming a caldera, a typically circular volcanic basin many kilometers in diameter. Some of these eruptions were truly enormous (Best et al., 1989), with eruptions producing 1000 to 5000 km³ of volcanic materials. (For comparison, the Mount St. Helens eruption of 1980 erupted 1 km³ of material.) Larson and Evanoff (1998) studied the mineralogy, geochemistry, and age relations of White River tuffs in Wyoming, Nebraska, and Colorado, and they determined that the vast majority of the volcanic sediments (called volcaniclasts) came from volcanic eruptions in the Great Basin of modern Nevada and Utah (Fig. 2.12). The White River Group has about 40

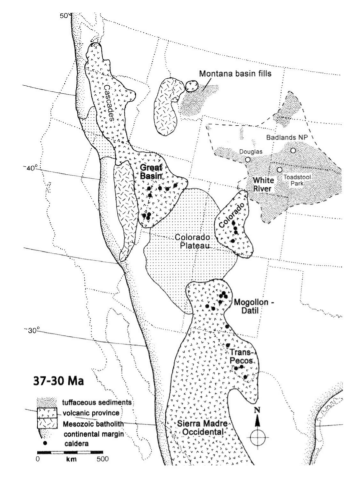

2.12. Palinspastic reconstruction of the west-central portion of North America showing the active volcanic provinces during the deposition of the White River sequence. Modified from Larson and Evanoff (1998).

percent volcaniclasts at it base and as much as 80 percent volcaniclasts at its top (Swinehart et al., 1985). Given an average of about 60 percent volcaniclasts for the White River rocks, their original distribution over ~400,000 km² of North Dakota, South Dakota, Nebraska, Colorado, and Wyoming, and the regional variation of their thicknesses, the total amount of volcaniclastic sediment in the White River Group is about 25,000 km³ (Larson and Evanoff, 1998). It was truly an ashy, dusty time.

How did this volcanic material get transported to the Badlands? Most of it did not come in from direct ashfalls through the atmosphere. Most of the volcaniclasts were erupted by volcanoes and carried by winds as ash to sites west of the Badlands, where winds picked up the ash and local epiclasts to make dust that fell to the east (Fig. 2.13A). In the earlier and wetter times of White River deposition in the Badlands, this dust fell and was weathered into clay or was picked up by streams and deposited as fine-grained overbank deposits during floods. Evidence of this process is preserved as ash shards mixed with muds in the overbank deposits of rivers

in the White River Group (Evanoff, 1990b; Lukens, 2013; Lukens and Terry, 2013). Eventually the climate dried out and the dust accumulated into thick dust deposits, called loess. Occasionally there were enormous eruptions that spread ash clouds thousands of kilometers downwind over the Great Plains, where the ash fell onto the land, forming volcanic ash deposits (Fig. 2.13B). The South Dakota Badlands were far downwind from most of the volcanic eruptions, and as a result there are only two distinctive tuffs known in the White River Group in the South Dakota Badlands.

Much of the original sediment of the Chamberlain Pass Formation, Chadron Formation, and Scenic Member of the Brule Formation was deposited by rivers and streams (fluvial processes). Approximately a third of sediments that made up the Poleslide Member of the Brule Formation are fluvial deposits. The coarse-grained channel deposits occur in two groups: as long, linear deposit with lenticular cross sections called sandstone ribbons and as sandstone blankets that are made of stacked thin sandstone sheets. The deposits of ancient stream channels rarely preserve the form and distribution of an individual channel in which water flowed at any one time. Over time, the channel swings from side to side, modifying preexisting channel deposits and creating a channel belt, or a broad band of internested individual channel deposits. For example, if the channel has limited movement from side to side, it will have a narrow channel belt that will be preserved as a ribbon sandstone body. The outcrop of a channel deposit that is preserved today actually reflects the geometry of the channel belt, not the individual channel in which water flowed at any one time. The two types of channel deposits in the White River Group reflect two kinds of fluvial depositional environments that are primarily related to the permanence of the flowing water and to the interconnectedness of the fluvial system.

Isolated sandstone ribbon channel deposits are characteristic of the Chadron Formation, and they are locally common in the Poleslide Member of the South Unit of Badlands National Park (Fig. 2.14A). Sandstone ribbons also rarely occur in the lower mudstones of the Scenic Member (Fig. 2.14B). They include a central ribbon filled with basal gravels, coarse sandstone in the middle, and muddy sandstones at the top. The lower coarse sandstones in the ribbons contain thick cross bed sets indicating stream flow from the west or northwest toward the east or southeast. These channel deposits typically are attached to one or more "wings," or tabular sandstone and muddy sandstone sheets that laterally extend from the tops of the ribbons and pinch out into the surrounding distal, well-bedded mudstones. Water flowed in the central channel throughout much of the year, but during floods, the rising waters dumped sand into broad sheets lateral to the channel to form splay deposits, and mud was deposited far from the channel as overbank deposits. Many of the channels have coarse-grained sediment derived from the Black Hills, but some isolated sandstone ribbons contain mudstone pebbles and fine-grained sand derived from sources within the White River Group. These sandstone ribbons containing intraformational clasts represent streams draining the Great Plains that were tributaries to the main trunk streams that carried sediment from the Black Hills. The areal geometry of the ribbons is typically a broken-stick pattern of long, straight reaches and short, abrupt bends. The bedding in the straight reaches is typically a series of thick internested cuts and fills, with the total thickness and width of the ribbon along the straight reaches less than in the bends. The thick and wide lenticular cross section of the ribbon at a bend contains lateral accretion deposits inclined toward the outside curve of the bend (Fig. 2.14B). These lateral accretion deposits represent the building of a point bar on the inside of the bend. The coarser basal gravels and sandstones of the ribbons are more resistant to erosion than the finer-grained upper channel and muddy overbank deposits, so after erosion the basal deposits will be preserved as long, sinuous ridges in badlands outcrops. The isolated sandstone ribbons contain a variety of aquatic animal remains, including the bones of freshwater fish, turtles, and alligators and the shells of freshwater unionid mussels. Locally, the channel deposits will contain bone beds in isolated pockets at the base of the channel where bones were concentrated as large sediment clasts.

The fluvial deposits of the middle Scenic Member and the Poleslide Member of the Brule Formation are dominated by sandstone blanket deposits. These sandstone blankets are thick sequences of interbedded sandstone and muddy sandstone sheets. These sandstone blankets are separated by wide, relatively thin sheets of mudstone or claystone that do not typically extend more than a kilometer along the Badlands outcrops. These muddy sediments were deposited as ponded overbank deposits in lows on the sand sheet surfaces, as indicated by the remnants of laminated bedding within them (Clark, Beerbower, and Kietzke, 1967). The alternating light gray sandstone blankets and brown to red mudstone/claystone sheets produce the striking banded outcrops of the middle Scenic Member in Badlands National Park (Plate 4, and the rocks between the marker beds in Fig. 2.5). Sandstone ribbons are scattered within these sandstone blankets, but they have different bedding features than those of the isolated sandstone ribbons. They are typically broad ribbons with very thin beds of sandstone that are horizontal or inclined at a low angle. They show abundant parting lineation, long linear streaks of grains on the bedding surfaces, and rare cross beds. The sandstone blankets are typically stacked, and the lateral

A. Volcanic eruptions in Nevada and wind transportation to Great Plains

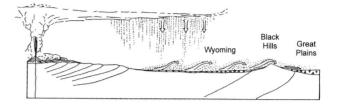

Wyoming   Black Hills   Great Plains

B. Caldera eruption in Nevada, fall of volcanic ash in the Great Plains

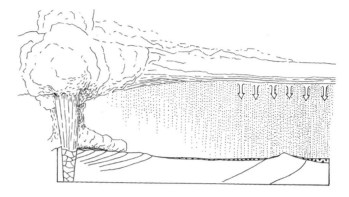

2.13. Origin of the White River sediments. (A) Normal deposition during White River time. Fine-grained volcaniclastic sediment originates from volcanic eruptions located in modern-day Nevada and Utah. Surface winds transport the volcaniclastic sediments as dust that mixes with fine-grained epiclastic material derived from local mountains. The dust falls into the Great Plains, where it either is reworked by stream or accumulates as loess. (B) An eruption of a huge caldera in modern Nevada or western Utah pushes fine-grained volcaniclastic material into the upper atmosphere. Winds carry this volcanic ash to the Great Plains, where it is deposited as a volcanic ash.

extent of the stacked blankets can be quite remarkable. The middle Scenic blankets occur in three stacked sequences separated by the Hay Butte and Saddle Pass marker beds. The lower stack of sandstone blankets below the Hay Butte marker is limited by the distribution of basal Scenic paleovalleys, but the middle and upper stacks of sandstone blankets extend along the entire Badlands Wall of the North Unit of Badlands National Park, a north–south distance of 22 km and an east–west distance of 58 km. The edge of the upper stack of sandstone blankets above the Saddle Pass marker occurs in the vicinity of the Ben Reifel Visitor Center, where the sandstone blanket thins and grades into tan mudstone beds that contain massive carbonate nodules. The fossil site at the Ben Reifel Visitor Center (the Saber Site) is in these tan mudstone beds.

The well-bedded sandstone beds in the lower and upper Poleslide sequences are similar stacked sandstone blankets but contain increasing amounts of siltstone sheets between the sandstones sheets in progressively higher fluvial deposits. Bone accumulations are rare in the Poleslide sandstone blanket deposits and occur only in the coarse sandstone ribbons

2.14. Cross sections of ribbon sandstone bodies in the White River Group. (A) Middle Poleslide channel-belt deposit showing the coarse basal conglomerate beds and the upper sandstone to muddy sandstone beds in the background as in the midground. The coarse basal conglomerate is filled with sediment derived from the Black Hills, indicating that the stream flowed from the Black Hills. Medium- to small-scale cross beds in these deposits indicate that the flow was toward the viewer. (B) Cross-sectional view of a channel-belt deposit in the lower Scenic brown mudstone interval. The outcrop was at a bend in the channel belt, so there are large, low-angle cross beds that represent lateral accretion deposits in a point bar. The channel migrated to the outside of the bed (to the left), and the point bar expanded toward the left. The final channel in the channel belt is represented by the sandstone fill on the left side of the channel. This stream channel occasionally flooded, producing sheet sands lateral to the top of the channel-belt deposits. These lateral sheet sandstones are labeled "wings" in the diagram. This channel-belt deposit has no basal coarse conglomerate and is filled with sandstone and muddy sandstone beds, indicating that this channel had its headwaters in the Great Plains, not the Black Hills. Photos by the authors.

within the sandstone blankets (for example, the *Protoceras* channels of Sheep Mountain Table).

The two kinds of fluvial deposits found in the White River Group in the Badlands reflect the kinds of sediments that are transported and the persistence of the stream flow in the rivers and streams. All of the streams that deposited the White River fluvial sediments were mixed-load streams, or streams that carried both gravel and coarse-grained sand as a bed load, and fine silt and clay grains as suspended load. The source of the coarse bed-load material was mainly from the Black Hills. The muddy suspended load was from erosion of older clay-rich rocks such as the Pierre Shale or, in the case of the lower brown mudstones of the Scenic Member, the claystone hills of the eroded Chadron Formation. Most of the suspended load in the streams of the White River Group was from dust derived from volcanic and epiclastic sources to the west.

The isolated ribbon sandstones of the Chadron Formation, the lower Scenic brown mudstones, and the *Protoceras* channels were deposited by rivers and streams with water flow throughout the year (perennial streams). This is indicated by their relatively coarse-grained, fairly well-sorted channel deposits and associated fossils of aquatic organisms. The freshwater unionid mussels in these channel deposits indicate that these streams had constant flow throughout the year, had well-oxygenated waters, and were part of an extensive interconnected river system. Freshwater mussels are dispersed by fish that carry their larvae as parasites attached to their gills and fins. The clam larvae drop off the fish at a certain point in their development and grow to become filter feeders on the stream floor. Fish will not inhabit streams that are dry for much of the year and cannot get to water bodies not connected by streams. Therefore, freshwater mussels will not live in intermittent streams or isolated water bodies, such as groundwater-fed lakes and ponds. The east-to-southeast current flow directions and the alligator fossils in the channels of the Chadron Formation indicate connections with an ancient Mississippi River system.

The fluvial deposits represented by the sandstone blankets were deposited in a completely different kind of fluvial system. The coarser sediment in the sandstone blankets are derived from the Black Hills (Clark, Beerbower, and Kietzke, 1967), so they had sources similar to those of the isolated sandstone ribbons. However, while the streams of the isolated ribbons transferred sediment along discrete channels that had flow throughout the year, the streams of the sandstone blankets flowed only during floods, eventually spreading the water and the sediment in extremely broad sheets across the plains. The alternating wet and dry conditions for these deposits are indicated by features in the interbedded mudstone and claystone sheets. Some of these mudstone deposits have desiccation cracks (mud cracks) that show drying of originally wet muddy sediment. Some discrete sandstone ribbons occurred within the sandstone blankets, and these have been traced for many kilometers in the Badlands (Ritter and Wolfe, 1958). These broad, thin-bedded ribbons with abundant parting lineation structures are typical of flash flow in ephemeral streams. Ephemeral streams are those that are dry for most of the year and only flow during local storms or as a result of floods derived in the headwaters. These streams eventually lost their water downstream as it percolated into the muddy and silty sediments of the Great Plains. These streams probably did not flow beyond the Great Plains. The large carbonate masses found in the mudstone beds lateral to the upper sandstone blankets near the Ben Reifel Visitor Center were derived from carbonate-rich groundwater flowing from the edges of the river system into the lateral fine-grained mud deposits. That these rivers did not flow all year or connect with other river systems is indicated by the lack of aquatic animal fossils, such as fish, turtles, and alligator bones, and freshwater unionid mussel shells. Such river systems, called distributive river systems (Hartley et al., 2010; Weissmann et al., 2010), or rivers that break into multiple channels that fan outward onto a plain, drop their sediment into broad sheets, eventually forming stacked blankets of sand. A modern analog to most of the river systems of the Brule Formation would be in the Pampas of South America, where streams flowing out of the Andes lose their water into the porous loess deposits of the Pampas and as a result make broad fans of isolated channels and broad sheet sand deposits (Arthur Bloom, pers. comm., 1989).

Lacustrine (lake) strata of the White River Group are preserved at various stratigraphic levels across the region and are preserved as two main end members: those dominated by muds and clays (siliciclastics) and those dominated by calcium carbonate (limestone). These lakes formed as a result of geologic and possibly paleoclimatic factors. In Badlands National Park, particular intervals of lacustrine sedimentation include siliciclastic lakes in the middle of the Crazy Johnson Member of the Chadron Formation and limestone-dominated lakes at the contact between the Chadron and Brule formations and within the Poleslide Member of the Brule Formation. Siliciclastic lake deposits from the Crazy Johnson Member are interbedded with ancient river deposits (Terry and Spence, 1997; Terry, 1998). The lake deposits appear similar to normal clays and silts within this part of the stratigraphic section, but upon closer examination, particular types of fossils (stromatolites) are preserved within these deposits that confirm their origin as lacustrine strata. Stromatolites are interlayered masses of photosynthetic

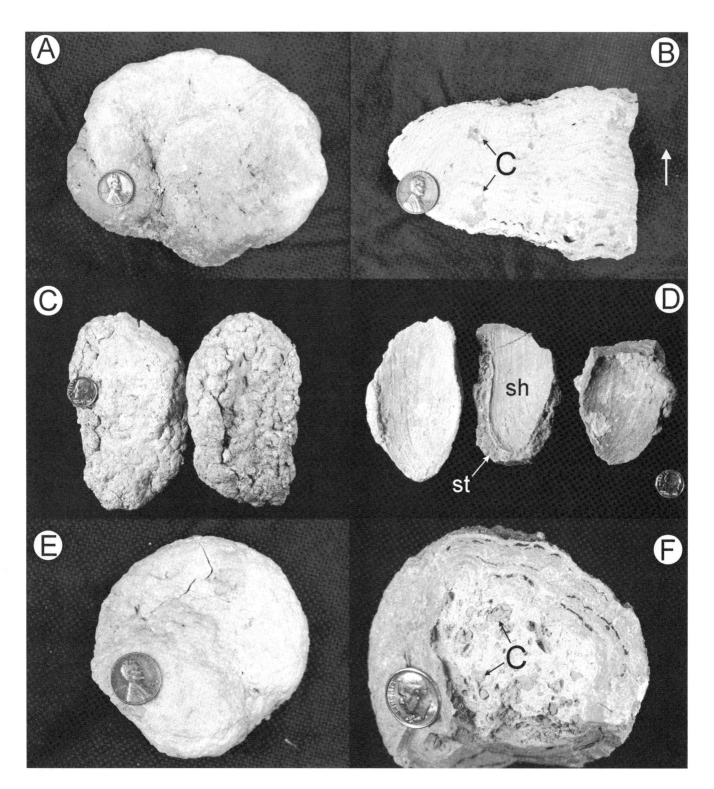

2.15. Photographs of lacustrine stromatolites. (A) Top view of stromatolite. (B) Cross section of stromatolite showing laminations and clay clasts (c) that were incorporated during growth. White arrow indicates original upward direction of growth. (C) Fossil bivalve shells encrusted by stromatolitic growth. (D) Interior view of bivalve shells (sh) encrusted by stromatolitic growth (st). (E) Exterior of oncolite. (F) Interior of oncolite showing laminations and clay clasts (c) that were incorporated during growth. Coin for scale in (A), (B), and (E) is 1.9 cm wide. Coin for scale in (C), (D), and (F) is 1.8 cm wide. Photos by the authors.

cyanobacteria and clay that form through the periodic addition of a thin layer of mud and clay to the surface of the cyanobacteria (Fig. 2.15). The bacteria, in an attempt to survive, grow through and recolonize the surface of the clay and mud. This process repeats, and over time, large stromatolitic bodies can be formed. The shape of the stromatolite body is a function of the water depth within which the cyanobacteria are growing. Very shallow waters and edges of lakes will have nearly flat layers of cyanobacteria. As water depth increases, the stromatolitic bodies become more domelike in shape (hemispheroids) as the photosynthetic cyanobacteria attempt to grow vertically in order to reach the light.

Stromatolites of the Crazy Johnson Member occur as laminar and hemispheroid masses. Some hemispheroids are up to 50 cm in diameter and 30 to 40 cm tall, although most are smaller (Fig. 2.15A). The large size of some of these stromatolites suggests that this was a stable perennial lake that existed for quite some time before eventually being buried by fluvial deposits. Some stromatolites show evidence of grazing, possibly by freshwater snails, which are known to be the primary predator of stromatolitic bodies. In addition to stromatolites, clams are also preserved in these deposits (Fig. 2.15D). Quite commonly the clamshells are covered with a layer of stromatolitic material that helped to preserve the shell (Terry and Spence, 1997; Terry, 1998).

In nearby ancient river deposits, rounded bodies of stromatolitic material are found as part of the coarser bed-load fraction of the channels. These rounded bodies, referred to as oncolites (Fig. 2.15F), represent the periodic movement of cyanobacterial masses by the river. As the oncolite rolls, its new resting place and orientation promotes the growth of new material in an upward direction on the oncolite mass. When the mass is moved again, the new orientation again promotes additional growth. Over time, this has a snowball-like effect on the growth of the oncolite. Some of these oncolites grew to the size of baseballs.

The contact of the Chadron and Brule formations is marked by occasional lacustrine limestone deposits. These deposits are easily seen in outcrops and can be recognized by their white color and resistance to erosion compared to the mudstones of the Badlands (Fig. 2.16). This difference in susceptibility to erosion has resulted in numerous table-top buttes across the region, with the limestone layer acting as a protective layer to the mudstones beneath it. Specific examples of these lacustrine tabletop buttes can be found just south of the NPS Minuteman Missile Site at exit 116 on Interstate 90, and to the north of Interstate 90 between exits 114 and 116 (Fig. P.1).

The predominance of limestone versus siliciclastics as the primary lacustrine sediment is related to the conditions

2.16. Photographs of lacustrine limestone caprock. (A) Massive lacustrine limestone representing a deeper part of an ancient lake. (B) Resistant layer of lacustrine limestone, which changes from massive at its base to progressively more laminated upward. Laminated sediments are indicative of shallower lake conditions. (C) Close-up of laminated lacustrine limestone. Pick for scale is approximately 60 cm long. Photos by the authors.

required for the formation of calcium carbonate in lakes. Limestone-dominated lakes require little influx of clastic sediment. Too much sediment and the clays and muds mix with calcium carbonate that is precipitating out of the water to form marls, in addition to inhibiting the growth of various forms of algae that biomineralize calcium carbonate. Lakes rich in limestone deposits generally form on stable landscapes away from rivers. It is also common for lakes that

2.17. Photographs of lacustrine limestone in the Poleslide Member of the Brule Formation (A) Multiple layers of lacustrine limestone (L) separated by mudstone. Pick for scale is approximately 60 cm long. (B) Close-up of lacustrine limestone (L). (C) Lacustrine gastropods (G) in limestone. (D) Stromatolitic structure. Photos by the authors.

are dominated by limestone to be the result of carbonate-enriched groundwaters exposed at the surface, or springs. As these waters interact with the atmosphere, carbon dioxide dissolved in the water can escape into the atmosphere, thus making the waters less acidic and allowing calcium carbonate to precipitate.

The lacustrine limestones south of the Minuteman Missile Silo are referred to as the Bloom Basin limestone beds (Evans and Welzenbach, 1998). These limestones range from 40 cm to meters in thickness and contain a rich diversity of aquatic life, including ostracods (a microscopic shelled arthropod), fish, turtles, charophytes (calcareous aquatic algae), and tracks of shorebirds (Evans and Welzenbach, 1998). These limestones, along with several other examples near the contact with the overlying Brule Formation, suggest that this time was geomorphologically stable across the region. This period of stability eventually gave way to active downcutting and erosion. Just northeast of Scenic, South Dakota,

just north of the old Chamberlain Pass along Highway 44 (Fig. P.1), lacustrine limestones of the Chadron Formation are dissected and surrounded by sediments of the overlying Brule Formation. This downcutting created a complex paleotopography regionally across the top of the Chadron Formation that was eventually filled in by sediments of the Brule Formation (Benton et al., 2007; Evanoff et al., 2010).

Lacustrine strata are also preserved within the middle of the lower Poleslide sequence (Unit 4, shown in Fig. 2.17). This depositional environment is unique within the Poleslide Member, which is dominated by eolian environments. The lake deposits range from relatively deeper water deposits in the center of the lake to shallow lake margin settings that interfinger with terrestrial environments (Fig. 2.18). At least two distinct episodes of lacustrine sedimentation are preserved. The lateral extent of these deposits is unknown because they are truncated by modern erosion. The central part of the lower lake level is preserved as a massive, chalky white micrite

Lacustrine limestone in the Poleslide Member

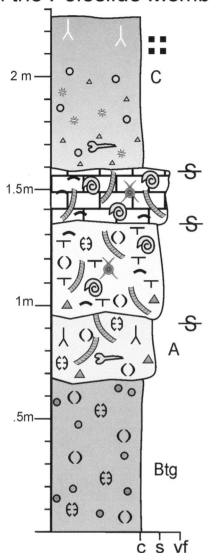

2.18. Measured section and outcrop photograph of lacustrine limestone (LS) in the Poleslide Member. Limestone in the measured section is concentrated around 1.5 m level. See Fig. 3.6 for explanation of symbols. Photos by the authors.

limestone. It rests on bluish-green, smectite-rich mudstones that weather similarly to the Peanut Peak Member of the Chadron Formation (Figs. 2.17, 2.18). The limestone is in turn capped by reddish marl that changes upsection to siliciclastics. Invertebrates and calcareous algae are the most common fossils within this part of the lake and include small gastropods, ostracods, and charophytes. Fish fossils are also present but are rare (Fig. 2.19). The upper lake level appears more laminated, possibly as a result of stromatolite growth. Domal stromatolites are found as float at the base of these exposures. Gastropods are also seen at this level.

Lake margin settings are marked by a transition to siliciclastic-dominated environments, but the nature of this transition is different depending on the individual lake level studied. The lowermost lake level preserves a lateral change to a more siliciclastic-dominated setting, whereas the uppermost lake level shows expansion of the lake system and the overlap of laminar, stromatolitic lacustrine limestone on top of terrestrial mudstone with fossil dung balls and sweat bee pupae. These strata represent a period of landscape stability. Because carbonate-dominated lakes form in response to low siliciclastic input, this suggests that this lake was set far away from an active channel system, that eolian influx was minor, or both.

These lacustrine limestones throughout the White River Group comprise a small part of the overall depositional

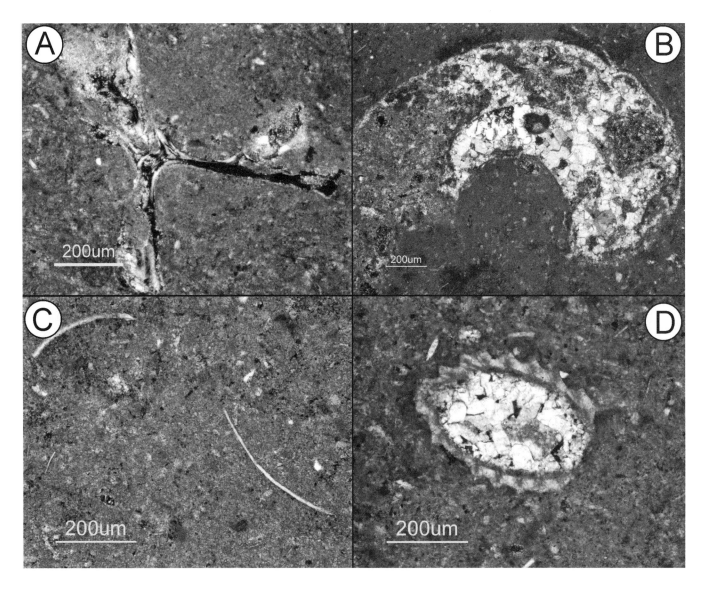

2.19. Photomicrographs of fossils from the lacustrine limestone in the Poleslide Member. (A) Fish vertebra. (B) Gastropod shell. (C) Ostracod shells. (D) Part of a charophyte. Photos by the authors.

history of these strata, but their paleoclimatic and paleogeomorphic requirements to form such deposits are important clues to the geologic history of the Badlands. The carbonate-rich nature of these lakes in the upper Eocene Chadron Formation suggests that climates by the end of the late Eocene were at times dry, that siliciclastic influx was at a minimum during stable geomorphic periods, or both. From a geologic standpoint, the source of these carbonate-rich waters is also interesting. It is likely that many of these lakes were fed by springs. Evans and Terry (1994) report the presence of tufa pipes (conduits for upward groundwater flow) that crosscut some of the lowest strata of the White River Group near the Bloom Basin limestone beds of Evans and Welzenbach (1998). The occurrence of these tufas was mapped by Evans and Welzenbach (1998) and Evans (1999), who postulated that these groundwater conduits could be related to the

structural development of the Badlands. In essence, these tufa conduits followed faults to emerge at the land surface. The carbonate-rich nature of the groundwater could be explained as the result of hydrologic recharge from the ancient Black Hills as groundwater passed through various carbonate rock units, such as the Pahasapa Limestone.

The transition from stacks of mudstone beds in the upper Scenic Member into stacks of thick, massive siltstone beds in the Poleslide Member represents a change from predominantly wet conditions during Scenic deposition to the accumulation of thick eolian dust deposits, or loess, during the dry conditions of Poleslide deposition. The loessites, or lithified loess deposits, are typically tan to gray, massive thick siltstone beds that cover large areas with no associated channel deposits. Fossils of the fluvial Scenic deposits tend to occur in concentrations at local areas, called bone beds, but the fossils are typically scattered throughout the loessites. The fossil fauna of the siltstone beds are dominated by animals that prefer drier conditions, such as oreodonts, camels, running rhinos, and land tortoises. The dust was derived from sources to the west (Fig. 2.13) and includes a mixture of volcaniclastic and epiclastic silt-size particles that were picked up and transported by winds. Studies of modern dryland loesses (Pye and Tsoar, 1987) show that dust is derived from arid areas that have little vegetation to protect the silt from wind erosion. It accumulates in semiarid conditions that have enough vegetation to trap the dust but that are not wet enough to weather the dust or transport it by water. Subhumid and humid areas can have dust accumulation, but the dust weathers into clays or is redistributed by streams and rivers. Therefore, the transition from the mudstones of the Scenic Member to the siltstones of the Poleslide Member indicates a transition from wetter, probably subhumid conditions during the deposition of the Scenic Member into drier, probably semiarid conditions during deposition of the Poleslide Member. All of the eolian deposits of the Badlands are fine-grained loessites; no eolian sand dunes are known from the White River Group.

SEDIMENTATION HISTORY OF THE WHITE RIVER GROUP IN THE BIG BADLANDS

A broad erosional surface extended into the Great Plains from the Black Hills at the start of the deposition of the White River Group during the late Eocene. This surface was under intense weathering in warm and humid climates over a long period of time, forming the deep red soil of the Interior Zone. Gravel that had been deposited to the east of the Black Hills in the earlier Eocene was involved with this intense weathering, resulting in preserved gravels composed of only quartz, chert, and quartzite. As the land surface was slowly eroded, much of this gravel was spread across the land as lag gravels. Streams flowed across this surface and concentrated these quartz-rich gravels at the base of their channels. We know little of the geometry of these channels; they occur as isolated outcrops within the Chamberlain Pass Formation.

A major stream flowing out of the Black Hills and into the west side of the Badlands cut a major valley through the Interior Zone to a depth of about 27.4 m. This valley started to fill approximately 36 Ma as an influx of volcanic sediment from volcanic eruptions far to the west started to fall into the area as dust and ash, initiating the deposition of the Chadron Formation. This volcanic material was reworked and mixed with local epiclastic sediment by streams coming out of the Black Hills. These streams flowed throughout the year in channels that stayed in a narrow channel belt that

are preserved as isolated sandstone ribbons. The freshwater fauna of these channels and east-to-southeast flow features indicate that these streams were part of a well-integrated river system that flowed into an ancient Mississippi drainage. The climate was still humid enough so that the volcanic dust that continued to fall into the area was either transported and deposited by the streams as overbank deposits or was weathered into clay where the dust fell but was not transported by streams. Locally, ponds and lakes contained aquatic animals and deposited sheets of lime mud. Eventually the deposition overtopped the valley walls and spread across the old erosion surface of the Interior Zone. Most of the deposits that lie on the old Interior Zone surface are the massive green-gray claystones of the Peanut Peak Member of the Chadron Formation, but broad channel complexes, such as the one in the vicinity of the town of Scenic and the modern Cain Creek drainage, were interbedded with these widespread massive clay beds. Locally, there were ponds and lakes that deposited sheets of lime muds, eventually becoming limestone sheets in the upper Chadron sequence.

Deposition of the upper Eocene Chadron sediments ended before the end of the period, and erosion carved valleys and hills into the Chadron Formation. The relief on this topography was at least 27.4 m with at least 1.05 million years of time missing across the Eocene–Oligocene boundary. A ridge of hills cut into the Chadron deposits in the Pinnacles–Dillon Pass area of Badlands National Park separated two broad paleovalleys at the start of Brule deposition. Deposition resumed with the accumulation of the basal brown mudstone beds in the lower part of the Scenic Member of the Brule Formation. These basal brown muds were derived from the eroded Chadron hills, the continued influx through the air of volcanic dust, and eventually muddy overbank deposition from streams. The brown mudstones blanketed the topography, and the valleys started to fill with sheets of sand from multiple channels and sheet floods. These sheet sands accumulated as widespread sand blankets that were separated by thin, often laminated sheets of muds and clay representing local overbank deposits. The ribbon sands within the sand blankets contain bedding features that indicate they were deposited by ephemeral streams that flowed only during flood events. The fluvial systems at this time (early Oligocene) were distributive channels flowing from the Black Hills and splitting into distributary channels on the Great Plains, depositing sand in broad but relatively thin fan-shaped sequences of stacked blankets of sand. Waters in these distributary channels eventually sank into the muds of the Great Plains, as indicated by the thick groundwater-derived carbonate nodules in mudstone sequences lateral to the sandstone blankets. The lack of aquatic animal remains in these sand blankets and associated channel deposits indicate the presence of ephemeral water flow and the lack of downstream continuity of the fluvial system. The thick stacked sand blankets are separated by two widespread mudstone marker beds, the Hay Butte and Saddle Pass markers. These represent accumulation and weathering of volcanic dust into widespread series of paleosols that occur within these markers, either from a shift in the deposition centers of the distributive fluvial systems or climatic cycles that shut down the deposition of the distributive river systems on the Great Plains. The widespread deposition of mud layers ended the aggradation of the Scenic Member. These muds resulted from the weathering of volcanic-rich dust still falling onto the Great Plains and from overbank deposits from rare stream channels. The highest mudstone beds in the Scenic Member in the eastern Badlands grade into siltstone beds of the lowest Poleslide Member in the western Badlands.

The Poleslide Member is characterized by the accumulation of dust deposits as dryland loess. The loess accumulated in semiarid environments where vegetation trapped dust derived from the west in more arid environments, and the local environment was dry enough to preserve the volcanic glass grains. The lower Poleslide sequence includes interbedded thick units of siltstone, sandstone blankets with almost as wide a distribution as those in the middle Scenic Member, and one widespread mudstone sequence that contains the last of the lacustrine limestone sheets. The middle Poleslide sequence in the North Unit of Badlands National Park is composed of massive siltstone beds that represent stacked blankets of loess. In the South Unit of Badlands National Park there are a few isolated sandstone ribbons in the middle Poleslide that indicate continually flowing streams that were connected to a larger river system. These sandstone ribbons and their associated overbank deposits are interbedded with thick, massive siltstone beds representing loessites. The upper Poleslide sequence is another thick fluvial sequence of alternating sandstone and siltstone sheets that are interbedded with broad coarse-grained sandstone ribbons, called the *Protoceras* channels. These well-bedded sandstone sheets are the last evidence of broad fluvial complexes in the White River Group in the Badlands. These upper Poleslide beds are found at the top of Sheep Mountain Table, where they are capped by the thick Rockyford Ash at the base of the Sharps Formation, a unit that is largely composed of sandy loessites. To the east, massive silty loess deposits continued to accumulate in the Cedar Pass area during the deposition of the upper Poleslide fluvial sequence. A solitary fluvial sandstone blanket marks the base of the upper Poleslide in the Cedar Pass area.

The buttes along the Badlands Wall, including those in the Cedar Pass area, are capped by valley-fill deposits of basal siltstone–pebble conglomerates and overlying alternating fine-grained sandstone and siltstone sheets. Sandstone beds are abundant in these upper valley fills, and the fauna is of late Oligocene (late Arikareean) age, indicating that these are valleys were filled with sediments of the Arikaree Group. Erosion has removed the upper levels of these Arikaree valley fills and the rocks through which they cut.

CLUES TO ANCIENT PALEOENVIRONMENTAL AND paleoclimatic conditions in the Badlands can be found in many sources, including fossils, sediments that make up the buttes and spires, and the enigmatic stripes of red, brown, and beige that cut across the Badlands. These razor-sharp lines of color are actually ancient soils (paleosols) representing former landscapes that have been buried and lithified (turned to rock). In order to utilize paleosols to reconstruct past climates and environments, we need a working knowledge of modern soils and the environmental conditions under which they form.

WHAT IS A SOIL?

Soils are zones of physical, chemical, and biological activity at the surface of the Earth that modify geological materials into a more stable form. Physical weathering is the process by which geological materials are mechanically broken down into progressively smaller pieces. Chemical weathering is the interaction of geological materials with water and organic acids and alters the minerals that comprise the rock. These various forms of weathering and modification represent a quest for equilibrium between the forces of weathering and the stability of the geological material that is being weathered. As a soil develops, the downward movement of soil water and dissolved/suspended materials will create distinct accumulations (soil horizons) that reflect the particular physical, chemical, and biological processes regulated by five primary factors of soil formation: climate, organisms, relief, parent material, and time (CLORPT) (Jenny, 1941). Variations in these five factors have generated over 18,000 different soils in the United States alone (Soil Survey Staff, 1999).

In terms of soil formation, climate can be thought of as the predominant temperature and precipitation conditions for any given area. Is it hot or cold? Is it rainy most of the time, or is it dry like a desert? Warm, humid environments will promote faster rates of chemical weathering, whereas extremely cold climates promote physical weathering by freeze–thaw processes. Humid conditions favor the downward movement

(eluviation) and accumulation (illuviation) of clays, whereas more arid conditions favor the accumulation of calcium carbonate by evaporation of soil water. The influence of organisms can be both physical and chemical in nature, and includes the smallest microbes to the largest animals as well as all varieties of plants. Relief is simply the shape of the landscape, otherwise known as topography. Changes in relief, if significant enough, will influence rates of soil erosion and stability, with areas of higher relief experiencing greater erosion. Parent material refers to the physical and chemical makeup of the geologic material that is being altered by soil formation. The last factor of soil formation is time. In the simplest sense, the more time that is available for weathering, the greater the degree of physical and chemical change that can be induced. Persistent deposition of new material or erosion of existing soil resets the time clock.

Soils are classified into 12 distinct orders (Soil Survey Staff, 1999). These soil orders are three-dimensional bodies of material that are composed of distinct combinations of horizons (Table 3.1). The types of horizons present in the soil are the direct result of CLORPT conditions. The first level of horizon classification is the master horizon (Table 3.1). O horizons are accumulations of purely organic matter at the land surface and can vary from a thin layer of leaf litter or a thick mat of marsh and swamp vegetation. A horizons are zones at the top of soil profiles (topsoil) that are defined by a mixture of geological materials and organic matter. B horizons are the primary zones of accumulation that form from the downward percolation of soil water and deposition of finer materials in pores and fractures, or the precipitation of minerals that were dissolved in the soil water. The C horizon represents unaltered parent material. These master horizons can be further classified on the basis of the presence of particular types of mineral and organic matter that accumulate within them (Table 3.1). Common diagnostic subsurface soil horizons include B horizons that are enriched in clay (argillic horizons, Bt) or calcium carbonate (calcic horizons, Bk). These horizons are stacked vertically and are referred to as a soil profile (Fig. 3.1).

Spheres of Influence

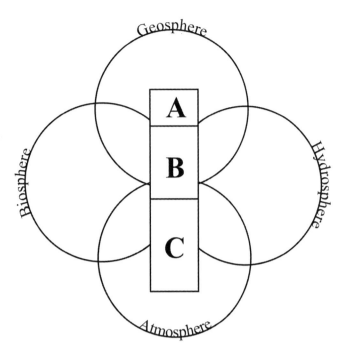

3.1. Primary factors affecting the development of a soil profile.

PALEOPEDOLOGY AND THE RECOGNITION OF PALEOSOLS

Paleopedology is the study of fossilized soils (paleosols). This is a relatively new field within geology and has been steadily gaining in popularity over the past couple of decades. Today, the study of paleosols is used in a wide variety of geological investigations including paleoclimatology, paleoenvironmental analysis, geoarcheology, and vertebrate paleontology. The utility of paleopedology is grounded in the philosophy of modern soil genesis. The clues to ancient environmental and climatic conditions (paleo-CLORPT) that are preserved in the paleosol are extracted by a combination of field and laboratory techniques.

Early paleosol research in the Badlands was focused mainly on the intense zone of soil formation at the base of the White River Group within the Interior Zone (Plate 2; Ward, 1922; Wanless, 1923; Pettyjohn, 1966; Hickey, 1977). The first in-depth research on paleosols in Badlands National Park was conducted by Retallack (1983b) in which he recognized 87 superimposed paleosols representing 10 different types of soils within 143 m of section in the Pinnacles area (Fig. P.1). The most obvious clues to the presence of paleosols include horizonation, soil structure, and roots (Plates 6, 7). Horizonation is manifested as prominent brownish-red striping seen within outcrops. Particular horizons, such as accumulations

Table 3.1. Soil orders, master soil horizons, and common subordinate indicators

Characteristic	Name	Description
Soil order	Entisol	Least developed soils composed of A/C profiles.
	Inceptisol	Weakly developed soils with the inception of a cambic (Bw) horizon.
	Vertisol	Extreme mixing of soil profile by expandable clays (smectites).
	Aridisol	Soils of extremely dry climates, commonly rich in carbonates and salts.
	Mollisol	Grassland soils with thick, dark, organic surface horizon (mollic epipedon).
	Spodosol	Pine forest soils with illuvial iron, aluminum and organics in B horizon.
	Alfisol	Hardwood soils with illuvial accumulation of clay (Bt horizon).
	Ultisol	Similar to alfisols, but with <35% base saturation.
	Oxisol	Highly leached soils of humid tropical regions. Rich in iron and aluminum.
	Histosol	Organic soils of swamps and marshes.
	Andisol	Slightly weathered soils formed in volcanic ash.
	Gelisol	Frozen soils of permafrost zones.
Master horizon	O horizon	Organic matter at the top of the soil profile.
	A horizon	Mixture of mineral and organic matter at the top of profile. This is a zone of loss as materials are washed into deeper parts of the profile.
	E horizon	Zone of intense leaching and loss. Mineral grains are stripped of clay and oxide coatings and commonly have a whitish appearance.
	B horizon	Zone of accumulation formed by the downward movement and collection of materials from higher in the profile. Common materials include clays, carbonates, and organics.
	C horizon	Unconsolidated parent material.
	K horizon	Intense accumulation of pedogenic calcium carbonate formed by downward percolation and precipitation of carbonate-rich soil waters.
	R horizon	Bedrock parent material.

Note: Modified from Soil Survey Staff (1999).

of calcium carbonate, can form resistant ledges. The different colors are clues to ancient soil environmental conditions. Red and brown colors within paleosols are usually indicative of former well-drained soil conditions, allowing more oxygen to react with iron minerals, whereas green, black, and gray colors are commonly associated with former waterlogged and swampy conditions and less available oxygen. Depending on the type of paleosol that is preserved, individual paleosol profiles will contain horizons of differing colors, which is responsible for the barber pole striping so prevalent throughout the park (Plate 6).

Upon closer examination, smaller-scale features of paleosols can be detected and related to current soil forming conditions. In the Badlands, these smaller-scale features (nodules, concretions, and peds) are most commonly exposed by digging into the rock outcrop because the highly erosive nature of Badlands rock often obscures these features. Nodules and concretions are cemented bodies of particular types of minerals, most commonly calcium carbonate or iron oxide, which can sometimes be seen standing out in erosional relief in the outcrop (Plate 6). Soil scientists refer to these bodies as glaebules, and they form as the result of precipitation of minerals out of the soil water as it percolates through the soil profile. Glaebules of calcium carbonate are associated with drier soil conditions, and those composed of iron oxides are commonly associated with wetter soil conditions.

Peds are three-dimensional blocks of soil structure that form by the repetitive action of wetting and drying of a soil profile as a result of rain events, the burrowing action of organisms, and the downward penetration and branching of roots (Plate 6). Peds can be blocky, angular, crumb, granular, platy, columnar, or prismatic in shape. The different shapes depend on different types of soil conditions and position within a soil profile, and they represent the physical modification of the originally deposited sediments into distinct bodies. Crumb and granular peds are usually several millimeters in size, are generally rounded to slightly angular, and are commonly associated with dense accumulations of fine root hairs in the A horizon. Anyone who has tried to strip sod out of a yard is familiar with the little clumps of dirt that cling to the roots. These are the peds. Larger angular, blocky, and columnar/prismatic ped structures tend to be found deeper in soil profiles, commonly in the B horizon. Platy peds are most commonly associated with the original texture of sediments that were deposited by geological forces, such as flooding, and have not been altered to any great degree by soil formation. This type of ped structure is most commonly seen in the C horizon of soil profiles. Peds may sometimes display striated surfaces (slickensides) created by the expansion and contraction of soil materials as they repeatedly saturate and dry out as a result of rain events (Plate 6). Slickensides are common in soils that contain large proportions of smectite, a clay mineral that expands and contracts with moisture.

The most obvious clue to the presence of ancient soils in the Badlands should be fossilized plants and roots. The orientation, size, and mode of preservation of roots and plant material in a paleosol are important clues to ancient soil conditions. However, the soil conditions that were active during the formation of the Badlands did not favor the preservation of fossilized plant material. The most favorable environment for the preservation of plant material in the geologic record is in a marsh or swamp. The acidic nature of swamp and marsh waters helps to retard the decomposition of plant material. Over time, and with burial, thick accumulations of organic materials in swamps can be converted to coal. Under typical soil-forming conditions, water does not collect in the soil but drains through to the water table. This well-drained and aerated setting promotes the decay of plant material. This latter set of soil conditions was dominant during the accumulation of sediments in the Badlands.

Nevertheless, we do have evidence of ancient plant life in the Badlands. Within paleosols of the Badlands, plants, seeds, and roots are preserved as mineralized bodies or as clay or mineral infills of hollow spaces left by former roots that have since decayed, greenish ghosts formed by the decay of roots in low oxygen (anaerobic) conditions, or sheathes of calcium carbonate or iron oxide that precipitated around former roots (Plates 6, 7). Permineralization is the replacement of original plant materials on a microscopic scale, usually by quartz but by other minerals as well, and produces fossils most commonly referred to as petrified wood. This process produces extremely detailed preservation of original plant structures to such a degree that original cellular structures and tree rings can be preserved. Scientists can use this detailed preservation to classify plant remains into modern taxonomic groups, and they analyze variations in tree ring morphologies to interpret paleoclimatic conditions (Falcon-Lang, 2005).

Under well-drained and aerated soil conditions, which were the dominant conditions in the paleosols of the Badlands, plant material will decay rather quickly. Plant remains on the land surface will break down and become recycled. Roots will also decay and leave voids in the soil profile. These voids are commonly filled in with either clay or silt (a cast) and sometimes with minerals that grow in these voids upon burial in the geologic record (Plate 7). In other instances, original root material can persist with burial of the landscape and then be broken down by microbial decay of the root under oxygen-poor conditions. This process produces a greenish, ghostlike halo around the former position of the root (drab-haloed root trace), but the central void of the root will persist. This central zone can then be filled in by minerals (Plate 7).

Last, roots can self-fossilize while they are still alive. As roots absorb nutrients from the soil, they modify the chemistry of soil near the root. With periods of rainfall and infiltration, the acidity of the soil can change, as can the potential to release or fix iron. These repetitive episodes of wetting and drying, in conjunction with chemical changes in the root microenvironment, will promote the precipitation of calcium carbonate or iron oxides around the root body. This lithified mass, referred to as a rhizolith (root rock), can then be easily preserved upon burial in the geologic record. Carbonate rhizoliths are associated with drier climates, whereas iron oxide rhizoliths tend to be associated with soils that are subjected to periodic waterlogging.

In addition to the mode of preservation, the size and orientation of fossilized roots in paleosols of the Badlands provide insight into the general environment of soil formation and clues to overall paleoclimatic conditions. The size of fossil roots is directly related to the type of plant community that was present on an ancient land surface. Large roots several centimeters across represent trees, whereas hairline root structures indicate former grasslands. Roots that penetrate downward represent well-drained soil conditions, whereas roots that are more horizontal suggest waterlogged or shallow water table conditions. Roots and plants preserved as carbonized traces or coal, which have yet to be identified in the Badlands, represent former waterlogged, marshy, and swampy conditions. Clay- and mineral-infilled root voids, drab-haloed root traces, and rhizoliths represent soils that formed under well-drained conditions.

Burrows of small animals and insects are also common features within soils, which raises the question of how one would tell them apart in the geologic record. In brief, both can branch in a downward direction, but roots will taper as they branch whereas burrows will remain constant in size (Plate 7). However, similar to roots, burrows can be an indicator of soil conditions. Deep, penetrating burrows will be found in soils that are well drained. Burrowing animals that make their dens below ground will not dig deeper than the water table. Common insect burrows in the strata of the Badlands include those made by dung-ball beetles and subterranean bees (Plate 7). Dens or burrows of mammals have yet to be identified in the Big Badlands.

WHAT CAN WE LEARN FROM THE PALEOSOLS IN BADLANDS NATIONAL PARK?

As we look at the strata of the Badlands, we can recognize a vertical change in the appearance of various rock formations (Plate 4). The base of the Badlands is composed of brightly colored orange and red strata, which are in turn covered by greenish-gray rock that weathers into haystack shapes, and finally beige and light brown cliff- and ridge-forming strata defined by vibrant stripes of color that cut across the Badlands. This vertical change in color and resistance to erosion is a function of the types of ancient environments that were responsible for the deposition of sediments that formed the Badlands, the types of sediments that were deposited, and the soil-forming environments that modified the sediments after deposition.

Paleosols of the Interior Zone

After the retreat of the Western Interior Seaway approximately 68 Ma, this entire region, from North Dakota to Colorado, was subjected to 30 million years of intense soil formation and erosion (Pettyjohn, 1966). Within Badlands National Park and the surrounding region, this episode of intense soil formation is represented as the bright yellow, purple, and red interval that underlies the White River Group Badlands and is referred to by various names, including the Interior Zone, Interior Paleosol, or Interior Formation, for exposures of this interval in and around the town of Interior, South Dakota (Plate 2; Fig. 3.2). Within Badlands National Park, this interval is best exposed in Dillon Pass and along the Sage Creek Rim Road (Fig. P.1).

From a distance, this brightly colored interval appears as a gradual transition from bright reddish rock at its top to progressively yellow strata farther down (Plate 2). In actuality, this brightly colored zone represents several different episodes of ancient soil formation that modified distinctly different sedimentary units before in turn being buried by the next depositional event (Retallack, 1983b). The lowermost of these distinct episodes of pedogenesis is referred to as the Yellow Mounds Paleosol Series and is named for exposures in Dillon Pass (Retallack, 1983b). The term *series* is used by modern soil scientists to describe a discrete package of soil with particular physical and chemical characteristics that can be compared to other nearby soils across a landscape. The Yellow Mounds Paleosol Series modified the underlying Cretaceous marine Fox Hills Formation and underlying Pierre Shale, which changed the Fox Hills Formation into a more yellowish-orange color with intervals of red and purple, and the Pierre Shale from its original dark grayish-black to bright yellow and orange (Plates 1, 2; Fig. 3.2). Within Dillon Pass, these colors extend and lessen downward for tens of meters, and in the lowest part of some stream drainages, the original black color of the Pierre Shale can be seen.

According to Retallack (1983b), the Yellow Mounds Paleosol Series represents an intense period of ancient soil formation under humid subtropical conditions, similar to what would be found today in Georgia or South Carolina. Within modern soil classifications (Soil Survey Staff, 1999), the Yellow Mounds paleosol represents an ultisol, a soil that has accumulated clays in its B horizon (A–Bt–C) and had a large proportion of its unstable, easily weathered minerals removed to leave behind only the most resistant minerals. Roots preserved within the Yellow Mounds Paleosol Series are suggestive of forested vegetation. With increasing depth into the Yellow Mounds Paleosol Series, the original unstable mineralogies of the Fox Hills Formation and Pierre Shale survived pedogenesis as the strength of soil formation waned.

Paleosols of the Chamberlain Pass Formation

Approximately 37 Ma, during the late Eocene, the rivers that had been flowing across the forested strata of the Pierre Shale and Fox Hills Formation began to deposit sediments. These sediments, referred to as the Chamberlain Pass Formation (Evans and Terry, 1994), are represented by the bright red stripe of mudstone just above the Yellow Mounds Paleosol Series in Dillon Pass and occasional bright white lenses of sandstone, which are seen more frequently to the south near the town of Interior (Plate 2) and to the north near the Minuteman Missile Site (Fig. P.1). Before the recognition of the Chamberlain Pass Formation, Retallack (1983b) recognized that this bright red stripe of mudstone capping the Yellow Mounds Paleosol Series was modified by a separate episode of ancient soil formation and that it not only modified the mudstone but also penetrated through this layer to partially overprint the underlying Yellow Mounds Paleosol Series. Retallack (1983b) named this distinct period of soil formation the Interior Paleosol Series (Fig. 3.2). According to Retallack (1983b), the Interior Paleosol Series represents pedogenesis in conditions similar to the underlying Yellow Mounds Paleosol Series. Within soil taxonomy, the Interior Paleosol Series is similar to modern alfisols, soils that form under forests and accumulate clays in their B horizons (A–Bt–C), but not to the extent of weathering and soil formation of an ultisol.

Lateral tracing of the Interior Paleosol Series reveals that paleosols next (proximal) to ancient stream channels of the Chamberlain Pass Formation formed under different environmental conditions. These proximal paleosols show evidence of hydromorphy, the effect of rising and falling water tables. As the position of the water table fluctuates, the chemistry of the soil will change from oxidizing to reducing conditions as the soil profile experiences alternating wet and dry conditions. Over time, this results in the buildup of iron oxide masses (glaebules) and discolors the soil profile into a mottled collection of reddish, greenish, and gray colors. Terry and Evans (1994) classified these particular soils as the Weta Paleosol Series for exposures east of Badlands National Park near the town of Weta, just southwest of Kadoka, South Dakota. Taken together, the Interior and Weta paleosol series represent an ancient soil catena, a laterally linked association of distinct soil types across a landscape.

After deposition of the Chamberlain Pass Formation, the rivers flowing across this region began to cut down through the strata of the Chamberlain Pass and Fox Hills formations, through the Yellow Mounds Paleosol Series, and eventually stopped downcutting after reaching unaltered Pierre Shale. This episode of river downcutting, which formed the Red River Valley of Clark, Beerbower, and Kietzke (1967),

Interior Zone Nomenclature in South Dakota (Badlands National Park area)

Yellow	Red	Interior Zone outcrop color
Pierre Shale		Toepelman (1922)
Interior Phase		
Post Pierre Shale/Pre-Titanotherium Beds		Wanless (1922)
Interior Formation		
Pierre Shale/Fox Hills Formation		Ward (1922a,b)
Interior Phase/Formation		
Pierre Shale and Fox Hills Fm.		Pettyjohn (1966)
Eocene Paleosol		
Pierre Shale		Clark & others (1967)
Interior Zone		
Pierre Shale		Harksen & MacDonald (1969a, b)
Interior Paleosol		
Pierre Shale	Chadron Fm.	Retallack (1983)
Yellow Mounds Paleosol Series	Interior Paleosol Series	
Post Pierre Shale/Pre-White River Group		Martin (1987)
Interior Paleosol		
Pierre Shale	Chamberlain Pass Fm.	Terry and Evans (1994)
Yellow Mounds Paleosol Series	Interior Paleosol Series	
Fox Hills Fm.	Chamberlain Pass Fm.	Stoffer (2003)
Yellow Mounds Paleosol Series	Interior Paleosol Series	
Pierre Shale and Fox Hills Fm.	Chamberlain Pass Fm.	This volume
Yellow Mounds Paleosol Series	Interior Paleosol Series	
Yellow	Red	Interior Zone outcrop color

3.2. History of nomenclature for the Interior Zone.

increased the percolation and drainage of precipitation through the paleosols of the Chamberlain Pass Formation and created highly weathered soils. The sediments in the former river channels of the Chamberlain Pass Formation, which were abandoned when the downcutting began, only preserve the most resistant of minerals, such as quartz (Terry and Evans, 1994), while the other minerals in these former stream channels were converted to kaolin clay by hydrolysis, which explains their blazing white color (Plate 2). The soils of the Interior Paleosol Series were so well drained and stable when the Red River Paleovalley was cut that the majority of mammal bones were destroyed before fossilization, a situation atypical for the majority of deposits in the Badlands. Fossils that have been recovered from the channel sandstones of the Chamberlain Pass Formation suggest a Chadronian age, and possibly as old as the Duchesnean (Clark, Beerbower, and Kietzke, 1967; LaGarry, LaGarry, and Terry, 1996; Terry, 1998).

Paleosols of the Chadron Formation

The sediments that filled the valleys that cut through the Chamberlain Pass Formation were deposited by ancient rivers to produce the Chadron Formation. The Chadron Formation is represented by the Peanut Peak Member across the majority of the region, but within one of the paleovalleys to the west (Plate 3), the Ahearn, Crazy Johnson, and Peanut Peak members are all present (Clark, Beerbower, and Kietzke, 1967; Terry, 1998). As with rivers today, periodic flooding buried ancient floodplain soils, and the paleovalleys were eventually filled, preserving numerous paleosols and their associated vertebrate fossils.

According to Retallack (1983b), the Chadron Formation (Peanut Peak Member) in the Dillon Pass area is dominated by soils similar to modern Alfisols (forest soils: A–Bt–C horizons) that would have formed under subhumid to humid, warm conditions (Plates 2, 4). Retallack (1983b) named these the Gleska Paleosol Series and recognized that on occasion these soils contained concentrations of pedogenic carbonate lower down in their horizons (A–Btk–Ck), which collected under drier soil forming conditions. To the west, within the paleovalleys filled by the Chadron Formation, permineralized roots from the Ahearn Member are several centimeters across and indicate forested vegetation (Plate 6; Terry, 1998).

Paleosols of the Brule Formation

The overlying Brule Formation represents a change to different soil-forming conditions than those that occurred during the deposition of the Chadron Formation. This change in paleo-CLORPT conditions manifests as a change in the types of sediments that were deposited as well as the types of paleosols that formed. The boundary between the Chadron and Brule formations straddles the transition from the Eocene to Oligocene and represents the local response to a significant worldwide climatic shift from warm and humid to cooler and drier climates: the Hothouse to Icehouse Transition.

Paleosols of the Scenic Member

Retallack (1983b) recognized five distinct types of paleosols over 38 m within the Scenic Member in the Pinnacles area (Fig. P.1). In stratigraphic order, these include the Conata, Gleska, Zisa, Ohaka, and Ogi soils. The profile identifications initially proposed by Retallack (1983b) have become outdated; they now fall within new categories in USDA soil taxonomy (Soil Survey Staff, 1999). The soils are described here as they were originally interpreted by Retallack (1983b), and reference to the old versus new classifications is made as needed.

The Conata Series represents soils that formed away from active river channels on terraces or floodplains. They are inceptisols (A–B–C profiles) and are characterized by clay enrichment in the B horizon (BE), moderately weathered volcanic ash shards and feldspars, medium-size greenish root traces, and an overall calcareous nature. The Gleska Series is still present in the Scenic Member, indicating that forested conditions were still present on these landscapes. The Zisa Series represents weakly developed soils that formed along active river channels, possibly within swales and on proximal levee deposits. They are entisols (A–C profiles), which are characterized by little development and small rootlets. The Ohaka Series represents weakly developed, possibly waterlogged soils, which formed along the edge of sloping floodplains or active rivers, or on stream-side bars. They are entisols in modern soil taxonomy (A–C profiles) on the basis of the descriptions of Retallack (1983b), although he lists them as inceptisols. They are characterized by surficial concentration of rootlets and reworked soil material from nearby Gleska Series soils. The Ogi Series represents thin, weakly developed, possibly waterlogged soils that formed in interstream depressions. They are also entisols by modern classification standards (A–C profiles), instead of inceptisols as reported by Retallack (1983b), and are characterized by poor development, colors suggestive of poor drainage, and roots ranging from small to large in size.

Additional research involving paleosols of the Scenic Member includes Terry (1996a, 1996b), who used paleopedology to interpret the genesis of the Big Pig Dig, an extremely concentrated Orellan vertebrate assemblage bone bed near

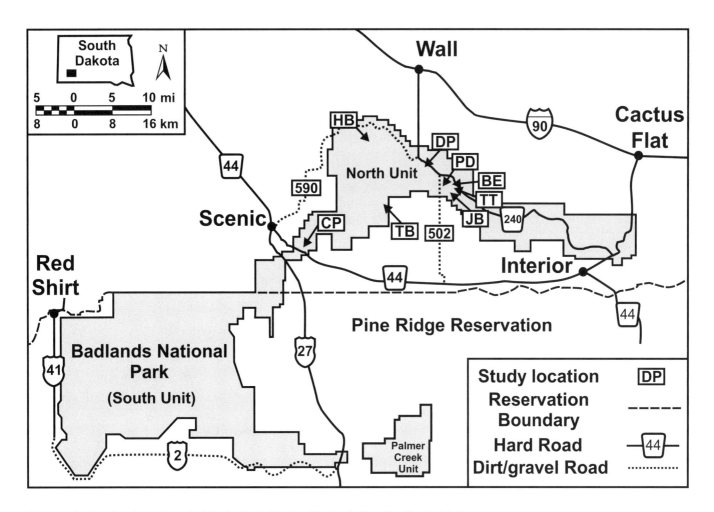

3.3. Map of paleosol study locations of within the Scenic Member. BE = Bessie Site; CP = Chamberlain Pass; DP = Dillon Pass; HB = Hay Butte; JB = Jerry's Bone bed; PD = Big Pig Dig; TB = Tyree Basin; TT = Tilted Turtle.

the Conata Picnic Ground. Other, more recent researchers have used paleopedology to interpret vertebrate taphonomy within the lower Scenic Member (Factor, 2002; Factor and Terry, 2002; Terry in Benton et al., 2007), including Mc-Coy (2002), who studied the paleopedology of the Hay Butte marker bed of Evanoff (in Benton et al., 2007), and Metzger, Terry, and Grandstaff (2004), who evaluated the effects of paleosols on the process of fossilization. This combination of research on the paleosols of the Scenic Member across the Badlands has revealed a pattern in which the variety of ancient soil types appears to be responding to changes in the environmental conditions of soil formation within the Scenic Member (Fig. 3.3). The majority of these observations fall within the lower part of the Scenic Member, below and within the Hay Butte marker of Evanoff (in Benton et al., 2007).

The paleosols of the Scenic Member are generally weakly to moderately developed (entisols and inceptisols) and were strongly influenced by their position on the ancient landscape during deposition of the Scenic Member. Those

paleosols closer to ancient streams were weaker (entisols) as a result of the more common episodes of overbank flooding and burial of the ancient landscape, and they sometimes show evidence of waterlogging (Fig. 3.4). Positions along the landscape further away from active rivers would have benefited from longer periods of time between flooding events and thus could have developed into better formed soils (inceptisols). Fossil roots within the Scenic Member are smaller overall in comparison to those preserved in the underlying Chadron Formation, 0.5 to 1 cm diameter, and suggest more open landscapes instead of forested ecosystems. Direct preservation of plant material is extremely rare in the Scenic Member, but a few locations in the park preserve silicified trees, still in life position (Plate 7).

To the west near Scenic, South Dakota, paleosols of the basal Scenic Member are most commonly weakly developed, and vertebrate fossil preservation is excellent throughout the area (Figs. 3.3, 3.5, 3.6). Along with preservation within paleosol profiles, fossil bone is also preserved as bed load, the coarse material found at the bottom of river channels. The

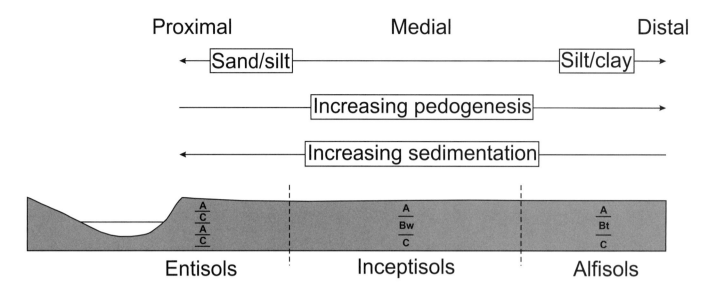

3.4. Model of soil formation on a floodplain from proximal to distal.

degree of abrasion of the bone varies from rounded bone chips to complete limb bones. From a facies perspective, this area represents a proximal floodplain position characterized by repetitive flooding events, now preserved as individual upward-fining packages of siltstone and mudstone, which buried former land surfaces and promoted fossil preservation. From a paleopedological perspective, this region is composed of a series of stacked entisols (A–C profiles) and inceptisols (A–Bw–C profiles) heavily influenced by river processes and minimal pedogenic modification (Figs. 3.4, 3.5). The individual profiles represent pedogenic modification of the discrete upward-fining overbank flood deposits. The fossils preserved with these deposits likely represent a combination of attritional accumulation on the landscape and animals caught up in individual flooding events. Commonly the fossils are lying directly on top of a former land surface (within an overlying C horizon), as indicated by drab-haloed root traces directly beneath the fossils (Plate 7; Fig. 3.5). These paleosols are most similar to the Zisa and Conata series of Retallack (1983b).

In the Dillon Pass area of the park (Fig. 3.3), the paleosols of the Lower Scenic Member are dominated by a combination of simplistic A–C profiles (entisols) and slightly more developed A–Bw–C profiles (inceptisols). The Dillon Pass area is situated on the crest and flanks of the Sage Arch. Because of syndepositional uplift during deposition of the White River Group, individual stratigraphic units thin across the top of the arch and create a condensed section in which larger amounts of geologic time are represented by the thinner strata and associated paleosols compared to areas off the arch that experienced greater rates of sedimentation. The paleosols in this area were also influenced by the infilling of paleotopography on the underlying Chadron Formation. Those soils that formed in lower parts of this aggrading paleotopography experienced greater influence from fluctuations on the ancient water table compared to others that formed on strata deposited after the majority of the paleotopography had been filled. Particular examples of paleosols within Dillon Pass serve to demonstrate this degree of pedogenic variability.

Jerry's Bone Bed profile is located southeast of the Conata Picnic Grounds (Fig. 3.3). This site is noted for a large concentration of disarticulated bones that are commonly stained orange-red (Fig. 3.7). From a facies perspective, this site represents a proximal floodplain position dominated by interbedded siltstones and mudstones. From a paleopedological perspective, this site is composed of a series of stacked A–C profiles (entisols) with minimal pedogenesis and a strong influence from fluctuating water tables. The individual profiles are upward-fining fluvial packages that have been overprinted by pedogenesis. Their proximal position to a former stream channel creates a condition in which pedogenesis is commonly interrupted by overbank deposition. The bones associated with these deposits were transported as bed load. Although dominated by disarticulated bones, several complete turtle fossils are found lateral to these paleosols. Paleosols in this profile are most similar to the Zisa Series of Retallack (1983b).

The Tilted Turtle profile is located approximately 0.5 km northeast of the Conata Picnic Grounds (Fig. 3.3). This site is noted for the odd tilted orientation of an associated turtle (tortoise) carapace (Fig. 3.8). From a facies perspective, this site represents a proximal to medial floodplain position and is composed of an upward-fining sequence of overbank

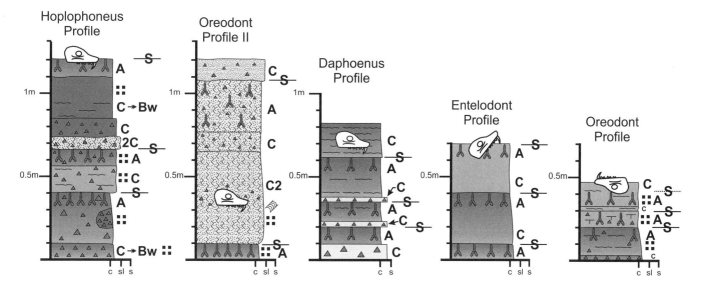

3.5. Soils of the Scenic Member in Chamberlain Pass. See Fig. 3.6 for explanation of symbols.

sediments, possibly a crevasse splay, given the siltstone at the base and the presence of claystone breccia that also fines upward. From a paleopedological perspective, this site is an A–Bw–C profile (inceptisol) that overlies an earlier soil and is in turn capped by an overlying profile. The proximal to medial floodplain position creates a condition in which pedogenesis was allowed sufficient time to begin modification of the profile to include a Bw (cambic) horizon (Fig. 3.4). The tortoise shell is in good condition and is buried within the uppermost horizon of the profile. On the basis of the presence of drab-haloed root traces, this was a stable land surface for a brief period of time before additional sediments accumulated and buried the profile. Vertical accretion was likely episodic given the presence of relict bedding and root traces in the uppermost A horizon. The entire profile was then removed out of the range of active pedogenesis by influx and burial of new sediments (uppermost C horizon). This paleosol is most similar to the Conata Series of Retallack (1983b).

The Bessie Site profile is located approximately 0.5 km north-northeast of the Conata Picnic Ground (Fig. 3.3). It is noteworthy for the articulated skull and mandibles of the oreodont, *Merycoidodon*, in the middle of the profile (Fig. 3.9). From a facies perspective, this profile is proximal, but not immediately next to, a former channel. From a paleopedological perspective, this is a simplistic A–C (entisol) soil that appears to have been truncated by erosion. The base of the profile is a claystone breccia-rich C horizon overlying a former A horizon of an underlying profile marked by drab-haloed root traces. This C horizon is partially eroded and capped by another claystone breccia-rich C horizon. Drab-haloed root traces penetrate this horizon from above, although no well-defined A horizon is present. This may

be due to frequent episodes of vertical accretion. Accretion eventually outpaced pedogenesis and buried the profile. The oreodont skull and mandibles are associated with the weak zone of drab-haloed root traces. The bones are in excellent condition except for the presence of gnaw marks on the sagittal crest area of the skull. This suggests that the skull was exposed briefly on the land surface before burial, but not long enough to incur damage from weathering or disarticulation by scavengers. The paleosols in this profile are most similar to the Zisa Series of Retallack (1983b).

Hay Butte Marker Bed

Although the majority of the paleosols and ancient river deposits suggest that the landscapes of the Lower Scenic Member predominantly supported weakly to moderately developed soils because of frequent flooding, one paleosol profile within the Scenic Member tells a different story. As we look at the various stripes of paleosol profiles that stretch across the buttes of the Badlands, we can occasionally find one that can be traced over great distances. Within the Scenic Member, one particular paleosol profile forms a prominent, regionally extensive marker (Evanoff in Benton et al., 2007).

The Hay Butte marker bed is a paleosol profile that formed across an ancient landscape of the Scenic Member (Plate 4; Fig. 3.10), but depending on where you look, the physical nature of the Hay Butte marker bed changes (McCoy, 2002). In Tyree Basin (Fig. 3.3), the Hay Butte marker has an extremely enriched horizon of percolated clay in the subsoil (argillic horizon; Bt). Clay accumulated to such a degree that it created films of clay (argillans) up to 0.5 cm thick around peds. Most argillans are no more than a millimeter

3.6. Legend of paleosol symbols used throughout Chapter 3.

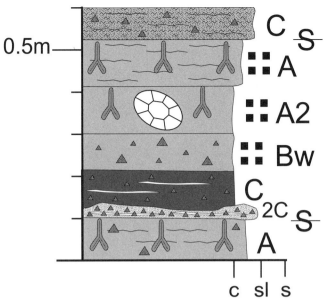

3.8. Titled Turtle Profile in the Scenic Member. See Fig. 3.6 for explanation of symbols.

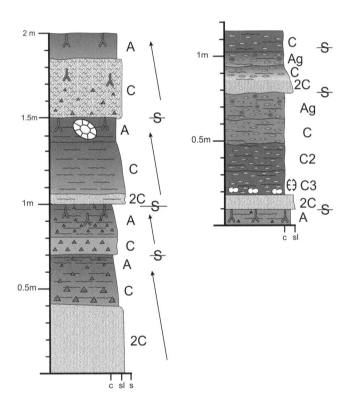

3.7. Jerry's Bone Bed paleosol profiles in the Scenic Member. These profiles are lateral to each other and demonstrate changing hydromorphic conditions. See Fig. 3.6 for explanation of symbols.

thick. Roots within the Hay Butte marker bed are quite small and suggest little to no influence by trees (similar to grassland soils of today: mollisols), and the sediments that make up the uppermost part of the Hay Butte marker bed are fine grained and massive, suggesting input via windblown (eolian) transport of silt. In Dillon Pass and at Hay Butte proper (Fig. 3.3), this interval is marked by thinner, less developed inceptisols.

The two most likely explanations for the regional variability across this ancient landscape (paleocatena) include tectonics and sedimentation. Whereas most paleosols of the Scenic Member are responding to their proximity to an ancient river and the effects of frequent versus infrequent flooding events, the Hay Butte marker bed represents a response to larger-scale tectonic forces. Dillon Pass rests on the Sage Arch, an uplifted area that acted as a highland during the formation of the Hay Butte marker. This relative difference in elevation from the "upland" of the Sage Arch to the "lowland" of the Tyree Basin area is reflected in the weaker soils on the flank and crest of the Sage Arch and the extremely well-developed paleosols of the Hay Butte marker in Tyree Basin (McCoy, 2002). The episode of pedogenesis that modified the Hay Butte marker was also influenced by a period of lower-than-normal sedimentation across the region that would have allowed the Hay Butte marker paleosols to develop into more mature profiles compared to those in other strata of the Scenic Member. The Hay Butte marker bed is capped by typical fluvial sedimentation found elsewhere in the Scenic Member, but the nature of sedimentation higher up in the Scenic Member suggests a greater influence of eolian sedimentation over time.

Paleosols of the Poleslide Member

The transition to the overlying Poleslide Member of the Brule Formation is marked by several distinctive features, and the paleosols within it record the influence of a combination of fluvial, lacustrine, and eolian environments, each of which is dominant at different times throughout the deposition of this unit. The Poleslide Member is distinguished from the underlying Scenic Member by an increase in the

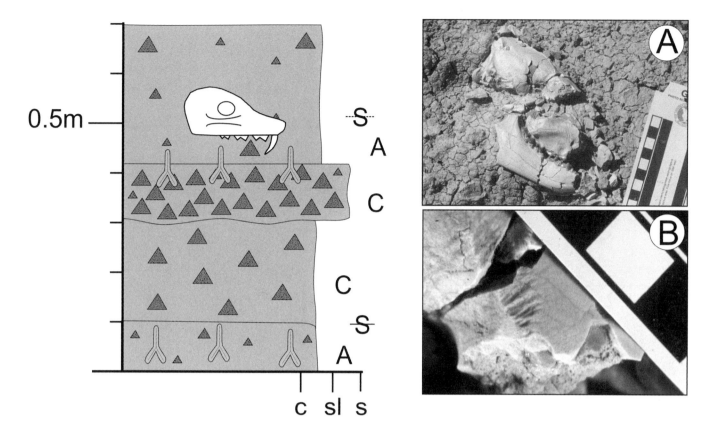

3.9. Bessie Profile in the Scenic Member, in situ skull (A), and ancient rodent gnaw marks along the top of the skull (B). See Fig. 3.6 for explanation of symbols. Photos by the authors.

amount of silt. This translates to a relatively greater resistance to erosion for the Poleslide Member, which is why this unit erodes into vertical cliffs (Plate 4). The other characteristic that coincidentally marks the boundary between the Scenic and Poleslide members is a silver-beige layer of volcanic ash, referred to as the Cactus Flat bentonite bed (more appropriately a marker bed) by Stinchcomb (2007), which extends across the northern part of the park.

Although there are stripes of color that represent paleosols that formed on landscapes of the Poleslide Member, some soils are different than those that formed earlier in the history of the Badlands and result in a more uniform color for the Poleslide Member (Plate 4). Retallack (1983b) recognized five distinct types of paleosols within 46 m of the Poleslide Member in the Pinnacles area of the park (Fig. P.1), two of which appear for the first time. In stratigraphic order, these include the Ogi, Wisange, Zisa, Gleska, and Pinnacles soils.

The Wisange Series represents weakly to moderately developed soils. They are inceptisols (A–B–C profiles) that formed on proximal floodplain positions, similar to the Ogi Series, but on slightly more elevated landscapes. They have slightly clay-enriched B horizons and roots ranging from small to large in size. The Pinnacles Series represents calcareous, ash-rich soils dominated by fine roots. They are aridisols with simplistic A–C horizonation that formed under dry

climates on distal floodplain settings. The soils and ancient environments described herein for the Poleslide Member are based on observations concentrated around the Door and Window trailheads, at Cedar Pass, and at Norbeck Pass (Fig. P.1).

Paleopedology of the Scenic/
Poleslide Member Contact

Pedogenesis associated with the Scenic/Poleslide contact is unique within the section. This is a moderately to well-developed, well-drained, and oxidized soil (Fig. 3.11). There are distinct soil horizons, root traces, ped structures, argillans, and soil fabric. Roots are abundant as branching, drab-haloed and clay-filled root traces up to 0.5 cm in diameter. The small root size suggests shrublike vegetation. This paleosol is similar to udalfs (humid climate alfisols; A–Bt–C profiles). According to Retallack (1983b), paleosols of the Scenic/Poleslide Boundary are classified as the Ogi Series (weak A–C profiles, occasionally with clay and carbonate accumulations) in the Pinnacles area. This paleosol is not an Ogi equivalent. The dissimilarity may be due in part to the revision of the boundary between the Scenic and Poleslide members by Evanoff (in Benton et al., 2009). This paleosol is most similar to the Gleska Series of Retallack (1983b), which

3.10. Paleosol profiles within the Hay Butte marker bed of the Scenic Member across the North Unit of Badlands National Park. See Fig. 3.3 for location of individual profiles and Plate 4 for outcrop picture. Modified from McCoy (2002). See Fig. 3.6 for explanation of symbols.

is found near the revised lithologic boundary. The base of the Poleslide Member is a platy, greenish, muddy siltstone layer with prominent relict bedding. This greenish siltstone is also rich in vertebrate fossils, ranging from isolated bones to articulated skeletons.

Basal fluvial soils The bottom 5 m of the Poleslide Member are characterized by a repetitive pattern of pedogenically modified nodular silty layers that are capped by finer muds and clays (Fig. 3.12). Small roots are common at the top of these individual 30 to 50 cm thick packages, which in turn are covered by the next silt to clay package. These soils are weakly developed entisols (A–C and A–Ck profiles; k = carbonate enriched) with little evidence of pedogenesis other than drab-haloed and clay-filled root traces, and occasional carbonate nodules. Fossils within these soils occur on top of upward-fining sequences as flat-lying, isolated bones,

commonly within carbonate nodules. The orientation of the bones, their association with weakly developed soils, the overall upward-fining nature of these individual packages of sediment, and the concentration of roots at the tops of these packages suggest a scenario in which bones lying on ancient landscape were covered by flooding events. These soils are most similar to the Ogi Series of Retallack (1983b).

Lacustrine influenced soils Lacustrine environments are dominant in the middle of the Poleslide Member in the Door and Window area. Lateral to this ancient lake, the paleosols show evidence of groundwater influence as mottles, but also zones of intense cementation by calcium carbonate. These zones of intense carbonate cementation manifest themselves as resistant ledges in outcrop and contain well-preserved traces of former soil ecosystems, including fossilized dung balls, larval chambers of subterranean bees,

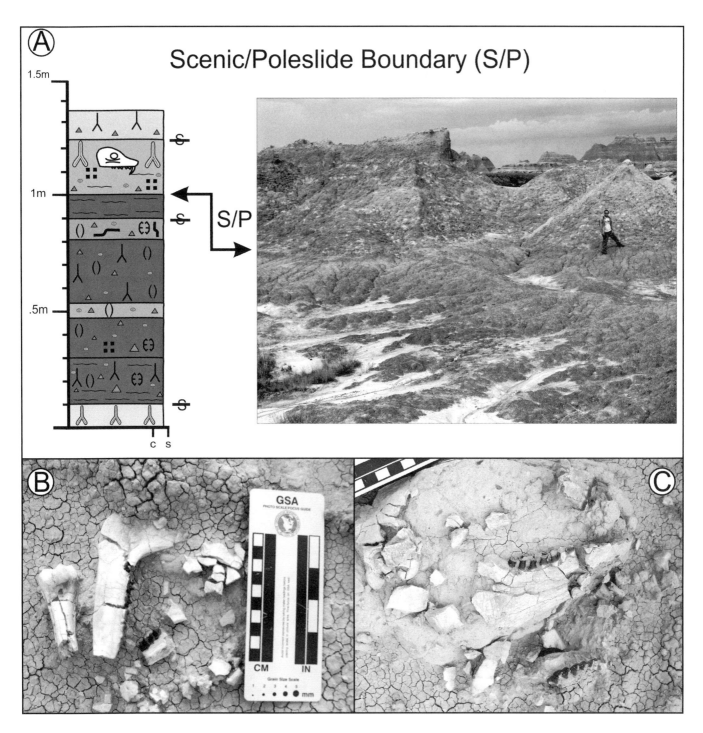

3.11. Paleosol profile and outcrop photograph across the Scenic/Poleslide Member contact near the Fossil Exhibit Trail (A). Person for scale is 1.8 m tall. (B), (C) In situ fossils associated with the boundary. See Fig. 3.6 for explanation of symbols. Photos by the authors.

burrows, and calcified root traces (Plate 7). On occasion, vertebrate fossils can be found within or directly on these resistant ledges, suggesting that they formed very near to, if not at, the ancient land surface.

The characteristics of these paleosols lateral to the ancient carbonate lake suggest a combination of wetter soils proximal to the lake and drier soils farther away (Mintz, 2007; Mintz, Terry, and Stinchcomb, 2007). The preservation of burrows and dung balls suggests soil conditions that were well drained because these features would not have been formed beneath the water table. The dense concentration of roots, dung balls, burrows, and larval chambers, along with a greater-than-normal background accumulation of vertebrate fossils, suggests that these were stable landscapes with little to no sedimentation. On the basis of root diameters (most 5 mm or less, with infrequent traces up to 1 cm), landscapes at this

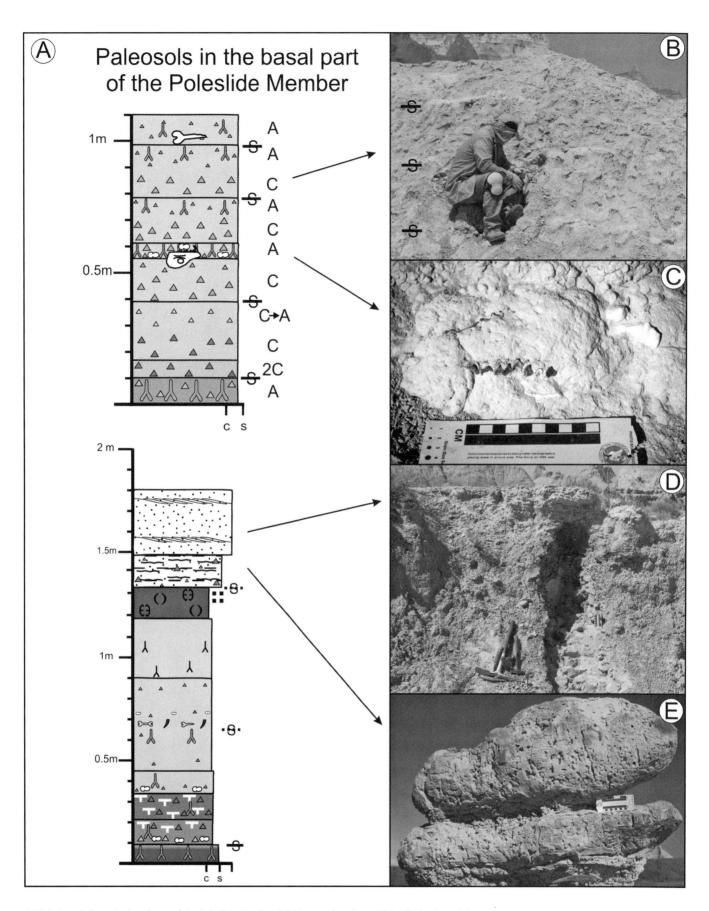

3.12. Paleosols from the basal part of the Poleslide Member. (A) Measured sections. (B) Stacked paleosols in outcrop. (C) Inverted skull on ancient land surface. (D) Fluvial sandstone capping paleosol profile. (E) Close-up of fluvial sandstone showing voids formed by ancient roots. See Fig. 3.6 for explanation of symbols. Photos by the authors.

Stable and aggradational landscapes of the Poleslide Member

3.13. Photograph of stable (A) and aggradational (B) landscapes and associated paleosol characteristics in the Poleslide Member at Door and Window Overlook. Note differences in erosional relief. See Fig. 3.6 for explanation of symbols. Photo by the authors.

period of time were open savanna-type environments with few to no trees. Within modern soil taxonomy, paleosols on these ancient landscapes would be similar to grassland soils of the Great Plains (mollisols) to those found in drier parts of the western United States (aridisols).

Eolian-dominated soils Landscapes dominated by eolian processes were present before and after this period of lacustrine-dominated sedimentation. The rate of eolian influx was not steady, resulting in dramatically different soil types (Fig. 3.13). During periods of relatively faster eolian influx, landscapes were progressively buried as soils kept pace with deposition, which created zones of massive, siltstone-dominated intervals. Pedogenic features, such as roots, are scattered throughout these zones of massive siltstone, and distinct paleosol horizons are absent (Fig. 3.14). Fossils appear to float within these massive intervals without reference to defined land surfaces. The lack of defined horizons and random distribution of fossils throughout these massive siltstones suggests cumulic soil-forming conditions during

which the landscape slowly aggraded by the addition of eolian materials. New material was incorporated into the developing soil, creating overly thickened soil profiles without defined horizons. These soils are most similar to the Pinnacles Series of Retallack (1983b). Conversely, periods of low influx of eolian sediments resulted in stable landscapes that allowed for concentrated accumulation of pedogenic features. These stable soils manifest as resistant grayish-white layers that contain a high concentration of hairline carbonate rhizoliths, dung balls, burrows, and sweat bee pupae (Plate 7; Figs. 3.13, 3.14). These resistant horizons grade downward to relatively less carbonate-rich zones. Vertebrate fossils are concentrated within and at the top of these resistant layers (Benton et al., 2009). The presence of calcium carbonate at such a shallow depth and the prominence of a thick mat of carbonate rhizolith root traces containing dung balls and sweat bee pupae suggest that these soils formed under arid conditions. There is no direct match with the paleosols of Retallack (1983b). They are most similar to the Wisange and

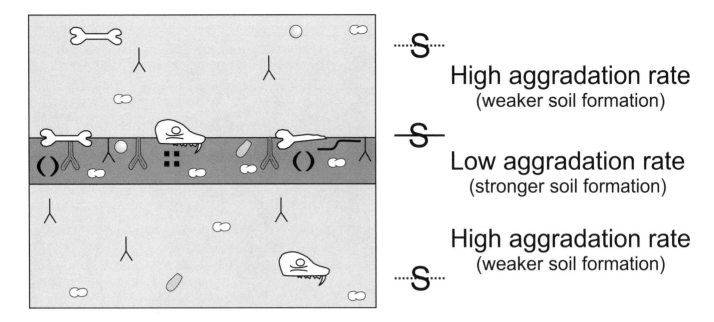

High aggradation rate
(weaker soil formation)

Low aggradation rate
(stronger soil formation)

High aggradation rate
(weaker soil formation)

3.14. Model of stable versus aggradational paleosol phases in the Poleslide Member. Note higher concentration of paleosol features during stable phases. See Fig. 3.6 for explanation of symbols.

Pinnacles series, although the degree of carbonate accumulation and overall profile development within the Wisange and Pinnacles series is less.

PALEOSOL PERSPECTIVE

Paleosols of the White River Group provide a unique perspective on paleoclimatic change across this region from the late Eocene into the Oligocene. On the basis of paleosols alone, the paleoclimates in this region of North America appear to have undergone a progressive drying trend that changed the landscape from a predominately forested habitat in the late Eocene Chadron Formation to open savannas of the Oligocene Brule Formation. These changes are recorded in the types of paleosols that formed. In particular, the most obvious clues to paleoclimatic change include a change from soil horizons enriched in translocated clays in the Chadron Formation to progressive enrichment of calcium carbonate in soils of the Brule Formation, and the change in the overall size of fossil roots from forested vegetation in the Chadron Formation to more open, prairielike vegetation in the Brule Formation. Across individual landscapes, the dynamics of sedimentation and relative degrees of time available to form these ancient soils resulted in weaker soils near ancient rivers, or during periods of rapid influx and aggradation of eolian materials. Geographic locations farther away from active river channels, or landscapes unaffected by rapid eolian influx, were able to generate better-developed soils. Some of the most distinct marker beds in these rock units, such as the Hay Butte marker, are the direct result of distinctive periods of pedogenic modification along a defined landscape. Although the data preserved within the paleosols suggest an overall change to progressively open landscapes from the late Eocene to the early Oligocene within the White River Group, the cause for this shift could have been paleoclimatic change, or possibly an increase in the rate of sediment delivery to this region. As with any research, more data are required in order to completely understand the dynamics of paleoecosystem change within these strata.

The other important consideration regarding the paleosols of the White River Group is the role they played in creating the lithologic units that we see today. The sediments that accumulated to form the White River Group were modified by pedogenesis before their eventual burial and preservation. The warm, humid conditions of the late Eocene resulted in soils enriched in the smectite clay formed by the breakdown of the large amounts of volcanic materials present across this region. By the time the Brule Formation was being deposited, the rate of weathering of volcanic material lessened, resulting in the preservation of original ash materials. This in turn created siltier sedimentary deposits with less smectite. These differences in ancient weathering during the deposition of the Chadron and Brule formations manifest today as the low, rounded, smectite-enriched hills of the Chadron Formation versus the silty cliff- and spire-forming Brule Formation (Plate 4).

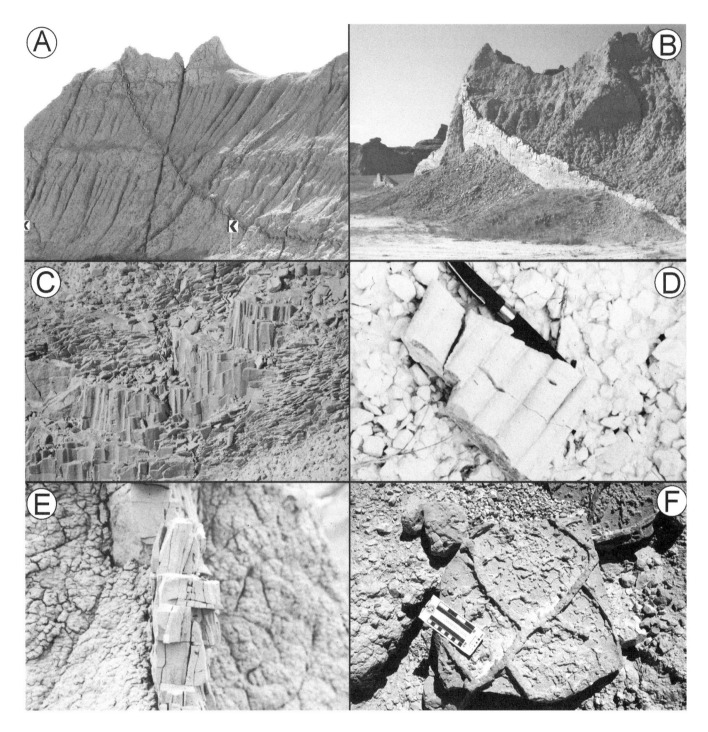

4.1. Photographs of clastic dikes. (A) Multiple crosscutting clastic dikes on Highway 240 near Pinnacles Overlook. (B) Clastic dike along the Old Northeast Road showing sheetlike morphology. (C), (D), (E) Close-up of clastic dikes showing crenulated texture. (F) Mud cracks from strata near Door and Window Overlook. Note size difference and fracture pattern compared to clastic dikes. Photos by the authors.

Postdepositional Processes and Erosion of the White River Badlands

4

THE BADLANDS THAT WE SEE TODAY ARE THE RESULT OF both depositional and postdepositional processes. Our understanding of these processes and their role in the formation of the Badlands is only possible as a result of the significant erosion that characterizes this region. We discuss postdepositional features first, with the understanding that it is erosion that has allowed us to study these features.

BURIAL AND LITHIFICATION

The rock strata that comprise the Badlands were once loose sand, silt, and clay, with the occasional wind-deposited tephra and freshwater limestone. Upon burial, these sediments were turned to rock by compaction and cementation during the process of lithification. Compaction is due to the enormous overlying weight of additional sediments that are brought into the basin. As a result, the thicker the accumulation of sediments, the greater the reduction of porosity (void space) between the grains. Cementation is the process by which individual grains of sediment are bound together by the precipitation of secondary minerals, commonly calcite, quartz, or iron oxide, out of groundwaters that have moved through the sediment. Geologists refer to lithification as just one form of diagenesis, the sum of physical and chemical changes that can be induced in sediment upon burial.

Additional chemical diagenetic changes include the wholesale change of certain minerals into new phases, such as the dewatering of iron oxyhydroxides to hematite and the conversion of smectite (swelling) clays to illite clay by the addition of potassium. Along with cementation, entirely new minerals can grow in available pore spaces. In the White River Group, the common secondary minerals include calcite, gypsum, quartz, chalcedony, barite, various uranium-bearing minerals, and zeolites (Retallack, 1983b; Terry and Evans, 1994). Many of these diagenetic changes require microscopes to detect small-scale mineralogical changes and specialized instruments to measure the chemical makeup of rock samples. Diagenetic changes that can be seen with the naked eye include the overall reddish appearance (due to the recrystallization of various yellowish-brown iron oxyhydroxides into hematite) that are seen in many of the stripes that cut across the Badlands, and secondary mineralization within fractures caused by tectonic forces that compressed and stretched rock units of the Badlands.

CLASTIC DIKES

Some of the more prominent postdepositional features of the Badlands are the numerous nearly vertical sheets of resistant material that crosscut the horizontal layering of the Badlands (Fig. 4.1). These features, referred to as either clastic dikes or chalcedony veins, depending on the type of material that is present, vary in size and in regional and stratigraphic distribution, and are formed by different processes.

Clastic dikes are composed of small particles of sediment that have been cemented to form ridgelike structures that cut across the Badlands. Their resistance to erosion is greater than that of typical Badlands rocks, which results in "fins" of clastic dike material that stand out in erosional relief and commonly help to support less resistant Badlands materials (Fig. 4.1). The appearance of these dikes can vary from light beige colors that are similar to typical Badlands strata to those that are stained with a greenish color. When viewed from the edge, these features range from paper-thin bodies to tens of centimeters thick that cut across tens of meters of solid rock. Some are relatively smooth along their surfaces, whereas others appear to be crenulated, similar to a wavy potato chip (Fig. 4.1). These dikes sound a glassy ring when pieces are struck together. When viewed along the surface of the ground or from the air, these sheets of clastic material are sometimes straight lines, but in other instances they form curvilinear features that seem to randomly run across the surface for tens to hundreds of meters. Models describing the formation of these dikes have included large-scale desiccation cracks that were filled with sediments (Retallack, 1983b), fractures that were formed (and later filled) in response to tectonic and structural strain (Smith, 1952), and injection of liquefied sediments from deeper in the Badlands strata (Whelan, Hamre, and Hardy 1996).

Modern-day desiccation cracks (mud cracks) form by the drying and shrinkage of clays and can be found in such settings as muddy stream banks or lakes that have dried. The resulting cracks decrease in width downward and tend to join other cracks to form polygonal patterns ranging in scale from several centimeters across to tens of meters. These cracks can be infilled with new sediment and preserved in the geologic record (Fig. 4.1). With respect to the clastic dikes in the Badlands, some of these features pinch out and narrow in a downward direction. The most common examples of this are found in association with green-colored clastic dikes.

Although some clastic dikes can be seen to narrow downward (which would support a desiccation model), when viewed from the air, these dikes do not create large-scale polygonal networks similar to modern mud cracks. In addition, shrinkage via desiccation only occurs in moist clay-size sediments, whereas these dikes crosscut tens of meters of solid rock of varying grain size. It is possible that these cracks were at one time open to an ancient surface sometime in the geologic past, but no such relationship has yet been documented. If these fractures were indeed open to an ancient land surface, such features as fossil bones from animals that fell into these fissures or evidence of younger sediments filling dikes that cut across older strata should be present. To date no such evidence has been documented, although no such study has yet been carried out in the Badlands. On one occasion, one of the authors (Terry) observed a fossil leaf impression in clastic dike material from strata equivalent to the Badlands at the Flagstaff Rim locality southwest of Casper, Wyoming. On another occasion the same author observed a clastic dike composed of pea-size gravel cutting through the Chadron Formation in the Big Corral Draw area in the South Unit of Badlands National Park (Terry and Spence, 1997). Both of these observations would suggest that at least some of these dikes were indeed open to the surface in the past, but they do not support the model of large-scale desiccation cracks.

Fractures in rock are commonly caused by geological forces of stress and strain acting over broad regions. In order to form large, open fractures, rock strata must be stretched. This induces tensional forces that can eventually fracture the rock to create open fissures. These fissures tend to be vertical and parallel to each other. Such forces were created by the uplift of the Black Hills, but this occurred before the majority of the Badlands were deposited. It is possible that some fracturing is related to the formation of the asymmetric basin that contains most of the Badlands strata in this region (Fig. 4.2). This basin is hinged on the northeast edge and drops downward to the south and west. This type of basin requires crustal thinning (tensional forces) and could be a consequence of aftereffects of the uplift of the Black Hills as the surrounding strata on the Great Plains could again relax after uplift.

If this model is correct, the majority of clastic dikes should show a roughly southeast-to-northwest orientation, parallel to the major structural axes of the Badlands. Although many dikes do follow this orientation, many more run southwest to northeast, suggesting that either a secondary stress field of extension was present or that fracturing was due to compressional forces, which creates sets of fractures that cut across each other. Tensional fracturing also does not explain the exquisite curvilinear nature of many of these fractures when viewed from the air.

Our final hypothesis to explain these features is injection of clastic materials from below. Under high pressures, sediments can become suspended in a fluid and injected into overlying strata. Such conditions are common during earthquakes, during which the intense shaking of the ground liquefies unconsolidated sediment as groundwater is forcefully ejected to the land surface (liquefaction). This process can result in localized sand volcanoes or small-scale dikes that are injected into surface soils. These dikes can be traced downward over several meters and can be seen to connect with laterally extensive beds of sandy material that were originally deposited by normal geologic processes, such as a former river.

In deeper geologic settings the fluidization of sediments and subsequent injection is more complicated. Certain minerals are more susceptible to breakdown after burial. In the case of the Badlands, the large volume of volcanic ash that was brought into this region by wind and rivers would have been easy to modify upon burial. This modification, referred to as desilicification and dewatering, creates a situation in which masses of sediment at depth become liquefied as the ash breaks down. The overlying layers of sediment exert significant downward pressure on these liquefied beds, which forces the liquid up and along any fracture that can be exploited.

Evidence of injection is prominent along the Old Northeast Road just north of Cedar Pass (Fig. P.1). These features were described by Whelan, Hamre, and Hardy (1996) and include upward-radiating fingers of dike material that branch off thicker dikes, and dikes that divide and rejoin. Upon closer examination, the dike material appears to be laminated. In some instances the dike fill appears to be cross-bedded in an upward direction. Both of these features are indicators of flow. Clastic dikes in this area also manifest a crenulated appearance, some of which may be the result of sculpting and scouring during fluid flow (Fig. 4.1). None of these features would be present if the dike fill was the result of passive

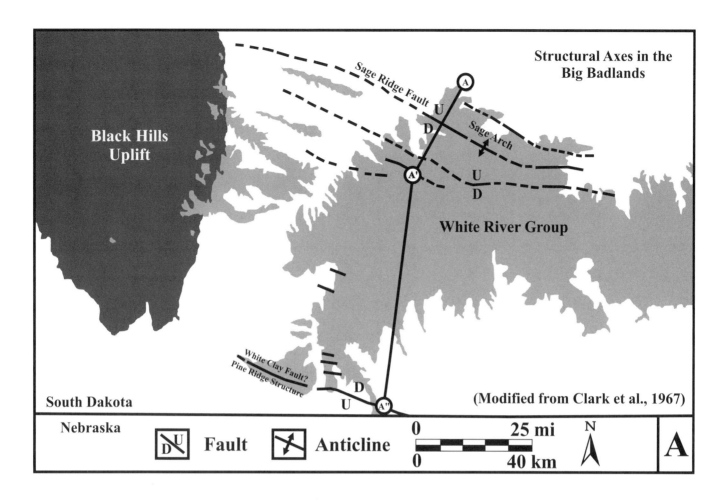

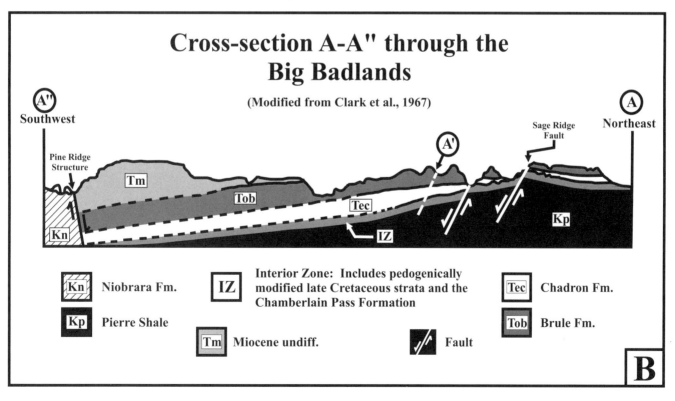

4.2. Regional geologic map (A) and cross section (B) showing the geographic and structural relationships between the deposits of the White River Group, the Black Hills, and the bounding faults that created this basin.

infilling of a fracture open at the land surface. In addition, some of these dikes show multiple generations of dike formation, which appear as variations in the color of the dike material or grain size.

One other possible explanation is the conversion of opaline silica to more stable forms by recrystallization (Davies et al., 2006). This process produces excess water and a reduction in sediment volume, creating overpressurized conditions and eventual fluidization of sandy materials and upward injection into overlying strata. This phenomenon has been documented at shallow depths (less than 50 m) in biosiliceous-enriched marine sediments. Although biosiliceous deposits are not known within White River Group, the underlying Cretaceous Pierre Shale, which would have a greater abundance of biosiliceous materials, could possibly be responding to such a recrystallization mechanism. If this is indeed the case, clastic dikes should be more widespread throughout the region because the Pierre Shale underlies the entirety of South Dakota.

CHALCEDONY VEINS AND OTHER SILICEOUS MASSES

At the other end of the spectrum are the numerous veins of chalcedony that can be seen in certain parts of the park. The chalcedony veins are commonly various shades of blue, olive, yellow-gray, and occasionally red, and they can vary in width from knife-edge to several centimeters. Although the veins are most commonly composed of chalcedony, calcite and gypsum are sometimes present, as are pockets of typical Badlands mudstone (Fig. 4.3).

The chalcedony veins appear to be restricted both stratigraphically and regionally throughout the park. The most prominent zone of chalcedony veins is primarily restricted to the upper part of the Chadron Formation and ends just below, or slightly within, the overlying Scenic Member of the Brule Formation. They can be easily seen along Highway 44 headed toward Scenic, South Dakota, in the narrow neck of NPS property that crosses the road (Fig. P.1). The veins stand out in erosional relief as sharp, low-standing ridges of material approximately 5 to 10 cm above the bedrock. Over time, these sheets of material break off and cover the surrounding slopes, which give the Badlands a slightly blue tint. On a regional scale, the chalcedony veins appear to be concentrated toward the southwest part of the park, especially in and around Chamberlain Pass, Sheep Mountain Table, and the White River Visitor Center (Fig. P.1). Veins identical in mineralogy but with a much deeper blue color, referred to as blue agate, can be found in Toadstool Geologic Park north of Crawford, Nebraska.

The genesis of the chalcedony veins is enigmatic. Unlike the clastic dikes, which are primarily vertical features, chalcedony veins are both vertical and horizontal, although vertical orientations are most prominent. At scales of 10 m² the chalcedony veins appear as rough polygonal patterns reminiscent of large desiccation cracks (Fig. 4.3). Upon closer examination, the veins show evidence of multiple episodes of mineral precipitation, sometimes as alternating colors of chalcedony and other times as alternating bands of gypsum and chalcedony. Some veins also show evidence of fracture and subsequent healing. In some instances, such as near the White River Visitor Center, chalcedony veins experienced ptygmatic folding (Fig. 4.3), suggesting compaction of strata surrounding the vein and subsequent deformation. On the very small scale, chalcedony occasionally manifests as three-dimensional boxworklike structures that pinch out over tens of centimeters in all directions. In other instances, isolated bodies of chalcedony can be found that display bladed forms, similar to desert roses (Fig. 4.3), which suggests replacement of an original calcite or barite precursor by silica. Amorphous, vesicular masses of chalcedony are sometimes found in association with bladed morphologies (Fig. 4.3).

The underlying cause behind the genesis of these veins is still a mystery. Any proposed hypothesis to explain the formation of these veins must be able to account for the production and availability of a large volume of free silica. The White River Group is enriched in volcanic ash, the hydrolysis of which produces smectite clays and free silica. The late Eocene Chadron Formation is abundant in smectite, which gives this unit the thick popcorn-textured weathered surface and rounded, haystacklike hills compared to the overlying Oligocene Brule Formation. The smectite enrichment in the Chadron Formation is thought to represent ancient weathering and soil formation under humid, forested conditions that gave way to drier conditions (and hence less hydrolysis) in the Oligocene Brule Formation (Retallack, 1983b; Terry, 1998, 2001). It is unlikely that a greater amount of hydrolytic weathering during the late Eocene is responsible for

4.3. Photographs of chalcedony veins and other siliceous accumulations. (A) Top view of vein showing multiple generations of vein fill and stress-related shearing (white lines and arrows). (B) Polished section of chalcedony vein showing clay masses (cl) inside of chalcedony (ch). (C, D) Patch of chalcedony veins just south of Sheep Mountain Table. Note curvilinear nature and rough polygonal outline. (E) Vertical vein exposure with ptygmatic folding (P). (F) Boxwork structure of chalcedony ranging from knife-edge to centimeters in thickness. (G) Bladed chalcedony pseudomorph from north end of Sheep Mountain Table. Original mineralogy is unknown. Coin for scale is 1.8 mm wide. (H) Frothy mass of chalcedony from north end of Sheep Mountain Table. Void spaces originally filled with mudstone. Coin for scale is 1.8 mm wide. Photos by the authors.

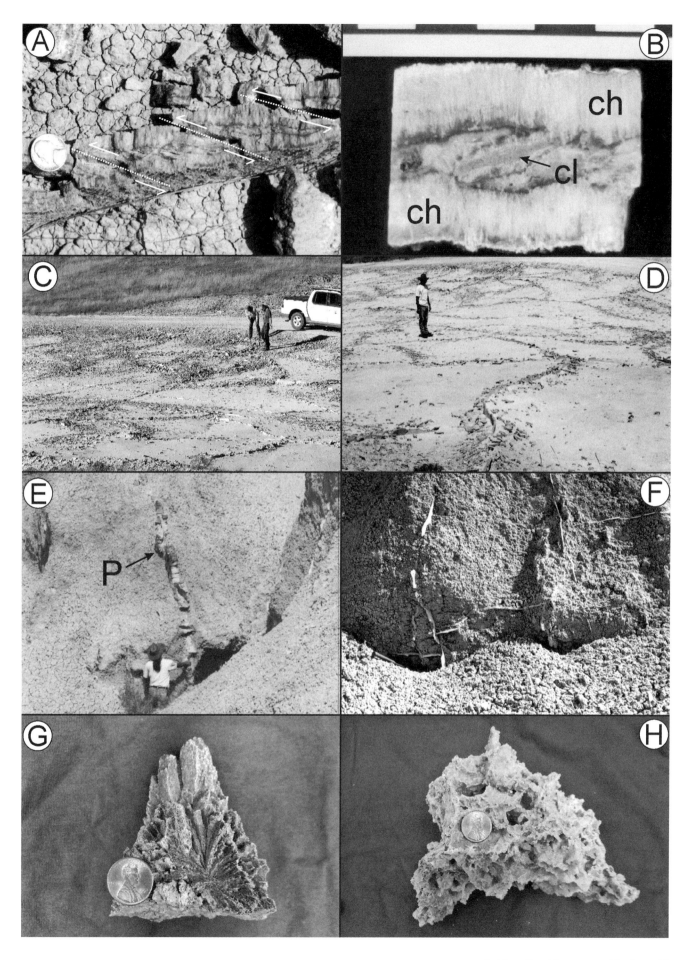

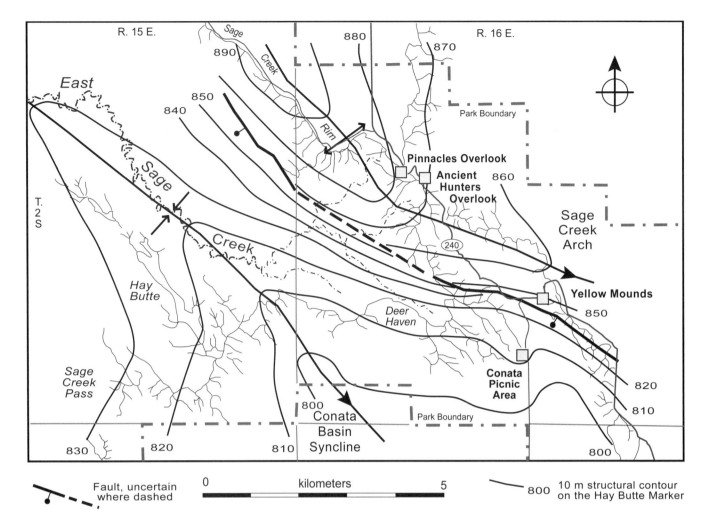

4.4. Dillon Pass fault and the structural features in the north central area of Badlands National Park. The structural contours are on the elevations of the Hay Butte marker. The fine lines are ridge lines along the Badland Wall and below the Sage Creek Rim.

contributing silica for these veins because the veins crosscut lithified sediments of the Chadron Formation and extend up into the lowermost Brule Formation.

Emplacement of these veins would have required water capable of carrying large amounts of dissolved silicon, calcium, and sulfur in order to produce chalcedony and gypsum. Because the majority of the veins end within the lower part of the Brule Formation, it is most likely that the veins (and fluids) originated at depth and migrated upward. This is supported by the presence of veins within outcrops of the Cretaceous Pierre Shale along Sage Creek near Sage Creek Campground (Plate 1). Several towns in and around the Badlands make use of geothermal wells that produce water heated at depth. Most notable is the town of Hot Springs, South Dakota, on the southern tip of the Black Hills; Wall, South Dakota, just north of the Badlands; and Philip, South Dakota, to the northeast of the Badlands. Geothermal waters are commonly enriched in dissolved minerals that precipitate upon exposure to air (e.g., the mineral precipitates around geysers at Yellowstone National Park), but precipitation at

depth is also possible if fluids become oversaturated with dissolved solids. The repetitive and alternating episodes of chalcedony/gypsum mineralization within the same vein structure, healing of fractured veins, and limited areal extent of chalcedony veins support a geothermal model. According to Evans and Terry (1994) and Evans (1999), localized occurrences of tufas and tufa pipes can be found throughout the Badlands and are interpreted to be the result of regional groundwater flow and recharge off the eastern flank of the Black Hills. The distribution of these tufas and pipes may be influenced by structure, such as the numerous normal faults that radiate off the Black Hills.

According to Hoff et al. (2007), chalcedony veins are organized and consistent in their geometry and orientation on the small scale (100 m²), but over larger regions of southwest South Dakota and northwest Nebraska, the orientation of the veins becomes increasingly random. This suggests a lack of tectonic influence and supports a syneresis model of vein genesis. Maher and Shuster (2012) also describe localized areas of uniform vein orientations, especially in association

4.5. Faulting along Highway 240. Arrows show relative direction of displacement; X–X' denote offset beds. Photo by the authors.

with faults, with increasingly random orientations at larger scales. They noted that chalcedony veins are concentrated at multiple stratigraphic levels within the Badlands, and they can be seen to transition into clastic dike materials upsection. Many of the veins appear to have been subjected to compaction, as assessed by their striated outer textures, suggesting that sedimentary volume loss associated with compaction and diagenetic phase changes in silica, smectite clays, or zeolite minerals, dewatering, and fluid migration are likely responsible for the genesis of these features.

FOLDS AND FAULTS

Postdepositional tectonic forces modified the strata of the Badlands with the formation of folds and faults. Folding is best exemplified by the Sage Arch, an area of uplift and warping in the Dillon Pass area of the park that is oriented northwest–southeast (Plate 8). Associated with this feature, especially in Dillon Pass and along the Sage Creek Rim Road, are numerous faults that were created by the tensional

thinning of the rock strata as a result of the uplift of the Sage Arch (Fig. 4.4). These faults, referred to by geologists as normal faults, are characterized by the downward slipping of overlying bodies of rock (headwall block) along a fault plane relative to the underlying footwall block (Plate 8). The movement along these faults can range from only a meter or so to tens of meters and are easily recognized by the juxtaposition of distinctly different rock strata or striping within the buttes. Normal faulting is common along the Badlands Wall from the Sage Creek Rim Road area to the southeast along the Highway 240 loop road, and through the Cedar Pass area of the park (Fig. 4.5). On a larger scale, these same tensional forces created the asymmetric basin that protects the Badlands strata from erosion (Fig. 4.2).

EROSION

The Badlands that we see today are the result of a shift from deposition to erosion that began about 660,000 years ago (Stamm et al., 2013). The last remnants of deposition in the park are preserved as patches of large gravels and cobbles, referred to as the Medicine Root Gravels, which overlie the lithified Badlands bedrock (Fig. 4.6). During this time, eastern South Dakota was repeatedly subjected to glacial advances and retreats of the Laurentide ice sheet, which generated the relatively flat terrain by deposition of glacial till, a mixture of grain sizes from clay to boulders carried by the glacial ice. The last of these glacial advances, the Wisconsinan glacial episode, pushed the Missouri River to its current position. East of the river, glacial deposits covered older bedrock.

West of the river, bedrock remained exposed at the surface, including deposits of the former Western Interior Seaway and the Badlands, and was subjected to extensive erosion primarily by rivers such as the Bad, White, and Cheyenne, but some erosion was due to wind that carried the lighter particles farther to the east. Erosion and downcutting by rivers is common, although the exact reason for this period of incision in western South Dakota is unknown. Possible triggers include a pulse of regional tectonic uplift, climate change, or a reduction in regional base level. The change from deposition to erosion was also enhanced by the actions of the Cheyenne River, which eroded headward and eventually captured the original drainages flowing east from the Black Hills and starved the Badlands of river-borne sediment. Eolian sedimentation was still possible, as documented by the accumulations of loess on the tops of tables across the region (Rawling, Fredlund, and Mahan, 2003).

In addition to the loss of new sediment from the streams that were captured by the Cheyenne River, the strata of the Badlands themselves contributed to a positive feedback of erosion and exposure that created more surface area to be eroded. The reason for this enhanced feedback lies in the composition of the strata in the Badlands. In particular, the secret lies in the smallest of sedimentary particles, a particular clay known as smectite. Smectite is formed by the weathering of volcanic ash via hydrolysis. Once converted to smectite, the clay is susceptible to swelling in the presence of water, similar to a sponge. Some forms of smectite can expand up to 50 percent of their original volume. Upon drying, the smectite will shrink back to its original size, until the next wetting event. The entire sequence of strata exposed in the Badlands, from the late Cretaceous Pierre Shale up through the Sharps Formation of the Arikaree Group, contains smectite, but some units contain more than others, which is why the stratigraphic units in the Badlands have such different appearances, such as the haystack mound shape of the Chadron Formation versus the cliff and spire morphology of the Brule Formation (Plates 4, 8). The reason for this difference in smectite content can be related to two different processes: original weathering of the volcanic ash grains when the sediments were first deposited and exposed to ancient soil-forming processes, and the amount of volcanic ash that was delivered to the Badlands region.

Within the Pierre Shale, discrete episodes of volcanic ash deposition from late Cretaceous eruptions in the west are preserved as thin whitish-yellow layers of bentonite, a particular type of smectite named for Benton, Wyoming, where it was first described. Windblown ash was also introduced as part of the background sedimentation into the Western Interior Seaway. Smectite is also present in the Fox Hills Formation, but the greater amounts of clastic grains brought in by the rivers that constructed the delta diluted the overall concentration of the smectite clay. This is seen by the change in erosional styles of the rolling hills of the Pierre Shale versus the slightly more cliff-forming strata of the Fox Hills Formation (Plate 1).

With the switch to nonmarine deposition of the White River Group, volcanic ash was incorporated into river and lake deposits and was weathered in the soils that were active at that time. In the late Eocene, climates were humid and tropical, which promoted hydrolysis of the ash into smectite. With the continued lateral migration of meandering rivers, smectites and partially weathered volcanic ash were recycled and redeposited onto new floodplains to continue hydrolysis. The end result was a series of stratigraphic units, the Chamberlain Pass and Chadron formations, which became enriched in smectite compared to the overlying Brule Formation, which contains a greater proportion of unweathered volcanic ash grains. The reason for this greater concentration in ash within the Brule Formation may lie in the hypothesis of climate change from warmer and wetter conditions during the late Eocene to cooler and drier conditions of the Oligocene. It is also possible that the rate of new ash introduced into the Badlands region was too great for hydrolysis to keep pace. Calculations of sedimentation rates within equivalent deposits from Wyoming, Nebraska, and the Badlands suggest that sedimentation rates doubled from the Eocene into the Oligocene (Zanazzi et al., 2007; Terry, 2010).

Regardless of the cause, the result of these differing proportions of smectite in the bedrock of the park explains why the Chadron Formation tends to weather into mounds whereas the Brule Formation is more cliff forming. Geologists refer to this variable resistance to weathering and erosion as differential weathering, which can manifest as differences between geologic units or between individual small-scale beds in a particular unit. Along with the general geomorphology of these two formations, the surface texture of each of these units is the result of their smectite content. The Chadron Formation weathers into a popcorn texture

4.6. Photographs of various features that postdate deposition of the Badlands strata. (A) Coarse unconsolidated gravel and cobble deposit resting unconformably on Badlands strata. This deposit is at the top of the hill directly east of the Pinnacles Overlook parking lot. Largest cobbles are 15 cm or more in length. (B) Surficial popcorn texture of the Chadron Formation. (C) Geodetic survey marker from the 1950s, now exposed approximately 30 cm above the land surface as a result of erosion of the surrounding bedrock. Photograph by Tyler Teuscher, courtesy of the National Park Service. (D) Dissection of the Badlands by water to create isolated tables (T). Over time, this process leads to large isolated tables, such as Sheep Mountain Table, southwest of Scenic, South Dakota. (E) Hoodoo at the base of Norbeck Pass along Highway 240. (F) Armored mud balls. Photo scale is 10 cm (4 inches). Photos by the authors.

as a result of the repeated shrink and swell of smectite (Fig. 4.6). When wet, the smectite is smooth, sticky, and slippery (a feature referred to locally as gumbo) and is almost impossible to walk or drive across. Upon drying, the popcorn texture becomes hard and abrasive, like sandpaper. This outer veneer of popcorn is gradually washed away by successive storm events and replaced by new popcorn that forms by the infiltration of water down to unaltered bedrock. This outer layer of popcorn can be quite thick – up to several meters in places near the base of some slopes. Beneath the popcorn surface of the Chadron Formation, the unaltered rock of the Badlands is hard, like any other rock, but a fragment of rock placed in a glass of water will break down in a matter of minutes, literally melting in place as the smectite absorbs the water, expands, and sloughs off the side of the rock fragment. The Brule Formation weathers by this same process, but the lesser amount of smectite translates to a thinner popcorn surface and a greater resistance to erosion, which is why the Brule Formation weathers into high spires and cliffs.

On average, Badlands National Park loses almost 2 cm (1 inch) per year to erosion. This rate of erosion was determined by calculating the amount of sediment that was removed from underneath U.S. Geodetic Survey markers that were originally installed flush with the ground surface in the 1950s. Many of these markers are now lying exposed and on their side (Fig. 4.6). This extremely high rate of erosion also explains why plants that would normally slow the pace of erosion find it hard to gain a foothold. It is this high rate of erosion and the lack of vegetation that give the Badlands its name. Many areas have badlands, but these exposures in Badlands National Park serve as the archetypical example of this geomorphic feature. The overall high rate of erosion gives the White River its name. The whitish color is created by the suspended silts and clays that have been eroded from the surrounding Badlands, which prompted an early settler in this region to describe the White River as "too thick to drink and too thin to plow."

The dissection of the Badlands has created particular geomorphic features. Some of these features are large in scale, whereas others are localized. The largest of these features are the numerous tabletop buttes throughout the park and the surrounding region, such as Sheep Mountain Table just southwest of the town of Scenic, South Dakota (Fig. P.1). From a distance, the tops of these tables, which in some cases are several hundred feet (50 to 100 m) above the surrounding prairie, can be visually reconnected from one to another (Fig. 4.6). The tops of these tables represent the last and highest stable geomorphic surface in this region before river incision became the dominant geologic process. The rivers did not incise at a steady rate. Along numerous rivers throughout this region are multiple terraces that represent periods of relative stability that allowed the river to laterally migrate back and forth within its valley for a period of time to form a small-scale floodplain before renewed downcutting lowered the level of the active river. Some rivers in and around the Badlands have three to four terrace levels above the active channel.

At a smaller scale within the strata of the Badlands, differential erosion exerts a powerful control on geomorphology. Channel sandstone bodies tend to be more resistant to erosion and can protect smectite-rich rocks below them. In special circumstances, blocks of sandstone can become isolated from the main body of the outcrop to form a hoodoo, a column of softer rock protected by a sandstone cap. Over time, these weaker columns of rock will erode, eventually toppling the sandstone cap. Hoodoos are found throughout the park, but they tend to be small-scale features (Fig. 4.6). At Toadstool Geologic Park in northwest Nebraska, large blocks of channel sandstone protect soft, smectite-rich mudstones to produce toadstools, which gave this park its name. The majority of the largest toadstools have since fallen, but new ones are presently forming.

The clastic dikes and chalcedony veins throughout the Badlands are more resistant to erosion than the surrounding rock. This increased resistance to erosion acts as an internal support to the surrounding strata and helps to hold up the ridges of the Badlands. When viewed from above, these dikes and veins have a preferred northwest–southeast and northeast–southwest orientation, which by default creates a preferred orientation in the buttes (Fig. 4.7). On a larger scale, this same orientation is expressed in river and stream drainages that dissect the Badlands. This symmetrical arrangement of rivers and streams is due to the exploitation of fractures in the bedrock that are less resistant to weathering and erosion.

Within the bedrock areas where dikes and veins are absent, the Badlands erode into small-scale gullies and rills with steep knife-edge ridge tops and slopes. Water is able to exploit smaller-scale fractures within the rock and, over time, develop networks of tubes and pipes under the popcorn weathered surface, similar to karst terrains (caves) that form in limestone. This piping erosion can, over time, wash away large areas of bedrock under the surface of the Badlands, which will eventually collapse in on itself and form a sinkhole.

Given the highly erosive nature of the rock in the Badlands, it is fortunate that this part of the United States only receives approximately 13 inches of rain per year. Otherwise the Badlands would have washed away long ago. The rain that does fall tends to come in storms and rapidly inundates the landscape. The smectites in the bedrock swell and act as a protective barrier to retard deep penetration of precipitation,

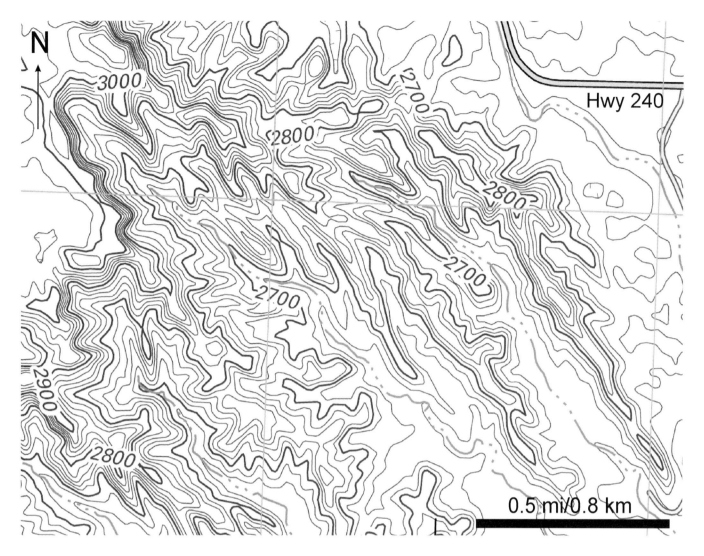

4.7. Portion of the U.S. Geological Survey Wall SW 7.5 Minute topographic map showing the preferred orientation of buttes and ephemeral streams.

instead forcing the water to flow over the saturated surface into the ephemeral streambeds and narrow gullies throughout the exposures. With rapid precipitation events, this can induce flash flooding as the water quickly collects and moves downslope. These flash floods can be quite powerful and are capable of moving large cobbles and chunks of more resistant rock. One of the more interesting features of these flash flood events is the creation of armored mud balls (Fig. 4.6). As the flooding proceeds, small pebbles are rolled along the bottom of the streambed and pick up bits and pieces of sticky clay. As this pebble moves along and grows in size, other, smaller pebbles can be picked up and incorporated, similar to rolling a snowball downhill. The farther the mud ball travels, the larger it will become, until the stream is no longer able to move it. Most mud balls are the size of a golf ball to the size of a softball, although some the size of basketballs are occasionally seen.

Not all of the water that falls on the Badlands is incorporated into gullies and streams. Some simply washes down the

sides of the buttes and onto the prairie. As it does so, this water carries clays and silts with it that are deposited as a broad sheet at the base of the slope (colluvium). Over thousands of years, these individual paper-thin colluvial deposits can accumulate into packages several meters thick. The prairie vegetation continues to grow upward as more sediment is added, which in turn protects the colluvium from erosion. Eventually these packages of individual sheet-wash storm events will be dissected by erosion to form sod tables, isolated patches of laminated sheet-wash deposits several meters high that are capped by prairie vegetation (Fig. 4.8). The prairie cap represents the former stable land surface before dissection and can commonly be traced from sod table to sod table.

Upon closer inspection, the laminae of colluvium that comprise the sod tables are sometimes disrupted by darker bands 10 to 20 cm thick. These bands represent periods of geomorphic stability during which dark, organic-rich surface horizons of prairie soils could form as a result of slower rates of sediment influx. With the return of increased sediment

4.8. Photographs of various features that postdate deposition of the Badlands strata. (A) Sod tables (S) formed by the dissection of sheet-wash deposits. The tops of the tables represent the former land surface. (B) Dark-colored colluvial sediments from the Pierre Shale interfingering with light-colored alluvium from stream flooding. (C) Rockfall in Norbeck Pass along Highway 240. (D) Earth flow in the Pierre Shale near the Sage Creek Campground. Photos by the authors.

input from the slope wash, the topsoil could not keep pace with deposition. In addition to buried soil horizons, sod tables also frequently display burrows and dens of prairie animals that were excavated before dissection of the landscape. Some sod tables preserve former campsites of indigenous populations that inhabited the Badlands thousands of years ago. In other parts of the park, where older bedrock of the Pierre Shale is exposed at the surface, dark-colored colluvium from the shale interfingers with light-colored river deposits coming out of the Badlands (Fig. 4.8).

The most dramatic types of erosion in the Badlands manifests as catastrophic mass wasting events, such as rockfalls, and slower-moving landslides, and can occur with or without the presence of heavy rains. Rockfalls can occur almost anywhere in the park but are most commonly associated with fractures in the bedrock that eventually break away or by undermining of cliff faces by rivers and streams during periods of heavy rain (Fig. 4.8). Other rockfalls are associated with faults, such as along the high cliff face of the Cliff Shelf Nature Trail in Cedar Pass, and form aprons of debris at the base of the cliffs. Highway 240 from the northeast entrance of the park descends from the upper prairie geomorphic surface to the lower prairie through Cedar Pass, just northeast of the Ben Reifel Visitor Center (Fig. P.1). This steep, twisting road is actually developed on several large landslide blocks that have been moving gradually downslope for decades. Evidence of this movement can be seen in the twisted stripes of rock along the highway and the constantly degrading road bed at the lip of Cedar Pass, which is split and offset by this constant movement. Periods of heavy rain tend to speed the rate of downhill movement. These landslide blocks were stabilized several years ago by installing artificial drainage systems to divert the water. On occasion, periods of heavy rain saturate less consolidated soils, causing

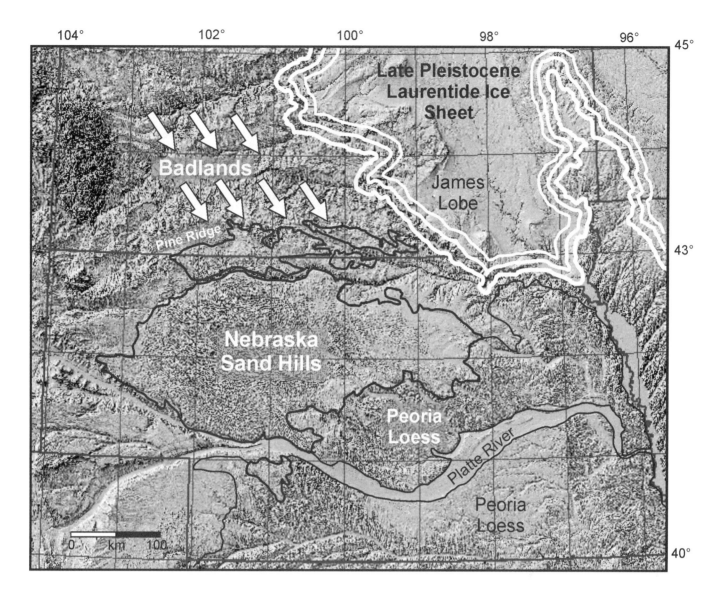

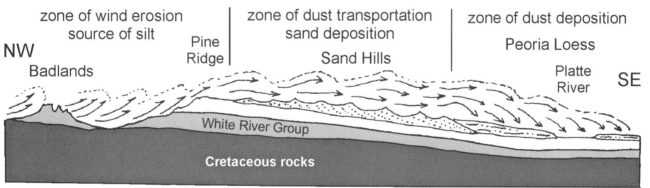

4.9. Physiographic map showing the late Pleistocene features of South Dakota and Nebraska. Winds blowing from the northwest (white arrows) through the Badlands and paralleling the Laurentide ice sheet margin picked up sand and silt grains. This sediment blew over the Pine Ridge and dropped sand-size grains in the dunes of the Sand Hills of southern South Dakota and central Nebraska (areas surrounded by a heavy line). The silt-size dust particles were transported across the Sand Hills and were deposited on the lee side of the dunes in eastern and southern Nebraska, forming the thick Peoria Loess. The cross-sectional model is modified from Muhs et al. (2008:fig. 23), after Mason (2001). The base map for this diagram is from the U.S. National Atlas Web site (http://nationalatlas.gov/mapmaker).

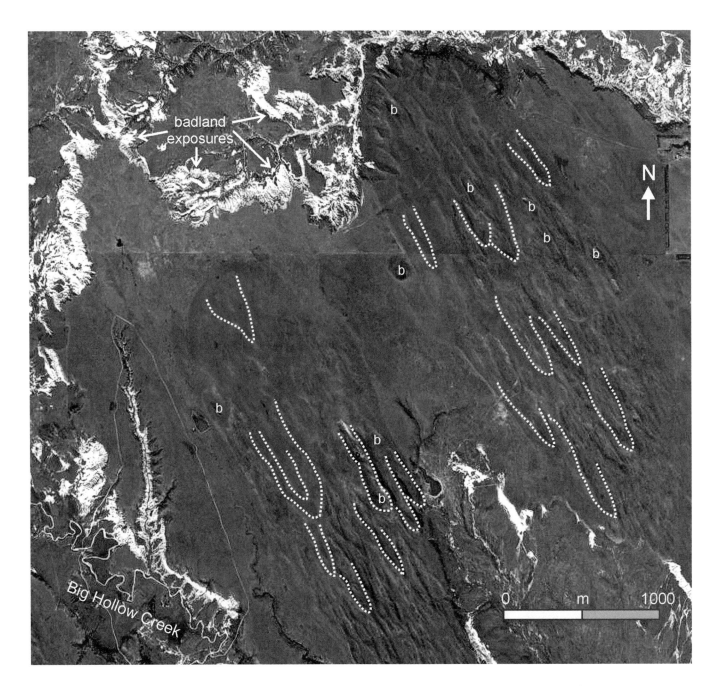

4.10. Aerial photograph of some of the Pleistocene dunes in the Big Badlands. The area shown is on the table lands between Big Hollow Creek and Cain Creek. The white dotted lines show the crests of some of the parabolic dunes in the area. The Pleistocene winds blew from the northwest, as shown by the alignment of the dunes with the "arms" of the dunes extending toward the northwest. Also shown are some of the blowouts (b), which are depressions formed by local erosion by winds. The base photo is from the U.S. Department of Agriculture, taken on September 25, 2011, and is available on Google Earth.

them to flow downhill (an earth flow) on top of more solid bedrock beneath. This is especially common in the Pierre Shale (Fig. 4.8).

Water is not the only agent of erosion in the Big Badlands. Windstorms can occur in the late spring, summer, and fall and pick up a great amount of fine-grained sediment, forming huge dust clouds. This is especially true during droughts, such as in the 1930s and as recently as the early 2000s, but a huge amount of wind erosion in the Badlands occurred during the last glacial maximum, between 14,000 and 25,000 years ago (Muhs et al., 2008). Glaciers did not occur in the Big Badlands during the Pleistocene, but during the greatest maximum extent of the Laurentide ice sheet, the margin of the continental glacier was 240 km to the east, where the Missouri River now flows (Fig. 4.9). Climatic conditions in the Badlands during the last glacial maximum were dry and exceptionally windy, with strong winds blowing from the northwest paralleling the margins of the continental glacier. The dry and cold conditions supported little vegetation, and silt grains eroded from the siltstones of the Brule and the Sharps formations were blown to the southeast across the Sand Hills of Nebraska, then deposited in eastern and southern Nebraska as a series of very thick loess deposits (Fig. 4.9). The thickest and most widespread of these dust deposits is the Peoria Loess, which reaches a maximum thickness of 40 m in central Nebraska. The Peoria Loess covers all of southern and eastern Nebraska, extending into Kansas, across Iowa, and into Illinois. Radiometric ages of detrital zircon grains and lead (Pb) isotopes within potassium feldspar grains in the Peoria Loess of Nebraska indicate a source from the White River Group and Sharps Formation in the Big Badlands (Aleinikoff et al., 2008). The ancient volcaniclastic loess deposits of the Brule and Sharps formations provided a source of silt grains for the Pleistocene loess of Nebraska, just like the ancient sand dunes of the Navajo and Entrada sandstones of Arches National Park are sources of modern sand in eolian dunes in eastern Utah. The thick Pleistocene loess deposits in Nebraska are some of the most productive agricultural lands in the world, and much of these sediments were derived from the Brule and Sharps formations in South Dakota.

Quaternary eolian deposits in the Big Badlands occur on the tops of the higher tables, such as on Sheep Mountain Table, Quinn Table, and Bouquet Table (Fig. P.1). These deposits include loess deposits and sand dunes that have been dated as late Pleistocene in age (Burkhart et al., 2008). The sand dunes are especially well preserved on the tables south of the Badlands Wall, such as on Imlay Table and to the west of Cain Creek (Fig. 4.10). These sand dunes are all parabolic dunes that have long ridges that point upwind and connect on the downwind side. These dunes are all stabilized by vegetation and did not move even during the Dust Bowl of the 1930s. Their orientation indicates winds blowing from the northwest. The presence of Pleistocene sand dunes in the Badlands region is significant, for silt either in the bedrock or reworked into stream deposits will not be picked up into the atmosphere as dust from only the direct action of winds. This is because of the weak cementation of the siltstone bedrock, and because the smoothness of any silt–sediment surfaces prevents the wind from lifting the silt into the air. It takes bouncing (saltating) sand grains during windstorms to hit the siltstone or silty sediments and launch the silt into the atmosphere as dust. Thus, saltating sand grains during late Pleistocene windstorms abraded the siltstone outcrops and silty stream sediments, sending up clouds of silt-size dust from the Badlands that eventually settled in Nebraska, forming thick loess deposits. The current limited outcrops of the Poleslide Member of the Brule Formation and the Sharps Formation in the Big Badlands is a result of not only stream erosion but also wind erosion.

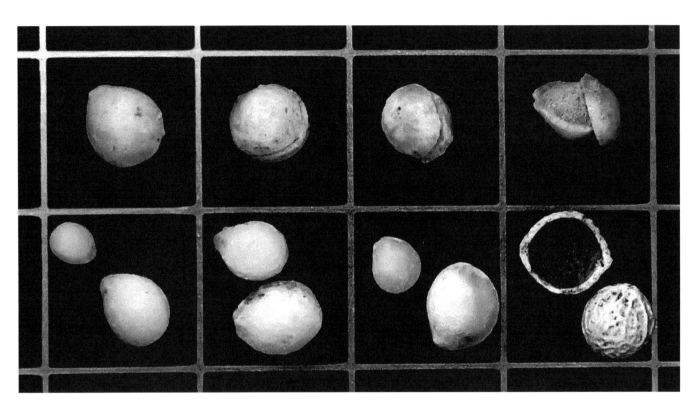

5.1. Hackberry endocarps, including specimens from the Brule Formation of Wyoming (*top row and lower left three squares on the bottom row*), and modern specimens (*lower right square*). The fossil endocarps are typically nearly smooth, with only a trace of the reticulation pattern seen in the modern specimens. The fossil endocarps occur in at least two size ranges. The original calcite shells have been recrystallized into coarse calcite crystals. The endocarps have two outer shells of calcium carbonate that can separate. The top row of fossils shows a progression from a complete endocarp (*upper left*) to an almost completely separated pair of shells (*upper right*). Grid in centimeters. Photo by the authors.

Bones That Turned to Stone: Systematics

5

INTRODUCTION

The sheer abundance and diversity of the paleontological record within the White River Badlands is indeed impressive. This has been well documented by the extensive amount of scientific literature that has been published on this region over the past 150 years. As a result, the fauna of the Big Badlands has played a key role in our understanding of how the North American biota has evolved and adapted in response to climatic change. The enormous depth of this topic has forced us to set some limits. The systematic discussions of paleontology are limited to taxa that have a published occurrence within a 100-mile radius of the Cedar Pass Area within Badlands National Park. However, many of the images featured in the systematics chapter come from areas outside the scope of this project but are of the same genus and species found in published records for Badlands National Park and the surrounding region.

During our research, we encountered some contradictory systematic classifications. This reflects the dynamics of the ongoing paleontological research with regard to the evolution, taxonomy, and systematics of the fossil taxa we include. We followed the most current publications whenever possible. For many taxa, there is a general agreement on the name that should be used for a taxon, but this is not always the case. We were thus forced to select a name. We emphasize that this is not a formal taxonomic revision and that our selection of names was often a matter of convenience to facilitate our ability to convey information. In those cases we have often selected historic names that are well established in the literature rather than more obscure names or newly proposed replacement names, which have not yet been fully accepted or vetted by the professional paleontological community. We realize that research is ongoing and that systematic classifications will continue to change in the coming years. To those specialists in different taxonomic groups who may disagree with our name choices, we apologize if our taxonomy does not exactly match your preferences.

PLANTAE

Because of the highly alkaline oxidizing soils, the plant record within the White River Badlands is poor (Retallack, 1983b). There is no record of fossil leaves or pollen. Instead, we find root traces, the endocarps of hackberry seeds (Chaney, 1925), petrified wood (Troxell, 1925; Berry, 1926; Wieland, 1935; Lemley, 1971), and partially digested plant material in herbivore coprolites (Stovall and Strain, 1936). Even with such a limited record, important interpretations about the vegetation communities on the landscape and how they reflect the paleoclimate can be made from these materials.

Fossil Root Traces

There is a broad variety of fossil root traces preserved within the White River Group (Plates 6, 7). Root traces are an important indicator of pedogenic activity (soil formation) and are discussed in detail in chapter 3. They can also be an important indicator of ancient climate. According to Retallack (1983b), within a root trace, there is no evidence of original organic material nor any details of anatomy or surface detail. They are simply an opening in the soil left by the root and later filled with soil material or crystalline calcite or chalcedony. The root traces taper and branch downward and are irregular in width and direction, depending on the root structure of the original plant. Some of the most striking root traces are large metaisotubules of reddish clay washed into the yellowish C horizon of the type Yellow Mounds silty clay loam paleosol of the Pierre Shale and Fox Hills Formation. These root traces are up to 3 cm in diameter and indicate the presence of trees or large shrubs. In contrast, fine root traces of 1 to 2 mm are abundant on the surface of all paleosols of the White River and Arikaree groups and are comparable to those of modern prairie grasses and forbes. They don't reach the density or deep development of root systems of a modern tallgrass prairie but are more similar to a mixed or short-grass prairie (Retallack, 1983b).

5.2. Permineralized (petrified) wood from the Chamberlain Pass Formation near Weta, South Dakota. Note the well-preserved ring structure, which is suggestive of seasonality, as well as the detailed preservation of wood grain. The nonuniform shape of this specimen suggests partial decomposition before burial and fossilization. (A) Cross-sectional view. (B) Long view. (C) Complete specimen. Scales in (A) and (B) are in centimeters. Penny for scale in (C). Photos by the authors.

Hackberry Seeds

Chaney (1925) was the first to report the presence of endocarps of hackberry seeds (*Celtis hatcheri*), which is a plant part much like a stone of a peach (Fig. 5.1). Hackberry seeds are common in many paleosols within the Chadron Formation and the Scenic and Poleslide members of the Brule Formation. Their preservation is due to the large amounts of biogenic calcium carbonate and silica deposited in the plant tissue (Retallack, 1983b). They are easy to identify because of the distinct wrinkled appearance of their outer surface.

Petrified Wood

The petrified wood found in the White River Badlands is mostly silicified and comprise the remains of dicotyledonous angiosperms (Troxell, 1925; Berry, 1926; Wieland, 1935; Lemley, 1971) and hackberry bushes (Fig. 5.2A) (Chaney, 1925). Stumps and fossil wood are rare but occur in the Chadron Formation and Scenic Member of the Brule Formation in the Badlands.

INVERTEBRATES

Animals that lack an internal skeleton composed of hydroxyapatite and lack a vertebral column are called invertebrates (Buchsbaum, 1977). The invertebrate fossil record within the White River Badlands is not well described because the fossils are poorly preserved and have a limited occurrence. However, a few valuable references can be found on the topic (Fig. 5.3) (Meek, 1876; White, 1883; Wanless, 1923). Mollusks, because of their hard shell of calcium carbonate, compose the majority of invertebrate fossils found within the White River Group. Other invertebrates are represented by tracks and traces.

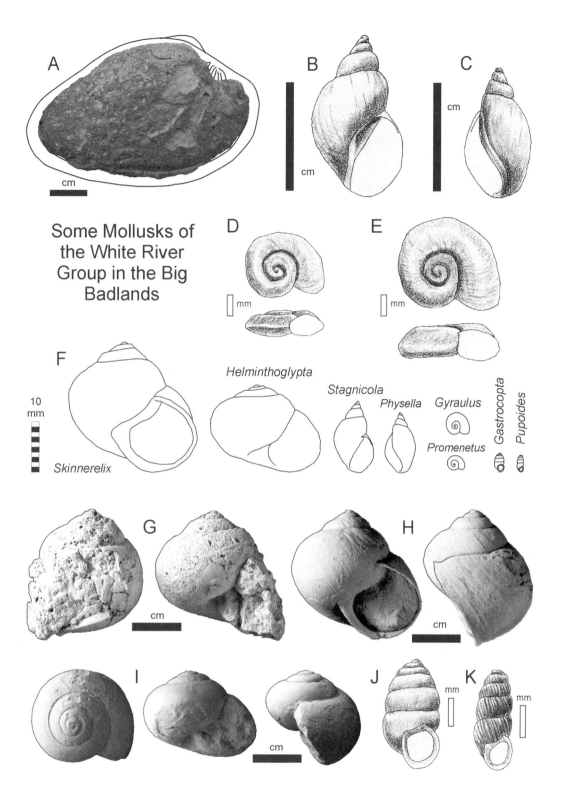

Some Mollusks of the White River Group in the Big Badlands

Helminthoglypta

Stagnicola   *Physella*   *Gyraulus*   *Gastrocopta*   *Pupoides*

Promenetus

Skinnerelix

10 mm

5.3. Freshwater mollusk shells (A–E) and land snails (G–K) known from the White River Group of the Big Badlands. (A) Steinkern of a unionid bivalve from a channel deposit in the Poleslide Member. An outline of the reconstructed left valve is shown. (B) Diagram of the shell of *Stagnicola shumardi* from the Brule Formation. (C) Diagram of the shell of *Physella secalina* from the Brule Formation. (D) Diagram of *Promenetus nebrascensis* from the Brule Formation. (E) Diagram of *Gyraulus* sp. from the Brule Formation. (F) Series of diagrams of fossil

gastropod shells showing the relative sizes of the various genera. (G) Side and apertural views of the type specimen of *"Helix" leidyi* Hall and Meek, 1855 (AMNH 11174/1, figured in Evanoff and Roth, 1992:figs. 1 and 2). This specimen was collected by Meek or Hayden in 1853 from the "turtle–*Oreodon* beds" (Scenic Member) in the headwaters of Bear Creek. It is the only known specimen from the Badlands of this species. (H) Apertural and side views of a nearly complete specimen of *Skinnerelix leidyi* from Chadronian rocks near Douglas, Wyoming (from Evanoff,

1990b:plate 1, fig. 5b and c; UCM 30736, University of Colorado Museum of Natural History, Boulder, Colorado, U.S.A.). (I) Spiral, apertural, and side views of *Helminthoglypta* sp. from the Brule Formation of northern Nebraska. (J) Diagram of *Gastrocopta* (*Albinula*) sp. from the Brule Formation. The apertural barriers are not known for this species. (K) Diagram of *Pupoides* (*Ischnopupoides*) sp. from the Brule Formation. Only well-preserved specimens of *Ischnopupoides* will show the fine radial ridges.

Phylum Mollusca

The Mollusca is the second largest phylum in the Kingdom Animalia on the basis of the number of living species. Mollusks live in all possible environments on the surface of the Earth, from the greatest depths of the sea to the highest mountains, and from the Arctic through the tropics to the Antarctic. It is one of the few phyla that have a greater number of fossil species than modern species. The fossil mollusks of the White River Group include the shells of freshwater mussels (clams), freshwater snails, and land snails. All three have relatively thin aragonite (mother-of-pearl) shells that either are recrystallized into calcite or are dissolved away, leaving only internal molds. There has not been a dedicated systematic analysis of the mollusks of the White River Group in the Badlands, but their fossils are locally common. The following taxa represent the most abundant or historically important fossil mollusks in the Badlands.

Class Bivalvia
Order Unionoida

Freshwater mussels are large clams with relatively thin shells of mother-of-pearl, containing complex hinge teeth with multiple grooves and ridges (Fig. 5.3A, B). As with all bivalves (clams and their kin), the hinge teeth lock the shells together when the clam closes its shell, preventing the shells' separation by twisting apart. Freshwater mussels thrive in clean, perennial, well-oxygenated running water. Their larvae disperse by attaching onto the gills and fins of fish that transport the larvae through the river system. The larvae eventually drop off the fish and settle on the bottom of the channel. When they are found in the rock record, they indicate that the stream or river that deposited the channel was perennial; had fairly clear water; and was connected with other drainages that allowed the fish (and thus the mussels) to disperse.

The White River freshwater mussels are typically found within the coarse-grained channel deposits of the White River Group. They indicate that the streams flowed throughout the year and were part of a connected tributary system. They are not found in all paleochannel deposits in the White River Group, and their absence indicates that these streams were intermittent and did not flow beyond the Great Plains. They can occur in fine-grained deposits next to the channels (Gries and Bishop, 1966) where floods transported the clams into the surrounding floodplains. The mussels also occur in some lacustrine limestone deposits along with fish fossils (Wanless, 1923). The lakes must have had connections with the river systems, allowing the fish with their attached mussel larvae to enter the lake.

Gries and Bishop (1966) described some internal molds (also called steinkerns) of unionid bivalves from the Chadron Formation in the Indian Creek drainage just outside Badlands National Park. The fossils are preserved in a calcareous claystone, possibly representing a "quiet backwater of a sluggish stream" (Cook and Mansfield, 1933:265), and were probably brought in from a river channel by floods (see Hanley, 1987, for a description of the process). The unionid bivalves described by Gries and Bishop (1966) have not been identified to genus because such an identification would require details of the shell not preserved in these specimens. At the time of this writing, all of the unionid bivalves found within the White River Group are internal molds, making a more specific identification impossible.

Class Gastropoda
Subclass Pulmonata

Pulmonates are lung-bearing snails that can breathe oxygen from the air and are extremely drought tolerant. The pulmonates include many freshwater snails (Basommatophora) and most land snails (Stylommatophora). Both groups have thin shells made of aragonite that are typically recrystallized to calcite or are dissolved away, leaving internal molds. Fossil pulmonates range from the Triassic to the Recent (Evanoff, Good, and Hanley, 1998).

Order Basommatophora

The freshwater pulmonate snails can live in waters containing low oxygen, turbid waters, or waters containing a high amount of suspended sediment, as well as standing water that seasonally dries (intermittent waters). The fossil freshwater snails of the Big Badlands are all pulmonates, including the high-spired shells of the lymnaeid *Stagnicola* and the extremely low-spired shells of the planorbids *Gyraulus* and *Promenetus*. All of these are easily transported to open waters by birds, where a single snail can start a population because these snails can self-fertilize by a process called hermaphroditism. Early Eocene snail faunas were dominated by gill-bearing snails (the prosobranchs), which require perennial, clean, well-oxygenated waters like that in the eastern United States and the Pacific Northwest. By the time of White River Group deposition, rivers and ponds were much more intermittent and turbid, an environment only the pulmonates could withstand. Turbidity was probably the greatest controlling factor in favoring the pulmonates during the late Eocene and early Oligocene. The snails probably lived in semipermanent ponds and longer-lasting small lakes on the floodplains away from river channels, living on abundant freshwater algae whose photosynthesis helped to precipitate calcium carbonate in the water. All the freshwater snails found in the White River Group in the Badlands have descendants living today in South Dakota (Beetle, 1976). The following descriptions of the modern geographic ranges are from Burch (1989), which is an excellent reference for modern freshwater snail shells. The fossil freshwater snails of the White River Group were summarized by Meek (1876) and White (1883).

Family Lymnaeidae
Stagnicola Leach (in Jeffreys), 1830

Systematics and Evolution The type species of *Stagnicola* is *Buccinum palustre* Müller, 1774.

Distinctive Characters The adult shells of the fossil taxa are medium to small, elongate, spindle-shaped shells with slightly inflated whorls; the shell surface is smooth or sculptured with microscopic spiral striations; the columella (the central shaft of the shell) has a well-developed twist or plait (Burch, 1989; Evanoff, 1990b).

Stratigraphic and Geographic Distribution *Stagnicola* is a geographically widespread genus that occurs throughout North America, Asia, and Europe. *Stagnicola elodes*, the marsh pond snail, lives in South Dakota today (Beetle, 1976). Fossils of *Stagnicola* can be found in rocks ranging in age from the Jurassic to Recent (Evanoff, pers. obs.).

Two species of *Stagnicola* have been described from limestone beds in the Chadron Member near Poeno Spring, north of Badlands National Park (Fig. 5.3C–F). The larger form is *S. meeki* (Evans and Shumard) 1854, and the smaller form is *S. shumardi* (Meek) 1876. Both are typically found in thin sheet limestone beds deposited in shallow lakes and ponds. *Stagnicola meeki* is only known from limestone beds of the Chadron Formation, while *S. shumardi* occurs in limestone beds extending from the Chadron through the lower Poleslide Member of the Brule Formation in South Dakota. *Stagnicola* is the most common freshwater snail within the White River Group.

Family Planorbidae
Gyraulus "Agassiz" Charpentier, 1837

Systematics and Evolution The type species for *Gyraulus* is the living *Planorbis albus* Müller, 1774.

Distinctive Characters The adult shells of *Gyraulus* are small discoidal shells with rapidly expanding whorls forming a distinct umbilicus away from the spire, and a shallow umbilicus on the spiral side. An umbilicus is a depression formed on the spiral axis of the shell by the whorls expanding away from the axis of coiling. The outer margin of the shell is rounded; the shells are typically smooth. The margin can be circular in cross section or can be slightly elongated laterally and away from the spiral side (Burch, 1989).

Stratigraphic and Geographic Distribution *Gyraulus* is geographically widespread, occurring throughout North America, Asia, and Europe. There are two species of *Gyraulus* that live in western South Dakota, *G. circumstriatus* and *G. parvus* (Beetle, 1976). Fossils of *Gyraulus* occur in rocks ranging in age from early Jurassic to Recent (Evanoff, Good, and Hanley, 1998). *Gyraulus leidyi* (Meek and Hayden, 1860) was described from limestone beds in the Chadron Formation near Poeno Spring, north of Badlands (Fig. 5.3G, H). This fossil snail occurs in pond and shallow lake limestone beds within the Chadron Formation through the lower Poleslide Member of the Brule Formation in Badlands National Park.

Promenetus F. C. Baker, 1935

Systematics and Evolution The type species of *Promenetus* is *Promenetus exacuous* (Say, 1821).

Distinctive Characters Adult shells of *Promenetus* are small lenticular–discoidal shells characterized by a distinct angular margin called a carina. The tip of the carina is not far below the midline of the coiled shell. The shell has an umbilicus on both sides of the shell, but the spire side has a shallower umbilicus. The shell is typically smooth (Burch, 1989; Evanoff, 1990b). *Promenetus nebrascensis* has a rounded, indistinct carina, while *P. vetulus* has a distinct, angular carina (similar to what differentiates the shells of *P. exacuous* from *P. umbilicatellus* today).

Stratigraphic and Geographic Distribution Fossils of *Promenetus* range from the early Eocene to the Recent (Hanley, 1976). *Promenetus nebrascensis* (Evans and Shumard) 1854 and *P. vetulus* (Meek and Hayden) 1860 are two fossil species that were first collected in the Chadron Formation near Poeno Spring north of Badlands National Park (Fig. 5.3I–L). *Promenetus* species occur in thin limestone beds deposited in pond and shallow lake environments ranging from the Chadron Formation through the lower Poleslide Member.

Natural History and Paleoecology *Promenetus* is widespread in quiet freshwater environments of the western United States, and two species, *Promenetus exacuous* and *P. umbilicatellus*, live in quiet freshwater environments of South Dakota (Beetle, 1976).

Superorder Stylommatophora
Order Sigmurethra

The fossil land snails in the Badlands are all pulmonates and include both snails with large shells (shell heights larger than 10 mm) and very small shells (microshells that are less than 5 mm high). All fossil land snails are identified by the form and size of their shells. The larger shells also have microsculpture on the shells, including small granulations and fine spiral grooves, that can help identify the snail to the family level. The microshelled snails typically have small knoblike barriers in the apertures that are unique for genus and species identifications. All of the known fossil lands snails from Badlands can be identified to the family level of snails now living in North America. Most can be classified to modern genera and in some cases to modern subgenera. All the known fossil land snails in the Badlands occur in the Brule Formation.

No systematic study of the fossil land snails of the White River Group in Badlands has been made, so the following list, based on the observations of snail shells recently collected during the various projects in the field, is rudimentary. Though fossil land snails can be used as biostratigraphic and climatic indicators for the late Eocene and early Oligocene (Evanoff, Prothero, and Lander, 1992), not enough is known about the fossil lands snails at Badlands to make such interpretations.

Family Humboldtianidae
Skinnerelix Evanoff and Roth, 1992

Systematics and Evolution The type species of *Skinnerelix* is *Helix leidyi* Hall and Meek, 1855.

Distinctive Characters *Skinnerelix* is a large land snail (Fig. 5.3M, N) characterized by a globose–conic shell with a narrow-flared aperture opening and a coarse growth lines with a coarse, granular microsculpture. *Skinnerelix* has a shell almost identical to the modern *Humboltiana* except *Humboltiana* does not have a flared aperture opening.

Stratigraphic and Geographic Distribution *Skinnerelix* is known almost exclusively from Chadronian (upper Eocene) rocks except for the type specimen. For a complete discussion on *Skinnerelix leidyi*, see Evanoff and Roth (1992).

Helix leidyi was the first fossil land snail species to be described from the western United States. A single slightly crushed specimen (AMNH 11174/1; Fig. 5.3) was collected during the Meek and Hayden expedition to the Badlands in 1853. The specimen came from the head of Bear Creek (in the Scenic basin) from the "turtle–*Oreodon*" layer (Hall and Meek, 1855) now known as the Scenic Member of the Brule Formation. The specimen is large and globose, and it retains some recrystallized shell just behind the crushed body whorl that has the characteristic coarse granular microsculpture. If this specimen came from the Scenic Member, then it is the only known specimen of *Skinnerelix* of early Oligocene age.

Unfortunately, no other specimens have been found in the Badlands of South Dakota.

Natural History and Paleoecology *Humboltiana* typically lives on a variety of substrates and woody vegetation, ranging from oak forests on limestone to high coniferous forests and mixed, scattered woodlands on volcanic rocks (Pilsbry, 1939). *Humboltiana* now has a geographic range from west Texas south to central Mexico.

Family Helminthoglyptidae
Helminthoglypta Ancey, 1887

Systematics and Evolution The type species of *Helminthoglypta* is the modern *Helix tudiculata* Binney, 1843.

Distinctive Characters The fossil shells of *Helminthoglypta* in the Badlands are medium to large size; typically with low spires and moderately inflated whorls; and with fine granular microsculpture oriented in rows oblique to the growth lines. The microsculpture can also include spiral incised lines; the umbilicus is quite narrow. The aperture is flared but narrow and is rarely preserved.

Stratigraphic and Geographic Distribution The geologic range of the genus *Helminthoglypta* is from Chadronian (late Eocene) to Recent (Evanoff, 1990b). In the Badlands, they are known from the Scenic through the lower Poleslide members of the Brule Formation.

Natural History and Paleoecology The 88 modern species of *Helminthoglypta* live in forests, shrub lands, and woodlands from southwestern Oregon south to northern Baja California and west of the Cascade and Sierra Nevada mountains. *Helminthoglypta* is the most abundant large fossil land snail from the Badlands. Their thin aragonite shells are always recrystallized into calcite, but most specimens are represented by internal molds. Typically mistaken for "*Helix leidyi*," these snail shells are distinct from *Skinnerelix leidyi* by their smaller size, lower spire, less inflated whorls, and fine granular and spiral groove microsculpture when the shells are preserved. The genus may be represented by several currently undescribed species that may be distinguished by size, form, and details of shell microsculpture.

Family Pupillidae
Gastrocopta (*Albinula*) Sterki, 1892

Systematics and Evolution The type species for *Gastrocopta* is the modern *Pupa armifera* Say, 1821.

Distinctive Characters Shells of *Gastrocopta*, subgenus *Albinula*, are small (less than 5 mm long) and oval, and have flared rounded apertures. *Gastrocopta* shells have many short knoblike barriers in the aperture, with the middle barrier on the spire side of the aperture characteristically long and split at the end (bifid). *Gastrocopta* (*Albinula*) has the largest shells of all the *Gastrocopta* species, with shell lengths as great as 4.5 mm (Metcalf and Smartt, 1997). The fossil *Gastrocopta* (*Albinula*) shells of the Brule Formation in Badlands is essentially identical in form and size to *Gastrocopta armifera*, but the apertural barrier forms and arrangements in the fossils are currently unknown.

Stratigraphic and Geographic Distribution *Pupa armifera* is widespread east of the Rocky Mountains and lives today in western South Dakota (Beetle, 1976). *Gastrocopta* (*Albinula*) shells occur in the Scenic and lower Poleslide members of the Brule Formation in the Badlands. The range of the genus and subgenus is Chadronian (late Eocene) to Recent (Evanoff, 1990b).

Pupoides (*Ischnopupoides*) Pilsbry, 1926

Systematics and Evolution The type species for *Pupoides* (*Ischnopupoides*) is *Pupa hordacea* Gabb, 1866.

Distinctive Characters Shells of *Pupoides*, subgenus *Ischnopupoides*, are all very small (less than 3.5 mm in length) and cylindrical, with an oval aperture that is thickened and expanded but not flared. The aperture contains no knob-like barriers. The outer whorls can be smooth or can have thin, fine ribs parallel to the growth lines. The shell of *Pupa hordacea* has fine ribs like most White River fossils of this subgenus (Evanoff, 1990b).

Stratigraphic and Geographic Distribution *Pupoides* (*Ischnopupoides*) now lives in the Colorado Plateau and eastern foothills of the Rocky Mountains as far north as Douglas, Wyoming (Evanoff, 1990b). They live in dry locations typically in association with shrubs (Evanoff, 1983). Its geologic range is Chadronian (late Eocene) to Recent (Evanoff, 1990b), but it is only known from the Scenic and lower Poleslide members of the Brule Formation in the Badlands.

VERTEBRATES

Class Osteichthyes

The majority of fish material from the White River Badlands is represented by catfish pectoral spines and some unidentified ganoid scales (Lundberg, 1975; Clark, Beerbower, and Kietzke, 1967). There are a few rare occurrences where cranial material has been preserved but none of this material has been identified to species (Lundberg, 1975). In North America, the majority of high-quality Tertiary fish fossils are limited to the quiet, low-energy lacustrine deposits of the Eocene and the Miocene through the Pleistocene (Lundberg, 1975), such as those found in the Eocene Green River Formation. Lacustrine deposits within the White River Badlands that might preserve whole fish skeletons are extremely rare.

Order Siluriformes
Family Ictaluridae
Ictalurus (*Ictalurus*) Rafinesque, 1820

Systematics and Evolution The type species of the catfish *Ictalurus* (*Ictalurus*) is *Silurus punctatus* Rafinesque, 1818, later transferred to *Ictalurus* by Rafinesque in 1820. Cope (1891) described three species of catfish on the basis of vertebrae from the lower Oligocene Calf Creek fauna in the Cypress Hills Formation of Saskatchewan: *Rhineastes rhaeas* (now *I. rhaeas* [Cope]), *Ameiurus maconelli*, and *A. cancellatus*, now all considered to be *I. rhaeas* (Lundberg, 1975). There are no specimens from Badlands National Park that are identified to species.

Distinctive Characters Modern ictalurids are distinguished from all other living siluriforms (catfish) by having a massive jaw adductor muscle originating on the skull roof (Lundberg, 1975:45). The oldest known ictalurid showing this character is from the Brule Formation, in Jackson County, South Dakota. Both of the known Oligocene fossil catfish, *Ictalurus rhaeas* and *Ameiurus pectinatus*, are assigned to their respective modern genera on the basis of features of the pectoral spine in the former and features of the ventral surface of the skull in the latter.

The pectoral spine shaft in large individuals is ornamented with prominent but narrow, subparallel ridges and deep grooves. On small individuals, the spine shafts are finely striate, and the anterior ridge is prominent and shows traces of regular spaced anterior dentitions. The posterior dentitions in small individuals are uniformly spaced, evenly retrorse, and single cusped (Lundberg, 1975).

Stratigraphic and Geographic Distribution Fossilized remains from *Ictalurus* can be found in deposits from the Paleocene to Recent (Lundberg, 1975). *Ictalurus* fossils are found in South Dakota, Saskatchewan, Kansas, Texas, Nebraska, Florida, Oklahoma, Ontario, Mexico, Montana, Colorado, Wyoming, Pennsylvania, Oregon, Idaho, and North Dakota. At Badlands National Park, *Ictalurus* specimens are found in the Poleslide member of the Brule Formation in the South Unit of the park.

Natural History and Paleoecology The fossil *Ictalurus* from the Big Badlands probably had a diet similar to modern catfish, including smaller fish, crayfish, mollusks, and plant material (Grande, 1984). Modern catfish also prefer fluvial environments such as streams and rivers, which do not usually have high-quality preservation or articulated remains. Today, native populations of fossil catfish in North America are only found in river drainages east of the continental divide, although fossils are known from the western side of the divide from the Eocene deposits of the Green River Formation in Wyoming and from the Pliocene Glenns Ferry

Formation in Idaho. Living catfish are temperature sensitive, and their northern limits are determined by the length of the growing season or, conversely, limited by the length of winters. Warm-water fishes like catfish living in temperate climates can tolerate temperatures close to 0°C for several months. They do, however, require a growing season sufficiently long enough and warm enough (measured in degree-days) to reach sufficient size, with sufficient stored reserves, to survive through the winter into the next growing season (Smith and Patterson, 1994). Their presence in the White River Group supports the idea of generally warmer temperatures throughout the year, or at least mild or short winters

Class Amphibia

Sediments in the White River Badlands indicative of swamp or marsh habitat are almost nonexistent. Because these areas often are the preferred environments for amphibians, their fossils are extremely rare members of the fauna. There is no record of the Order Caudata or fossil salamanders from the White River Group in western South Dakota. *Ambystoma tiheni* is present in the late Eocene Calf Creek Fauna in Saskatchewan, Canada (Holman, 1972). The record of *Ambystoma* continues into the Miocene in South Dakota, where they are fairly abundant. It is difficult to say whether fossil salamanders lived in the White River Badlands or if they were simply not preserved. Hutchison (1992) notes a reduction in the North American amphibian fauna during the Oligocene as a result of increasing aridity. *Ambystoma tigrinum* is found in South Dakota today.

Order Anura

The order Anura includes frogs and toads. The record of anurans in the White River Group is better than the salamanders but still quite limited. This group, especially frogs, would also be sensitive to increased aridity during the Oligocene, but perhaps not as extremely sensitive as the salamanders. The specialized modes of swimming and jumping in frogs and toads distinguish them from other amphibians, and this is reflected in the distinctive morphology of their skeleton, which makes them easier to identify. They also have feeding adaptations that allow them to swallow relatively large prey species whole (Holman, 2003).

Family Pelobatidae

Although fossil frogs are fairly rare, the pelobatid frogs are some of the most common frogs found in the fossil record (Estes, 1970b). The family is defined by the following characters: procoelous vertebrae, imbricate neural arches, arciferal pectoral girdle, single coccygeal condyle, prominent sternal style, wide dilation of sacral diapophyses, and long anterior and short posterior transverse processes of the vertebrae (Estes, 1970b).

Eopelobates Parker, 1929

Systematics and Evolution The type species for *Eopelobates* is *E. anthracinus* from middle Oligocene lignite beds near Bonn, Germany (Parker, 1929). Zweifel (1956) described an early Oligocene North American species, *E. grandis*, on the basis of Princeton University No. 16441 (now housed at the Peabody Museum of Natural History, Yale University, New Haven, Connecticut), a nearly complete skeleton lacking only some skull and distal limb bones (Fig. 5.4) from the early Oligocene middle part of the Ahern Member of Chadron Formation, Pennington County, South Dakota.

Distinctive Characters *Eopelobates* is defined by the following characters: prominent elongated external style, strong posterior projection of the ischium, spade absent, long, relatively slender limbs, urostyle either separate, partially or completely fused with the sacrum, sacral diapophyses strongly dilated, tibia longer than femur, approximately subequal orbit and temporal openings, dermal ossification well developed and fused to the skull roof, skull roof flat or concave dorsally, ethmoid wide and blunt anteriorly, squamosal–frontoparietal connection absent, and femur–tibia length approaching or exceeding head–body length (Estes, 1970b).

Stratigraphic and Geographic Distribution Fossilized remains from *Eopelobates* can be found in the late Cretaceous through the late Eocene. *Eopelobates* was believed to be Holarctic in its coverage by the early Eocene (Estes, 1970b). *Eopelobates grandis* was described from the late Eocene Chadron Formation from Shannon County, South Dakota (Clark, Beerbower, and Kietzke, 1967), and in Pennington County, South Dakota (Holman, 2003).

Natural History and Paleoecology The various species of the genus, *Eopelobates* have proven to serve as strong climatic indicators. As fossils, they are known from deposits from the humid, subtropical environment of the late Cretaceous to a warm, temperate environment of the late Eocene (Estes, 1964). *Eopelobates grandis* is found in late Eocene sediments (Chadronian) and is associated with a more temperate climate. The deterioration of climate during the Eocene–Oligocene transition not only caused the eventual extinction of *Eopelobates* but also gave rise to the more temperate spadefoot toad line (Estes, 1970b). Close relatives of this frog now inhabit the lowlands of Indonesia and southern China, and the upland forests of the southeastern Tibetan Rim (Clark, Beerbower, and Kietzke, 1967).

5.4. *Eopelobates grandis.* Skeleton courtesy of the Division of Vertebrate Paleontology, YPM VPPU 16441, holotype, Peabody Museum of Natural History, Yale University, New Haven, Connecticut, U.S.A. Scale in centimeters. Photo by Ethan France, Peabody Museum of Natural History.

Class Reptilia

The record for reptile vertebrate fossils within the White River Badlands is sparse. The only exception is the fossil turtles, which occur in great abundance. The majority of reptilian fossils occur as scattered bone or jaw fragments, while complete shells of turtles are often preserved.

Order Chelonia

The turtles within the White River Badlands are some of the most common fossils found. Hayden (1858) coined the term "turtle–*Oreodon* layer," which was characterized by the abundance of turtles and oreodont fossils at the base of the Scenic Member within the Brule Formation. Both land tortoises and aquatic turtles are found within the White River Group, with the former being much more abundant. Of the land tortoises, *Stylemys nebrascensis* is by far the most common (Fig. 5.5). Described by Joseph Leidy in 1851, it is the first fossil tortoise documented in North America (O'Harra, 1920; Hutchison, 1996). The thick, high-domed shell of *Stylemys*

lends itself well to preservation within the Badlands strata. Unfortunately, the skull and limb elements are rarely preserved. Shell sizes range from about 100 mm to over 530 mm in length. The aquatic turtles have flatter and more delicate shells and hence are not as well preserved.

Family Kinosternidae (Mud turtles)

Hutchison (1991) characterized the family as having moderately broad plastron lobes. The plastron is 67 percent to 78 percent of the carapace length. The carapace is distinctly tricarinate to smooth. It contains six neurals.

Xenochelys Hay, 1906

Systematics and Evolution The type species of *Xenochelys* is *X. formosa* Hay, 1906, from the Chadron Formation in Shannon County, South Dakota.

Distinctive Characters The carapace of *Xenochelys* can be as large as 208 mm in length. The plastron is about 82 percent of the carapace length. The plastron and carapace are relatively thin to moderately robust. The perimeter length of the nuchal is no longer than any peripheral. There is a well-developed caudal notch between the two halves of the xiphiplastron (Hutchison, 1996).

Stratigraphic and Geographic Distribution Fossilized remains from *Xenochelys* can be found in the Wasatchian through the Chadronian. Fossils are found in South Dakota and Wyoming.

Natural History and Paleoecology Clark, Beerbower, and Kietzke (1967) noted that the aquatic turtles are rare and not well preserved but were important climatic indicators. *Xenochelys* did not make it past the Eocene–Oligocene boundary, probably as a result of the cooling and drying trend found in the Oligocene.

Family Trionychidae (Soft-Shelled Turtles)

The trionychids, also known as the soft-shelled turtles, can be traced back to the late Cretaceous. They show a sharp increase in size in the Bridgerian and then begin to decrease in diversity. They disappear from the geologic record in the Badlands and western North America in the earliest Orellan, but the living form, *Apalone spinifera,* is found in the eastern United States. The family is broadly distributed today, and members of this family occur in Africa, Asia, North America, and Southeast Asia, indicating a southern or eastern refugium during the Oligocene and early Miocene (Hutchison, 1992; Estes, 1970a).

Apalone Cope, 1891

Systematics and Evolution The type species for *Apalone* is the living Florida soft-shelled turtle, originally *Trionyx ferox* (Schneider, 1783) but transferred to the new genus *Apalone* by Rafinesque (1832). The fossil species, *A. leucopotamica,* Cope, 1891, from the Cypress Hills, Saskatchewan, Canada, was originally placed in *Trionyx.* Meylan (1987) resurrected the genus *Apalone* for the North American forms, including the three extant species, and restricted *Trionyx* to certain soft-shelled species found mainly in Africa

Distinctive Characters The diagnosis for *A. leucopotamica* is as follows. The carapace can grow as much as 325 mm in length. The nuchal width ratio to carapace width ratio is about 0.6. The carapace length to width ratio is about 0.9. The costal margins are tapered, and the carapace is truncated posteriorly (Hutchison, 1998). Hay (1908:537) gave the diagnosis for *A. leucopotamica* as the only Oligocene species of *Apalone.* He also describes the carapace as "thin broader than long; nuchal 0.6 the width of the carapace; a fontanel on each side of first neural."

Stratigraphic and Geographic Distribution Fossilized remains from *Apalone* can be found in the Chadronian through the Orellan. Fossils are found in South Dakota, Nebraska, North Dakota, and Saskatchewan. In the White River Badlands, *Apalone* is found in the Chadron Formation. Today the genus is represented by a single species in North America, *Apalone spinifera,* with a distribution from central-eastern United States (western New York and southern Carolina) to Wisconsin and from Minnesota and southern Ontario as far south as Mexico.

Natural History and Paleoecology Modern soft-shelled turtles are entirely aquatic. An active predator, they are fast in the water. Because of a marked sexual and morphological variation in the carapace, Hutchison (1998) concludes that the group needs major taxonomic revision.

Family Emydidae (Pond Turtles)

The emydids (pond turtles) can be traced back to the early Eocene. They show a sharp increase in size in the Uintan, followed by a decrease. They also disappear from the geologic record in the Badlands in the earliest Orellan. The family appears to also be sensitive to cooling and drying trends, which has profoundly affected their distribution (Hutchison, 1992; Estes, 1970a). They are characterized by a low domed shell, the absence of a musk duct foramina, and a weak or absent caudal notch. There is only a slight nuchal overlap and little or no anterior constriction (Hutchison, 1996).

Pseudograptemys Hutchison, 1996

Systematics and Evolution The type species for *Pseudograptemys* is *P. inornata* (Loomis, 1904) and includes *Graptemys cordifera.* The type specimens were collected, respectively, in the Chadron Formation, from Pennington and Shannon counties, South Dakota.

Distinctive Characters The carapace is unsculptured except for traces of a low, rounded median keel. The posterior peripherals are notched at the marginal sulcus. The hypoplastron buttress is well developed and articulates with costal 5. The xiphiplastron ends are distinctly pointed (Hutchison, 1998).

Stratigraphic and Geographic Distribution Fossilized remains from *Pseudograptemys* can be found in the Chadronian. Fossils are found in South Dakota. The genus *Graptemys* is considered to be the closest living relative of the fossil form.

Natural History and Paleoecology The extant map turtle *Graptemys* provides the closest relative for comparison to the fossil genus *Pseudograptemys.* Map turtles are freshwater turtles restricted to river systems in the eastern United States (Wood, 1977). Most of the modern species have some temperature constraints and are only found in the warm rivers of the southern United States. A few species have a more northern distribution, occurring in rivers that regularly freeze in the winter. *Graptemys* is also limited to temperate regions. Of the nine living species, seven are described as carnivores. It appears that *Graptemys* and *Pseudograptemys* were influenced by climatic and physiographic changes throughout the geologic record (Wood, 1977).

Chrysemys Gray, 1844

Systematics and Evolution *Chrysemys antiqua* Clark, 1937, was found in the Chadron Formation in Pennington County, South Dakota. The type specimen of *C. antiqua* is YPM PU 13839.

Distinctive Characters The carapace is of moderate size (to 190 mm), is moderately domed with or without sculpture, lacks keels, and has a sharp but moderately incised sulci. The anterior nuchal margin is straight to moderately concave (Hutchison, 1996:343).

Stratigraphic and Geographic Distribution *Chrysemys antiqua* is found in the Chadron Formation in Pennington County, South Dakota. It is also found in Colorado, Nebraska, and North Dakota. *Chrysemys antiqua* is found in the Chadronian through the Whitneyan. The genus *Chrysemys* ranges from the Oligocene through the Recent.

Family Testudinidae (Tortoises)

The family Testudinidae includes the most common turtles within the White River Group (Hay, 1908; Hutchison, 1992, 1996), and they occur throughout the section. First appearing in the Wasatchian, they increase in size in the Bridgerian and then tend to decrease in size into the Orellan in response to lower precipitation levels and cooler temperatures (Hutchison, 1992). Four tortoise genera have classically been assigned to this family. Two of the genera, *Testudo* and *Geochelone*, are now restricted to non–North American taxa (Williams, 1952; Preston, 1979; Hutchison, 1996). *Hesperotestudo, Stylemys*, and *Gopherus* are still used for the tortoises from the White River Group and will be discussed here.

Hesperotestudo (Williams, 1950)

Systematics and Evolution The type species of *Hesperotestudo* is *H. brontops* Marsh, 1890. The type specimen was collected from the Chadron Formation in Pennington County, South Dakota.

Distinctive Characters *Hesperotestudo* is characterized by a reduced or absent premaxillary ridge. There is no symphyseal dentary groove. The costals alternate from narrow to wide laterally. *Hesperotestudo* has a moderate to short tail and caudal vertebrae with interpostzygapophyseal notches. The armor on the forelimbs is moderately to well-developed with a patch of sutured dermal ossicles on the thigh and perhaps tail in adults. The nuchal scale is longer than wide (Hutchison, 1996:345).

Stratigraphic and Geographic Distribution *Hesperotestudo* occurs in the Chadron and Brule formations of the White River Badlands of South Dakota.

Stylemys Leidy, 1851

Systematics and Evolution The type species of *Stylemys* is *S. nebrascensis* Leidy, 1851. The type specimen was collected from the Scenic Member, Brule Formation, in South Dakota.

Distinctive Characters Unfortunately, the majority of defining features for the genus *Stylemys* relate to the skull, dentary, and pes structures, which are rarely preserved. Hay (1908) proposed many shell features that later were linked to ontogenetic changes (Auffenberg, 1964). One shell feature used to define the genus *Stylemys* is that the costals (bones of the carapace derived from ribs) only slightly alternate between narrower and wider width laterally (Fig. 5.5), in contrast to *Gopherus*, in which they distinctly alternate between narrower and wider width laterally. Another shell feature is that the nuchal scale is longer than it is wide (Auffenberg, 1964, 1974; Hutchison, 1996). Some other defining features are as follows: premaxillary ridge present; a symphyseal dentary groove; a moderately long, unspecialized tail with some of the caudal vertebrae lacking interpostzygopophyseal notches; free proximal portion of the ribs long; and minimum of armor on forelimbs (Auffenberg, 1964, 1974; Hutchison, 1996).

Auffenberg (1962) described *Stylemys nebrascensis* as having a carapace as much as 530 mm or more in length; peripheral pits for reception of costal ribs; proportionally thicker and more rounded shell; weak lateral notch on xiphiplastron; anterior lobe of plastron wider than long; posterior length distinctly less than bridge length.

Stratigraphic and Geographic Distribution Fossilized remains from *Stylemys* can be found in the Chadronian through the Whitneyan. Fossils are found in Oregon, California, South Dakota, Nebraska, Utah, Texas, Wyoming, and Colorado.

Natural History and Paleoecology *Stylemys* was defined as an extinct Holarctic genus with a climatic range of temperate to subtropical (Brattstrom, 1961; Auffenberg, 1964). Fossil turtle eggs from the Oligocene of the Western Interior were

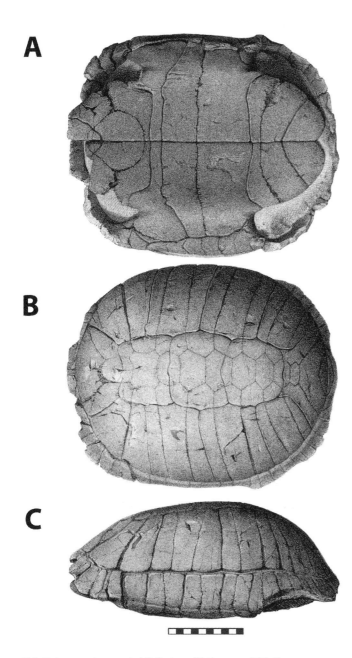

A

B

C

5.5. *Stylemys nebrascensis.* (A) Plastron. (B) Carapace. (C) Left lateral view (Leidy, 1853:plates 22, 24). Scale in centimeters.

described by Hay (1908:391) and on the basis of their macro-features were assigned to the genus *Stylemys*.

Gopherus Rafinesque, 1832

Systematics and Evolution The type species for *Gopherus* is the living species, *G. polyphemus*, originally *Testudo polyphemus* Daudin, 1802, that today lives in the southeastern United States. The fossil form found in the White River Group is *G. (Oligopherus) laticuneus* Cope, 1873, the type specimen of which was collected from Weld County, Colorado.

Distinctive Characters *Gopherus* shares many diagnostic features with *Stylemys*. Some of these include premaxillary ridge present and a symphyseal dentary groove

(Hutchison, 1996). In contrast with *Stylemys*, *Gopherus* has costals that distinctly alternate narrower with wider laterally, and the nuchal scale is wide or wider than long (Hutchison, 1996). Auffenberg (1974) includes the following additional features: short cervical vertebrae; flattened forelimbs adapted for digging; fourth vertebral scute usually wider than long.

Hutchison (1996) proposed the subgenus *Oligopherus*, characterized by the following features: the posterior epiplastron excavation is shallow; cervical vertebrae not appreciably shortened; pre- and postzygopophyses not elongated, widely separated; first dorsal vertebra with small zygopophyses and the neural arch structurally united with neural 1. The new subgenus included *G. laticuneus*, a better-known taxon.

Stratigraphic and Geographic Distribution Fossilized remains of *Gopherus* can be found throughout the White River Group in the Badlands. Beyond this area the time range of the genus is from Oligocene to Recent. The type of *G. (Oligopherus) laticuneus*, AMNH 1160, was found in Weld County, Colorado. Fossils are found in all of the Nearctic Region south of Canada and south throughout northern Mexico. Extant species are now limited to the southern United States and northern Mexico (Brattstrom, 1961; Auffenberg, 1974).

Natural History and Paleoecology The genus *Gopherus* includes the modern gopher tortoises, which are named for their ability to dig large, deep burrows. All four living species are found in xeric habitats.

Order Squamata

Fossil members of the order Squamata (lizards and snakes) from central North America are best preserved and have their greatest diversity in the Orellan. In contrast, the squamates are much less diverse in the Chadronian and the Whitneyan. Overall, the squamatofauna of the Great Plains in North America are more primitive than originally thought (Sullivan and Holman, 1996). All of the fossil squamata found within the White River Badlands discovered to date are fossorial or ground dwelling (Maddox and Wall, 1998), features of their ecology that facilitated their preservation as fossils. Hutchison (1992) concluded that the increase in aridity and accompanying decrease in aquatic habitats during the Eocene–Oligocene transition had a greater impact on herpetofauna species diversity than a change in temperature. The squamate fauna within the White River Badlands seem to support Hutchison's interpretation (Maddox and Wall, 1998).

Suborder Lacertilia (includes the lizards)
Family Anguidae
Peltosaurus Cope, 1873

Systematics and Evolution The type species for *Peltosaurus* is *P. granulosus* Cope, 1873. The type specimen was collected from the White River Formation in Logan County, Colorado.

Distinctive Characters *Peltosaurus* is characterized by seven teeth on the premaxillary and 10 teeth on the dentary. The parietal bone is broad and flat, and the frontals are greatly narrowed and united. The postorbital and postfrontal are coalesced, and the parietal is in contact with the squamosal (Fig. 5.6). The head and body are covered with unkeeled, finely granular scutes (Gilmore, 1928).

Stratigraphic and Geographic Distribution Fossilized remains from *Peltosaurus* can be found in the Orellan to Arikareean. Fossils are found in South Dakota, Nebraska, Florida, and Colorado. In the White River Badlands, *Peltosaurus* is found in the Brule Formation.

Natural History and Paleoecology *Peltosaurus* is the most common of all Oligocene "melanosaur" lizards and is known from numerous specimens (Sullivan and Holman, 1996).

Helodermoides Douglass, 1903

Systematics and Evolution The type species of *Helodermoides* is *H. tuberculatus* Douglass, 1903. The type specimen is Chadronian and was found in Jefferson County, Montana.

Distinctive Characters *Helodermoides* is characterized by distinct frontals and a bulbous cephalic osteoderms. There are numerous tubercles without definite arrangement. Sometimes they are arranged in a ring shape. Six or seven rows of cephalic osteoderms occur between the orbits. The teeth are subconical and the posterior ones are slightly recurved. The jugal blade is curved and the maxilla is straight. The dentary is moderately slender and the supratemporal fenestra is closed. The skull is highly vaulted (Sullivan, 1979).

Stratigraphic and Geographic Distribution Fossilized remains from *Helodermoides* can be found in the Chadronian to Orellan. Fossils are found in South Dakota (Maddox and Wall, 1998), Wyoming, Montana, and Nebraska. In the White River Badlands, *Helodermoides* is found in the Chadron and Brule formations.

Natural History and Paleoecology Gilmore (1928) synonymized *Helodermoides* with *Glyptosaurus*. Sullivan (1979) resurrected the genus *Helodermoides*, describing numerous characters that separated it from *Glyptosaurus* and other glyptosaurs.

(R)

5.6. *Peltosaurus granulosus*. Skull, left lateral view, AMNH 8138. Scale in centimeters. Photo by the authors of a specimen from the American Museum of Natural History, New York, New York, U.S.A.

Family Rhineuridae
Rhineura Cope, 1861

Systematics and Evolution The type species of *Rhineura* is *R. floridana* (Baird, 1859). *Rhineura hatcheri* (Baur, 1893) is the species found at Badlands National Park. The type specimen was found in Logan County, Colorado. According to Sullivan and Holman (1996), the various characters used by previous workers to diagnose species of *Rhineura* may not be valid for recognizing distinct genera. The size differences may be due to growth stages as opposed to character differences. The Oligocene *Rhineura* may be only known by one species, *R. hatcheri*. If *R. floridana* is a valid taxon, then the genus *Rhineura* would have a relatively long temporal range (Oligocene to Recent). It would be the only recent squamate genus represented in the early Oligocene squamatofauna (Sullivan and Holman, 1996).

Distinctive Characters *Rhineura* is characterized by a compact, well-ossified skull, with pleurodont teeth lacking the postorbital and postfrontal squamosal arches and epipterygoid (Gilmore, 1928).

Stratigraphic and Geographic Distribution Fossilized remains from *Rhineura* can be found in the Brule Formation of the White River Badlands. Fossils are found in South Dakota and Colorado.

Suborder Serpentes

The snake fauna of the White River Badlands is sparse and consists of a few taxa of boids, mostly small in size.

Family Boidae
Calamagras Cope, 1873

Systematics and Evolution The type species of *Calamagras* is *C. murivorus* Cope, 1873. The type of *Calamagras* is from the early Oligocene (Orellan) part of the White River Formation, Logan County, Colorado.

Distinctive Characters *Calamagras* has a short, thick neural spine on its vertebrae. The vertebral centra are less than 9 mm in length. The neural spine is less than one-half the total length of the centrum, but it is not tubular or dorsally swollen (Holman, 1979). Cope (1873) described a second species, *C. talpivorus* from the same locality as the type of *C. murivorus*, Cedar Creek, Logan County, Colorado. *Calamagras talpivorus* is considered to be the same species as *C. murivorus*. A third species of the genus from the same locality, also described by Cope (1873) and still considered valid, is *C. angulatus*.

Stratigraphic and Geographic Distribution Fossilized remains from *Calamagras* can be found in the late Bridgerian (Hecht, 1959) through the Arikareean. Fossils are most abundant in the Brule Formation of the White River Badlands of South Dakota (Maddox and Wall, 1998). *Calamagras* is found in the Chadronian in Saskatchewan. They are also found in Nebraska and Colorado. The genus is also known from the early Miocene (early Hemingfordian) of California and Delaware (Holman, 2000).

Natural History and Paleoecology Holman (1979) believed this poorly defined small boalike snake might have vestiges of hind limbs. *Calamagras* is a member of the Infraorder Henophidia, and general characteristics for this group include vaulted neural arches, round condyles and cotyles on the vertebrae, and a poorly developed hemal keel (Maddox and Wall, 1998).

Geringophis Holman, 1976

Systematics and Evolution The type species of *Geringophis* is *G. depressus* Holman, 1976. It is from the late Oligocene (early Arikareean) Gering Formation, Crawford locality, Dawes County, Nebraska. *Geringophis vetus* Holman, 1982, is believed to be found in the Brule Formation in the White River Badlands (Maddox and Wall, 1998). *Geringophis vetus* is the earliest occurrence of *Geringophis* in the fossil record (Sullivan and Holman, 1996). It is unknown whether the genus arose from *Cadurcoboa* from the Eocene of France and immigrated to North America from the Old World or whether it originated from an erycine boid with a flattened vertebral form such as *Calamagras angulatus* (Sullivan and Holman, 1996).

Distinctive Characters This erycine boid is distinct from other small boid genera found in the White River Group in that the vertebrae have a flattened shape and a long, high neural spine (Holman, 1982). Holman (1979) describes *Geringophis* as having the following unique characters; the vertebrae contained a reduced neural arch, a long, well-developed, dorsally expanded neural spine, and a well-developed hemal keel and subcentral ridges.

Stratigraphic and Geographic Distribution Fossilized remains from *Geringophis* occur from the Orellan through the late Barstovian. Fossils are found in South Dakota (Maddox and Wall, 1998), Colorado; the early and late Arikareean; medial and late Barstovian of Nebraska; and early Arikareean of Wyoming (Holman, 2000). *Geringophis vetus* was described from the early Oligocene (late Orellan) Toadstool Geologic Park in Sioux County, Nebraska.

Superfamily Booidea
Coprophis Parris and Holman, 1978

Systematics and Evolution The type species for *Coprophis* is *C. dakotaensis* Parris and Holman, 1978. The type specimen for *C. dakotaensis* was found in the Big Badlands of South Dakota (Pennington and Shannon Counties). Sullivan and Holman (1996) believe the genus should be assigned to the superfamily Booidea because of they have a wider, more indistinct hemal keel.

Distinctive Characters *Coprophis* is characterized by vertebra with a narrow, well-developed hemal keel (see Parris and Holman, 1978).

Stratigraphic and Geographic Distribution The type and only known record of *Coprophis* is from the early Oligocene (Orellan) Scenic Member of the Brule Formation. Fossils of this animal are only known from South Dakota.

Natural History and Paleoecology The vertebrae of *Coprophis* were discovered during a comprehensive study of the mammalian coprolites from the Brule Formation by D. C. Parris. The type specimen, which consists of vertebrae, are somewhat eroded because of their partially digested state. The coprolite is thought to be produced by a mammal because large nonmammalian carnivores are unknown in the Brule Formation. They were assigned to the new genus *Coprophis* on the basis of a combination of characters outlined in Parris and Holman (1978).

Order Crocodylia

The suborder Eusuchia includes the modern crocodiles, which have been known since the late Cretaceous. There are three living families of crocodiles: the Alligatoridae, the Crocodylidae, and the Gavialidae. Only members of the

5.7. Complete skeleton of *Alligator* cf. *prenasalis*. Courtesy of the Division of Vertebrate Paleontology; YPM VPPU 013799, Peabody Museum of Natural History, Yale University, New Haven, Connecticut, U.S.A.

Alligatoridae are found in the White River Badlands. The basic crocodilian features such as the massive skull and long snout have not changed since the late Triassic. Other features that characterize the order include a skull that is strongly buttressed and has a fenestrate appearance. Crocodilians also have a secondary palate that separates the nasal passages from the mouth (Carroll, 1988). The suborder Eusuchia is defined by procoelous vertebral centra, and the internal nares within the skull are completely surrounded by the pterygoid. The Eusuchians were much more diverse, numerous, and widespread in the early Tertiary than they are today. Their decline was due to the climatic cooling and drying that has occurred since the early Cenozoic.

Family Alligatoridae

The family Alligatoridae is characterized by a short and broad snout. The nasal bones usually reach the external narial aperture. The supratemporal fenestrae are usually smaller than the orbits and sometimes closed over secondarily. The mandibular symphysis is short. The first mandibular teeth and usually the fourth bite into pits in the palate. There are two or more rows of dorsal scutes (Mook, 1934).

Alligator Cuvier, 1807

Systematics and Evolution The type species of *Alligator* is the living *Crocodilus mississippiensis* Daudin, 1801 [1802]. *Alligator prenasalis* (Loomis, 1904) is the only fossil alligator found in the White River Badlands. Originally described as *Crocodilus prenasalis* by Loomis (1904), Matthew (1918) reassigned it to *Alligator* on the basis of more complete material. *Alligator prenasalis* is well described and has been extensively

collected in the White River Badlands. It is includes the oldest specimens assignable to *Alligator* (Brochu, 1999).

Distinctive Characters *Alligator prenasalis* has a broad and flat snout, which is considered primitive when compared to other species within the group (Fig. 5.7) (Brochu, 2004). The undivided nasal opening is located quite far forward on the rostrum and does not have a distinctive anterior border. The nostril opening would be directed to the front rather than upward on top of the snout (Loomis, 1904). There is a constriction on the maxilla to receive a tooth from the lower jaw. This occurs just behind the ninth superior tooth. The upper surface of the frontals is covered with large pits (Loomis, 1904). The animal was about 1.5 m long when alive (Fig. 5.7) (O'Harra, 1920).

Stratigraphic and Geographic Distribution Fossilized remains from *Alligator* can be found in the Chadronian to the Recent in North America (Brochu, 1999). In the White River Badlands, they are found in the Chadron Formation. Fossils of similar age are also found in Nebraska, Colorado, and Florida.

Natural History and Paleoecology Because alligators do not tolerate saltwater, it is thought that they only traveled to other continents via land bridges and were not as widely dispersed as crocodylids. Alligatorids occur exclusively in the Western Hemisphere and Asia during the Tertiary. Within the White River Badlands, alligator remains are found within the channel deposits in the Chadron Formation (Clark, Beerbower, and Kietzke, 1967).

Alligators today are restricted to the southeastern United States along the Gulf Coast through Florida and as far north along the Atlantic Coast to North Carolina. Alligators live in wetland habitat. They are apex predators. The living American alligators are less prone to cold than other crocodilians and can survive colder temperatures. This may explain their presence farther north in the fossil record: the living American alligators can spread farther north than the American crocodile. It is found farther from the equator and is more equipped to deal with cooler conditions than any other crocodilian. As apex predators, they help control the population of rodents and other animals that might overtax the marshland vegetation.

Climatic Interpretations from the Occurrence of Reptiles and Amphibians

Cenozoic herpetofaunas serve as good indicators for climatic change. A decrease in the diversity of reptiles and amphibians, primarily aquatic forms, begins to occur as early as the Uintan in North America (Hutchison, 1992). There is a significant decline in the diversity of aquatic amphibians and reptiles in faunas within the interval between the Chadronian and Orellan but only a modest decrease in diversity within the terrestrial herpetofauna (Hutchison, 1992). Hutchison (1992) also notes a change in maximum carapace length in tortoises from 80 cm in the Bridgerian to 50 cm between the Chadronian and Orellan. A significant modernization of amphibians and reptiles can be seen after the Eocene–Oligocene transition and through much of the Oligocene. The smaller, lower vertebrates from this time period are not well preserved and so are not well documented. In contrast, a much better record of the larger forms such as the turtles and crocodiles exists. Hay (1908) correlated the reduction in diversity of aquatic turtle faunas to a significant drying on the western plains. Tihen (1964) noted an abrupt transition between the archaic to more modern faunas during this time period. Hutchison (1982) attributed the significant decline in diversity to the reduction in the availability of permanent surface water. The large-scale changes involved the infilling of early Tertiary lacustrine basins, a general increase in aridity, and the decrease or even absence of permanent rivers and streams (Hutchison, 1992).

In addition to being strong climatic indicators, the herpetofauna of the Tertiary Great Plains sediments have provided important information on the transition between archaic and modern faunas from both a biostratigraphic and evolutionary perspective (Sullivan and Holman, 1996).

Class Aves

Compared to other vertebrates, the fossil record for birds is limited in the White River Badlands. Most avian fossils are from the limbs of birds. Complete skeletons or even skull fragments are rare (Adolphson, 1973). The skeletons of birds are lightweight and fragile and are easily destroyed.

Fossilization of avian skulls is a rare occurrence because of the delicate nature of this part of the skeleton. Braincases are often not found attached to the beak, as they can be easily separated at the hinge between these two parts of the skull. When a bird dies and floats on a body of water, it is subject to destruction by aquatic predators and scavengers. Its buoyancy usually prevents it from settling on the bottom where it can be buried by sediments. Most bird fossils are restricted to fine-grained, relatively undisturbed geologic sediments (Welty and Baptista, 1988).

The avian skeleton achieves strength with lightness and is constructed with the greatest possible economy of materials. Some bones common to other vertebrates are completely eliminated, and others are fused together. Many of the bones are pneumatized, or filled with airspaces instead of bone marrow (Welty and Baptista, 1988).

There is also a limited record of fossil bird eggs within the White River Badlands. The eggs can be identified on the basis of their general shape and shell texture.

Order Falconiformes
Family Accipitridae

The family Accipitridae includes the red-shouldered and red-tailed hawks. This family and the genus *Buteo* are widely spread over the world today (Adolphson, 1973).

Buteo Lacépède, 1799

Systematics and Evolution The type species of *Buteo* is *Falco buteo* Linnaeus. The fossil species, *B. grangeri* Wetmore and Case, 1934, extends the age of the genus *Buteo* back to the Orellan (Wetmore and Case, 1934). Hawks of a similar type are well represented in the Miocene and Pliocene sediments in Sioux County, Nebraska. The type specimen of *B. grangeri* (UMMP 14405) was found in the Brule Formation, Jackson County, South Dakota.

Distinctive Characters *Buteo grangeri* is similar to *Buteo melanoleucus*, the living black buzzard-eagle, but is slightly smaller, with the frontal relatively narrower between the orbits. The maxilla is stronger, heavier, and deeper. The palatine bones are somewhat more slender. The species is similar to the living red-tailed hawk, *Buteo borealis*, except that it is larger.

Stratigraphic and Geographic Distribution The range for the genus *Buteo* is from the Orellan to the Recent. *Buteo grangeri* fossils were found in South Dakota. The genus *Buteo* is found worldwide today.

Natural History and Paleoecology Because of their trophic level as predators, hawks and other birds of prey are even less common then other types of birds in the fossil

5.8. *Procrax brevipes,* holotype, SDSM 511. Skeleton is almost complete except for skull and vertebral column on limestone slab. Scale in centimeters. Photo by the authors of specimen from the Museum of Geology, South Dakota School of Mines and Technology, Rapid City, South Dakota, U.S.A.

record, and it is difficult to determine specifics about their paleoecology.

Order Galliformes
Family Cracidae

The family Cracidae includes the curassows, guans, and chachalacas; members of the family are found today from the lower Rio Grande Valley in Texas to south to Argentina. In North America, fossils from this family are known from the Cenozoic of Florida, Nebraska, and South Dakota. Features that define the family include the absence of the strongly developed intermetacarpal tuberosity on the metacarpal II, the absence of the prominent notch on the posterior palmer

surface of the carpal trochlea, and a small pollical facet on the metacarpal I. There is a flaring of the internal condyle into the shaft of the tibiotarsus, a less prominent internal condyle of the humerus, and a less ruggedly built coracoid (Tordoff and Macdonald, 1957).

The family now has a Neotropical distribution, but it originated in North America. It underwent a substantial adaptive radiation before it retreated southward. Living cracids are most abundant in the tropical zone, but they range also into the subtropical and even the temperate zones in the mountains. They are arboreal and live and nest in trees (Wetmore, 1956).

Procrax Tordoff and Macdonald, 1957

Systematics and Evolution The type species for *Procrax* is *P. brevipes*. The type specimen (SDSM 511) was collected from the Chadron Formation in Pennington County, South Dakota, and consists of an almost complete skeleton except for the skull and cervical vertebrae (Fig. 5.8). Five other fossil members of the family are known in North America. *Procrax brevipes* is the oldest known cracid.

Distinctive Characters *Procrax brevipes* has shorter legs and smaller feet and claws than its living relatives *Mitu*, *Crax*, *Penelope*, *Ortalis*, and *Penelopina*. *Procrax* most resembles *Pipile* but is smaller and has proportionally shorter and thicker claws. It is larger than the Oligocene *Palaeonossax* (Tordoff and Macdonald, 1957).

Stratigraphic and Geographic Distribution *Procrax brevipes* fossils were found only in South Dakota. The specimen is preserved in a fine-grained freshwater limestone (Fig. 5.8). These types of deposits are rare in the White River Badlands.

Natural History and Paleoecology *Procrax* is the largest of the fossil cracids and was believed to have a superior flying ability in comparison to modern forms. Because of its shorter legs and shorter and thicker claws, it was capable of rapid locomotion on the ground and through trees (Tordoff and Macdonald, 1957).

Palaeonossax Wetmore, 1956

Systematics and Evolution The type species of *Palaeonossax* is *P. senectus* Wetmore, 1956. The type specimen (SDSM 457) was collected from the Poleslide Member of the Brule Formation in Pennington County, South Dakota. The genus and species definitions are based on one fragmentary specimen, a distal end of a right humerus. However, Wetmore (1956) thought the features that were preserved were diagnostic enough to name a new genus and species.

Distinctive Characters The distal end of the humerus is similar to *Ortalis* but with the entepicondyle reduced in size and with a more definite separation from the internal condyle. The internal condyle is relatively smaller and more rounded. The external condyle is relatively shorter and slightly broader. The distal end is more delicate and less swollen (Wetmore, 1956).

Stratigraphic and Geographic Distribution Fossilized remains from *Palaeonossax senectus* can be found in the Poleslide Member of the Brule Formation. The distal end of the humerus of *Palaeonossax senectus* was found in Pennington County, South Dakota.

Order Gruiformes

The order Gruiformes includes cranes, rails, limpkins, and their relatives. This group arose late in the Mesozoic from the base of land bird radiation. Most of the order is specialized for life in or over the surface of the water. The order Gruiformes is the most primitive modern order within the assemblage. Most members of the order inhabit aquatic or swampy environments; however, there are few that live on dry land, and some even live in deserts (Carroll, 1988).

Suborder Cariamae

The Cariamae include one or more groups of giant, early Tertiary predators.

Family Bathornithidae

This is an extinct Gruiform family that was most abundant in the Oligocene. The center of its distribution was in the Northern Great Plains, especially in Colorado, Nebraska, South Dakota, and Wyoming. All of members of this family were flightless and were adapted to a cursorial way of life, similar to the cariamas in South America. Cracraft (1973) proposes that these morphological changes were an adaption to the overall cooling and drying trend in the Oligocene. There is a great diversity in size, ranging from birds 1 m to over 2 m tall (Cracraft, 1968). The fossil history of the family is one of the best-documented records for an avian family over such a short period of time.

Bathornis Wetmore, 1927

Systematics and Evolution The type species of *Bathornis* is *B. veredus* Wetmore, 1927. The type specimen (CMNH 805) is a lower portion of a right tarsometatarsus from the Chadron Formation in Weld County, Colorado. Three other species of *Bathornis* have been described, *B. celeripes*, *B. cursor*, and *B. geographicus*. The type specimen (SDSM 4030) of *B. geographicus* Wetmore, 1942, is an almost complete left tarsometatarsus from Shannon County, South Dakota. It was collected from the Poleslide Member of the Brule Formation.

Distinctive Characters The genus *Bathornis* is characterized by a tibiotarsus with condyles not as compressed lateromedially compared to *Eutreptornis*. The internal condyle with notch in distal border is slight or absent. Internal ligamental prominence is poorly developed. The supratendinal bridge narrows proximodistally. The anterior intercondylar fossa is deep. Condyles spread anteriorly. The tarsometatarsus has a short proximal hypotarsus. The intercotylar prominence is smaller than in *Eutreptornis*. The humerus has an

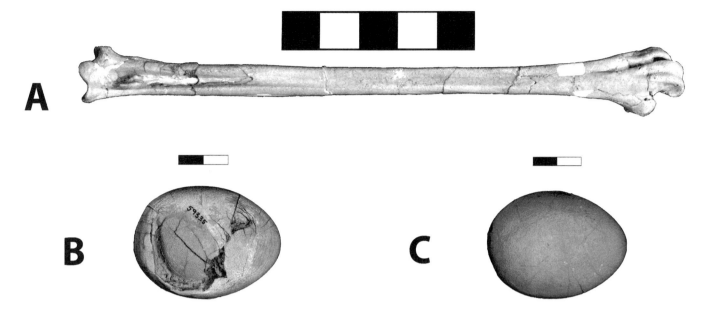

5.9. Fossil bird and bird egg from the White River Badlands. (A) *Badistornis aramus,* SDSM 3631, holotype, almost complete left tarsometatarsus. (B) Fossil bird egg, crushed side, DMNS 59335.

(C) Bird egg, uncrushed side, DMNS 59335. Scales in centimeters. Photos by the authors of specimens SDSM 3631, Museum of Geology, South Dakota School of Mines and Technology,

Rapid City, South Dakota, U.S.A., and DMNS 59335, Denver Museum of Nature and Science, Denver, Colorado, U.S.A. All rights reserved.

internal condyle that is raised somewhat distally relative to the external condyle in comparison to *Paracrax* (Cracraft, 1973). It is believed that many species of *Bathornis* coexisted at the same time. The tarsometatarsus of *B. veredus* is much larger than any known species within the genus.

Stratigraphic and Geographic Distribution *Bathornis* is found in Colorado, South Dakota, Nebraska, and Wyoming. Fossils are found from the late Eocene through the early Miocene.

Natural History and Paleoecology Species from the genus *Bathornis* range in a broad variety of sizes and are adapted to wetland or swampy types of environments.

Paracrax Brodkorb, 1964

Systematics and Evolution The type species of *Paracrax* is *Paracrax antiqua*. The type specimen (YPM 537) is the distal end of a right humerus and was collected from Weld County, Colorado. Three species are described: *P. wetmorei* Cracraft, 1968; *P. gigantea* Cracraft, 1968; and *P. antiqua* Cracraft, 1968. The type specimen (FAM 42998) for *P. wetmorei* Cracraft, 1968, is a complete right humerus from the Poleslide Member of the Brule Formation, Shannon County, South Dakota. The type specimen (FAM 42999) for *P. gigantea* is the distal end of a right humerus from the Poleslide Member of the Brule Formation, Jackson County, South Dakota.

Distinctive Characters The humerus resembles that of *Bathornis veredus* but with an internal condyle that is less distinctly raised relative to the external condyle. The

intercondylar furrow is less well marked; the entepicondyle is slightly less raised distally relative to the internal condyle; the brachial depression is slightly less deep; the distal end of the shaft is straighter, not curved when viewed from the side; the area of attachment of the anterior articular ligament is slightly less pronounced; and the external condyle is turned more internally as seen from palmar side. *Paracrax gigantea* is much larger than *P. wetmorei*, with all of the features proportionately more massive. *Paracrax gigantea* was truly a huge bird, standing at about 2 m (Cracraft, 1968).

Stratigraphic and Geographic Distribution *Paracrax* was found in South Dakota and Colorado in the Oligocene Brule Formation.

Natural History and Paleoecology *Paracrax* was considered to be a ground-dwelling predator.

Suborder Grues
Family Aramidae
Badistornis Wetmore, 1940

Systematics and Evolution The type species of *Badistornis* is *Badistornis aramus* Wetmore, 1940. The type specimen (SDSM 3631) is a complete left tarsometatarsus (Fig. 5.9A) collected in Pennington County, South Dakota.

Distinctive Characters *Badistornis* is characterized by a tarsometatarsus with the inner trochlea projecting distally only to the base of the middle trochlea and turned far posteriorly. In anterior view, the anteroposterior plane of the middle trochlea is inclined internally away from the longitudinal axis of the shaft (Cracraft, 1973:92). *Badistornis aramus* is the earliest record of a limpkin in North America (Brodkorb, 1967). The tarsometatarsus of *B. aramus* closely resembles that of the living *Aramus scolopaceus* (Wetmore, 1940). The tarsometatarsus of *B. aramus* is slightly longer than that of the limpkin *A. guarauna* (Wetmore, 1940). *Badistornis* is a distinctive genus within the suborder Grui and appears to be an aramid that became cranelike during the Orellan, and probably had similar locomotor habits. It shares intermediate features between cranes and limpkins even though these taxa do not have a common ancestor (Cracraft, 1973).

Chandler and Wall (2001) describe three fossilized bird eggs from Badlands National Park, which they believe to be from *B. aramus*. The eggs compare closest to the eggs from the living limpkin, *Aramus guarauna*. The overall egg geometry consists of average elongation, average bicone, and asymmetry. The eggs are elliptical in profile and similar to those of other members of the family Aramidae (Preston, 1968, 1969). They also describe the eggshell as porous. Figure 5.9B, C shows DMNS 59335, a partially crushed fossil bird egg from Converse County, Wyoming.

Stratigraphic and Geographic Distribution Fossilized remains from *Badistornis aramus* can be found in the Scenic Member of the Brule Formation, in South Dakota.

Natural History and Paleoecology Another published record of a fossilized bird egg, possibly from a duck, is from the White River Badlands; however, the collecting data are not precise (Farrington, 1899). It should be noted that no bones of any member of the Anseriformes (ducks, geese, swans) have been found in the Big Badlands, which makes the identification of this egg as a member of this order suspect.

Order Ardeiformes
Family Ardeidae
Gnotornis Wetmore, 1942

Systematics and Evolution The type species for *Gnotornis* is *Gnotornis aramielus* Wetmore, 1942. The type specimen for *Gnotornis aramielus* is SDSM 40158, distal left humerus from the Poleslide Member of the Brule Formation, Shannon County, South Dakota. This is the only species.

Distinctive Characters The humerus is similar to *Aramus* but differs in that the entepicondylar area projects only slightly laterally. The ectepicondylar area is relatively larger (Wetmore, 1942). The entepicondyle of *Gnotornis* projects more distally and that internal contour of the distal end of the shaft is straighter (Cracraft, 1973).

Stratigraphic and Geographic Distribution The type specimen was collected in the Poleslide Member of the Brule Formation in Shannon County, South Dakota.

Natural History and Paleoecology *Gnotornis aramielus* is described as a volant piscivore, meaning it was able to fly and it fed on fish. These fossils were collected from the *Protoceras* channel sandstones from the Poleslide Member of the Brule Formation, thus associating them with a riparian environment.

Class Mammalia

Mammals have been the dominant terrestrial vertebrates for the last 65 million years. The majority of mammalian groups are either marsupials or placentals. The late Eocene and Oligocene have often been referred to a period in Earth's history when mammals became quite abundant and diverse. Mammal skeletons from the White River Badlands are well preserved in the rock record, with the majority of remains consisting of skulls and teeth.

Order Marsupialia

Marsupials originated in North America during the Cretaceous (Clemens, 1979; Johanson, 1996; Cifelli and Muizon, 1997; Cifelli, 1999). They dispersed from North America and developed two major living assemblages, one in South America and one in Australia (Benton, 2000). The North American Tertiary marsupial record extends from the Paleocene into the early middle Miocene (Barstovian), when they became extinct (Slaughter, 1978). They reappear in North America in the middle Pleistocene (Irvingtonian) (Kurtén and Anderson, 1980) when the modern opossum, *Didelphis*, entered from South America after the formation of the Panamanian land bridge between the two continents.

North American marsupials were not very diverse, either morphologically or taxonomically. Even though they are represented in nearly all major faunas from the Paleocene through the Oligocene, they make up a small percentage of the sample (Korth, 2008). The marsupial fossil record consists mostly of dental elements, and the group becomes rare by the Miocene.

North American fossil marsupials have a postcranial morphology similar to the extant didelphid marsupials, suggesting that they were primarily terrestrial. Their cheek teeth are pointed and cuspate and have little modification from primitive forms. The lower molars can be distinguished from Eutherians by the close positioning of the entoconid and hypoconulid on the talonids, often referred to as twinning (Korth, 2008). Marsupials have a primitive dental formula of three premolars and four molars. This is in contrast to primitive placental mammals, which have a primitive dental formula of four premolars and three premolars.

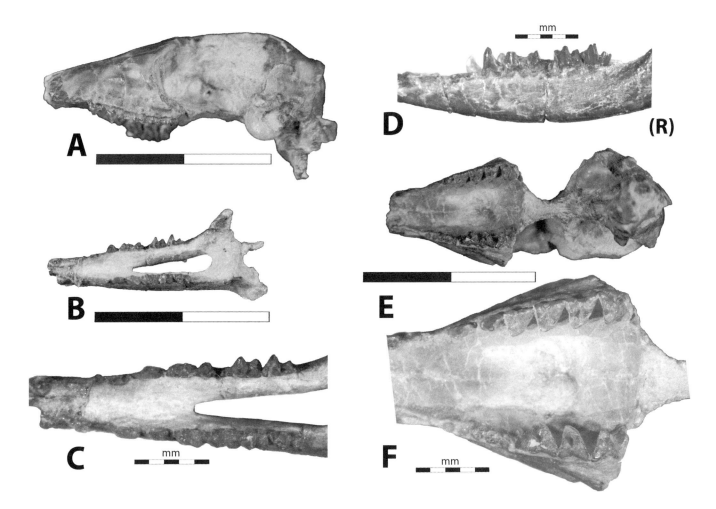

5.10. *Herpetotherium fugax,* SDSM 77332. (A) Skull, left lateral view. (B) Mandibles, occlusal view. (C) Mandibles, occlusal view (enlarged). (D) Mandibles, left lateral view (E) Skull, occlusal view. (F) Skull, occlusal view (enlarged). Scales in centimeters unless otherwise marked. Photos (A), (B), and (E) by the authors; and photos (C), (D), and (F) by Edward Welsh of specimens from the Museum of Geology, South Dakota School of Mines and Technology, Rapid City, South Dakota, U.S.A.

In contrast to modern placental mammals, marsupials are born in a relatively undeveloped or embryonic state, and in most species, the female carries the young in a fur-lined pouch (marsupium) until they are capable for caring for themselves (Jones et al., 1983). The pouch is supported by a pair of bones called the epipubic bones that are absent in placentals, except a few bats. It is difficult to determine whether the fossil taxa had similar habits and behavior as their modern relatives.

Family Didelphidae

The didelphid dentition is characterized by upper molars with a wide stylar shelf and more pronounced stylar cusps. The centrocrista is V shaped, and the conules are larger. The metacone is larger than the paracone, and the hypoconulid is posteriorly projecting. The first two lower incisors are greatly enlarged and procumbent. The size of the stylar cusps of the upper molars is reduced over time. The central stylar cusp becomes dominant. The angular process at the back base of the lower jaw (mandible) is bent inward (Korth, 2008).

Herpetotherium (including *Peratherium*) Cope, 1873

Systematics and Evolution The type species for *Herpetotherium* is *H. fugax* Cope, 1873. The type specimen (AMNH 5254) was collected from the White River Formation in Logan County, Colorado. Cope (1873) first named *Herpetotherium* but described it as an insectivore. Later Cope (1884) recognized that it was a marsupial and referred all species to the European genus *Peratherium* Aymard (1846). Several taxonomic revisions have been made over the years (Crochet, 1977, 1980; Fox, 1983; Krishtalka and Stucky, 1983; Reig, Kirsch, and Marshall, 1985; Korth, 1994a; Rothecker and Storer, 1996), followed by McKenna and Bell (1997), who placed all North American species in *Herpetotherium* and distinguished it from the European genus *Peratherium*.

Distinctive Characters There is a central dominant stylar cusp on the upper molars. On the upper M1–2, stylar cusps C and D progressively fuse, and stylar cusp B is progressively reduced (Fig. 5.10). The lower incisors 1 and 2 are enlarged and procumbent (Korth, 2008).

Stratigraphic and Geographic Distribution Fossilized remains from *Herpetotherium* can be found in the early Wasatchian through the late Hemingfordian. Fossils are found in Wyoming, Colorado, South Dakota, Nebraska, Saskatchewan, Montana, North Dakota, New Mexico, Texas, Florida, California, Oregon, and Utah.

Natural History and Paleoecology *Herpetotherium* is the only fossil marsupial found in Badlands National Park and is mostly known by isolated teeth and a jaw with fragmented teeth from the Peanut Peak Member of the Chadron Formation (Clark, Beerbower, and Kietzke, 1967). Like the living North American genus *Didelphis*, *Herpetotherium* was probably omnivorous.

INSECTIVOROUS MAMMALS

The "Order Insectivora" proposed by Simpson (1945) is considered a catchall term that included several morphologically disparate and unrelated groups of mammals and is not considered consistent or unified phylogenetically (Gunnell and Bloch, 2008). The term now commonly used is "insectivorous mammals," which contains several different orders, each listed in this section (Gunnell and Bloch, 2008), and which more accurately reflects similarities in diet and ecology rather than being indicative of a phylogenetic relationship. Insectivorous mammals are characterized by a relatively small body size, elongate snouts, small eyes, relatively small brains, five-toed and plantigrade feet, and external ears that are small or absent. Their dentition was specialized for eating invertebrate prey, and they have rudimentary endothermy (Eisenberg, 1980; Symonds, 2005). As an ecological category, they also have a relatively long history in the geologic record (Symonds, 2005). The rocks at Badlands National Park preserve a diverse record of insectivorous mammals, although the fossils are fairly rare.

Order Proteutheria

The order was set up to include extinct insectivorous mammals that had no clear relationship with living insectivorans. Little postcranial material has been found. More fossils need to be described to better document this group. Proteutherians tend to be characterized by a larger body size (Gunnell and Bloch, 2008).

Family Apatemyidae

The apatemyids are characterized by the lack of an ossified auditory bulla on the skull. The first lower incisor is enlarged and procumbent with a root extending posteriorly to the m3. The I1 lacks lateral enamel, and the canines are absent. The p3 is bladelike and single rooted. The upper molars are simple and lack mesostyles. Members of the family Apatemyidae range from the late Paleocene to the late Oligocene in North America (Gunnell et al., 2008).

Sinclairella Jepsen, 1934

Systematics and Evolution The type species for *Sinclairella* is *S. dakotensis* Jepsen, 1934. The specimen was collected from the upper part of Chadron Formation (about 11 feet below the limestone zone, which is usually considered as the boundary between the Chadron and Brule formations), near the head of the west fork of the east branch of Big Corral Draw, Shannon County, about 13 miles south-southwest of Scenic, South Dakota. The holotype and most complete specimen of *Sinclairella* have been lost for more than 30 years.

Distinctive Characters *Sinclairella* is characterized by its fairly large size compared to other insectivorous mammals. The upper molars are squared and have a large hypocone. The p4 is tiny, and the lower molars have reduced paraconids. The m3 is elongate with an expanded hypoconulid. The lower molar row is about four times as long as the lower premolar row (Gunnell et al., 2008).

Stratigraphic and Geographic Distribution Fossilized remains from *Sinclairella* can be found in the Chadronian through the Arikareean. Fossils are found in Wyoming, Colorado, South Dakota, Nebraska, Saskatchewan, and North Dakota.

Natural History and Paleoecology *Sinclairella* makes up a small part of the North American mammalian faunal sample until the end of the early Arikareean in the Great Plains. It is also the last surviving proteutherian in North America (Gunnell et al., 2008).

Family Pantolestidae Incertae Sedis

Two pantolestids, *Chadronia margaretae* Cook, 1954, and *Cymaprimadon kenni* Clark, 1968, were collected from north of Crawford, Nebraska, and Custer County, South Dakota, respectively. Both are from the White River Group but are beyond the area covered by this book. A new occurrence of *Chadronia* sp. has been documented in the Scenic Member of the Brule Formation in Badlands National Park, which

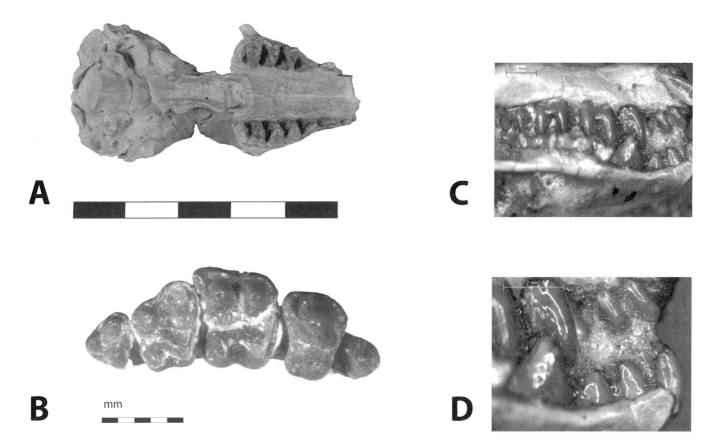

5.11. Insectivorous mammals from the White River Badlands (A) *Leptictis* sp., partial skull with right P2–M3 and left P1–M3, occlusal view, BADL 23756. (B) *Proterix bicuspis,* right P3–M3, BADL 11130/SDSM 7102 (C) *Proscalops* sp., skull, right lateral view, BADL 59754. (D) *Proscalops* sp., skull close-up showing antemolar region, right lateral view, BADL 59754. Scale bars for (A) and (B) are in centimeters; scale bars for (C) and (D) = 2 mm. Photo A by the authors; photo B by Edward Welsh; photos C and D by Mindy Householder. BADL 11130/SDSM 7102 from Badlands National Park, Interior, South Dakota, U.S.A., and the Museum of Geology, South Dakota School of Mines and Technology, Rapid City, South Dakota, U.S.A. BADL 59754 and BADL 23756 are from Badlands National Park, Interior, South Dakota, U.S.A. They are the property of the U.S. government.

extends the temporal range of pantolestids in North America (Boyd et al., 2014).

Order Leptictida

The Leptictida are believed to be ancestral or related to several other groups. These include the Lipotyphla, the Apatemyidae, primates, and the Erinaceomorpha (Gunnell, Bown, and Bloch, 2008). They first appeared in the late Cretaceous of North America and range from the early Paleocene through the late Oligocene. Members of this order were either insectivorous or omnivorous. They had an elongate and mobile snout (Novacek, 1986). The forelimbs were much shorter than the hind limbs, and they had a distally fused tibia–fibula. The humerus was robust and was designed for digging for either food or shelter (Gunnell, Bown, and Bloch, 2008). The Leptictida were relatively small mammals, with an estimated body mass ranging from 400 to 700 g. They were also terrestrial mammals capable of rapid running and quadrupedal jumping (Rose, 1999).

Family Leptictidae

The family is characterized by no more than five upper and lower premolars in total, often separated by a diastema. On the p5, the paraconid is large and the metaconid is reduced. The upper molars have district hypocones (Clemens, 1973; Novacek, 1986).

Leptictis Leidy, 1868

Systematics and Evolution The type species of *Leptictis* is *L. haydeni* Leidy, 1868. No type specimen was designated. *Leptictis* is the last of the order Leptictida in North America.

Distinctive Characters The skull is characterized by a twinned sagittal crest. The upper molars are transverse with narrow stylar shelves and large hypocones. The p4–p5 are similar in size. The P4 has a complex lingual moiety (Gunnell, Bown, and Bloch, 2008) (Fig. 5.11A).

Stratigraphic and Geographic Distribution Fossilized remains from *Leptictis* can be found in the Uintan through the early Arikareean. Fossils are found in South Dakota,

Nebraska, Saskatchewan, Montana, Wyoming, Texas, Colorado, and North Dakota. In the White River Group of South Dakota, *Leptictis* is common in the Scenic Member but appears to have been rare in the Poleslide Member of the Brule Formation.

Natural History and Paleoecology *Leptictis dakotensis* likely was a quadrupedal cursor with possibly some bipedal hopping. It had relatively short forelimbs that were adapted for digging, a common behavior in terrestrial insectivores. *Leptictis dakotensis* has been recovered from burrowlike structures (Sundell, 1997; Rose, 1999, 2006). Certain structures in the nasals suggest that *Leptictis* had increased snout mobility similar to modern moles (Meehan and Martin, 2012).

Order Lipotyphla

The Lipotyphla include the moles, hedgehogs, and shrews. Modern forms are pentadactyl, plantigrade, and quadrupedal. The forelimbs are shorter than the hind limbs. They are characterized by terrestrial, fossorial, or semiaquatic locomotor patterns. The skulls were low with laterally facing small orbits. They had a small braincase and a long snout. The skulls have elongate nasals, indicating that the animals relied on olfaction for locating insect prey (Gunnell et al., 2008). Most of the fossils from this group are only known from jaws and teeth. They had dental specializations for eating soft- and hard-bodied insects. The postcrania are not well known.

Family Erinaceidae

The family Erinaceidae includes the modern hedgehogs, which today are found in Europe, Asia, and Africa but which in the past extended into North America. Modern hedgehogs belong to a separate subfamily called Erinaceinae, which originated in Asia and Europe approximately 35 Ma and continues to thrive today (Rich and Rasmussen, 1973). They were believed to be part of a faunal interchange between the Old World and North America (Rich and Rasmussen, 1973).

The dental formula for this family is I2–3/3, C1/1, P3–4/2–4, M3/3. The incisors are often enlarged, and the lower molars are progressively reduced from the m1 to m3. The lower molars are also rectangular with small or absent hypoconulids and weak exodaenodonty. The m1 has a strong, anteriorly angled paraconid. The upper molars are quadrate and often bunodont. The M1–2 have well-developed hypocones, and the M3 is reduced in size and oval in outline. The zygomatic arch is complete on skulls from members of this family (Gunnell et al., 2008).

Proterix Matthew, 1903

Systematics and Evolution The type species for *Proterix* is *P. loomisi* Matthew, 1903. The type specimen for *Proterix* is AMNH 9756, collected from Indian Draw, Cheyenne River, South Dakota, in the upper *Oreodon* beds (Lower Poleslide Member, Brule Formation) according to Gawne (1968). *Proterix bicuspis* (Macdonald, 1951), originally *Apternodus bicuspis*, SDSM 4048, was transferred to *Proterix* by Gawne (1968:11); it came from the *Protoceras* channel sandstone, 7 miles east of Rockyford, Shannon County, South Dakota.

Distinctive Characters The anterior portion of the skull is short with a completely ossified palate. On top of the skull, there are well-developed supraorbital crests and postorbital processes. The temporal crests form a strong sagittal crest. The nasals are elongate and extend posteriorly to the postorbital process. There is a deep depression anterior to the orbits for snout musculature. The zygomatic arches are well developed. The auditory bullae are completely ossified and extremely large. The P3–M3 are three rooted. The M3 is small and triangular. The M1 is quadrate and large with a subequal paracone and metacone. The m1 is five cusped, and the m2 and m3 are four cusped with a relatively broad cingulum (Rich, 1981) (Fig. 5.11B).

Stratigraphic and Geographic Distribution Fossilized remains from *Proterix* can be found in the Whitneyan through the early Arikareean. Fossils are found in the Poleslide Member of the Brule Formation in the White River Badlands of South Dakota. They are also found in Nebraska.

Natural History and Paleoecology Erinaceids are present throughout the Eocene and most of the Oligocene at a rather low diversity, but the group does continue into the Miocene. Beginning in the late early Arikareean, they begin to diversify and continue through most of the Miocene (Gunnell et al., 2008).

Family Proscalopidae

The family Proscalopidae, including ancient moles, are the earliest known Lipotyphla to exhibit highly specialized burrowing adaptations. Unlike modern moles, which use their forearms to break the soil during burrowing, in the Proscalopidae the head is used as the primary means of burrowing. This is indicated by the fusion of the cranial bones in adults, and the rostrum is distinctly longer than the cranium. The cranium is broad and deep with occipital condyles placed ventrally and well below the plane of the dentaries. The premaxillae are marked with prominent lateral shelves that extend anteriorly to the nares. The tympanic bulla is ossified but only slightly inflated (Gunnell et al., 2008).

The anteriormost incisors are enlarged and piercing in most genera. There are seven or fewer antemolars with at least some premolars that are single rooted. The lower molar cingulids are incomplete labially, and the molars are brachydont to hypsodont in latest forms.

The postcranial skeleton in fossil forms is also well adapted for burrowing. The neck is shortened, and some cervical vertebrae co-ossify to form a rigid base. The deltoid crest of the scapula is greatly elevated and arches broadly over the glenoid. The humerus and forelimb are highly modified for rotary burrowing (Reed and Turnbull, 1965; Barnosky, 1981, 1982). The humeral head is laterally compressed, and the distal end is greatly expanded, with distinct fossa for insertion of a major muscular ligament. The olecranon process of the ulna is transversely expanded. The tibia and fibula are unfused.

Proscalops Matthew, 1901

Systematics and Evolution The type species for *Proscalops* is *P. miocaenus* Matthew, 1901. The type specimen for *Proscalops* (AMNH 8949a) was collected in Whitneyan rocks of northeast Colorado. The locality of the holotype of *P. tertius* (AMNH 19420) is uncertain. It was collected by G. L. Jepsen from the "White River Formation, Badlands South Dakota." Reed (1961) noted that it may have come from the Brule Formation, but no precise information is available (Gunnell et al., 2008).

Distinctive Characters The antemolar region is relatively short with six antemolar teeth. The molars are relatively high crowned. The coronoid process is high and spicular. The anterior lower incisor is gliriform, and the lower molars are brachyhypsodont to hypsodont. The upper molar metaconules and paracones are prominent (Fig. 5.11C, D) (Gunnell et al., 2008).

Stratigraphic and Geographic Distribution *Proscalops* is found in the Chadron and Brule formations of the White River Badlands in South Dakota. Fossils are also found in Wyoming, Colorado, Nebraska, Saskatchewan, and Montana.

Suborder Soricomorpha

Soricomorphs, or shrews, have a more derived and specialized dentition, which indicates a more strictly insectivorous diet. Many soricomorphs have molars with high, pointed cusps on tall trigonids, with either low and short talonids or well-developed broad talonids. Upper molars tend to be tritubercular with small or absent hypocones, and tall and sharply pointed cusps in more primitive forms. Skulls tend to be low with short infraorbital canals, enlarged lacrimal canals,

relatively short rostrae, and relatively globular braincases. Zygomatic arches are often absent or weakly formed, and there is no ossified auditory bulla. There are few postcrania of fossil soricomorphs known (Gunnell et al., 2008).

Family Geolabidae

The snout is long and tubular, abruptly narrowing above the P3 in later forms. The zygomatic arch is lost and the lacrimal foramen is large, opening into the orbit. The nasals are elongate, widest above the P3, and not fused to the other skull bones.

The dental formula for this family is I3/3, C1/1, P4/4, M3/3. The I1 is enlarged. The P1 is doubled rooted, and the P4 lacks a metacone. The M1–3 have a strong stylar shelf. The M3 has a strong parastyle and a weak metacone. The M1–2 have a separate hypoconal root. The lower premolars are double rooted, the p3 becomes progressively reduced, and the p4 becomes progressively simplified (McKenna, 1960).

Centetodon Marsh, 1872

Systematics and Evolution The type species for *Centetodon* is *C. pulcher* Marsh, 1872. The type specimen for *Centetodon* is YPM 13507, which was found in Sweetwater County, Wyoming. *Centetodon magnus* Clark, 1936, originally described as *Metacodon magnus*, was found in the Chadron Formation in the White River Badlands.

Distinctive Characters *Centetodon* is a small soricomorph. The snout abruptly narrows at the level of the P3. The palatal margin is elevated at the anterior to the midpoint of the P4 (Lillegraven, McKenna, and Krishtalka, 1981).

Stratigraphic and Geographic Distribution *Centetodon* has a long temporal range. Lillegraven, McKenna, and Krishtalka (1981) thoroughly reviewed the Bridgerian and younger *Centetodon*, making it one of the most completely studied early Tertiary insectivores. *Centetodon* is the last of the Geolabidae and survives almost to the end of the Arikareean. Fossilized remains from *Centetodon* can be found in the Wasatchian through the late Arikareean. Fossils are found in Texas, Utah, Wyoming, Colorado, California, Saskatchewan, North Dakota, South Dakota, Nebraska, Montana, and Florida.

Family Micropternodontidae

Members of this family are about the size of moles and were probably fossorial. The rostrum was long, moderately wide, and deep. The sutures tend to fuse easily. The upper cheek teeth are hypsibrachydont, and the ectolophs slant strongly lingually. The I1 is greatly enlarged, the P1 is absent, the P3 has an extremely reduced protocone, and the P4 is submolariform with a small hooklike cusp behind and above the mesostyle at the posterior end of the sectorial blade of the ectoloph. The M1–2 are much larger than the p2 and are somewhat molariform. The p4 and lower molars are high and have anteroposteriorly narrow trigonids with the upper half of the crown curving posteriorly (Stirton and Rensberger, 1964).

Clinopternodus Clark, 1937

Systematics and Evolution The type and only species of *Clinopternodus* is *C. gracilis* Clark, 1937. The type specimen is YPM 14197 and was found in the Chadron Formation in Shannon County, South Dakota.

Distinctive Characters *Clinopternodus* is one of the larger genera in this family. The lower canine and the crowns of the p3–m1 are relatively procumbent and lingually curved. The tips of the crown are not recumbent. The p2 is absent, and the p3 has a relatively low crown. The m1 talonid is equal to two-thirds of the height of the trigonid (Stirton and Rensberger, 1964).

Stratigraphic and Geographic Distribution Fossilized remains from *Clinopternodus* are only found in the Chadronian. Its geographic range is also limited. Fossils are found in South Dakota and Wyoming. *Clinopternodus* is only known at one locality where it has been identified to species level.

Family Apternodontidae

The anatomy of the Apternodontidae is well known (Schlaikjer, 1933; McDowell, 1958; McKenna, 1960; Reed and Turnbull, 1965; Asher et al., 2002). The family consists of relatively small soricomorph insectivores that, on the basis of scant postcranial material, were almost certainly fossorial.

The lacrimal foramen is enlarged and laterally facing, and the anterior margin of the infraorbital canal is concave. A sagittal crest is present and continuous with a strong nuchal crest. The posterolateral braincase is expanded into lambdoid plates, and the entoglenoid process is enlarged. The postglenoid process is absent. The zygomatic arch is incomplete, and there is no ossified auditory bulla (Asher et al., 2002).

The dental formula for the Apternodontidae is as follows; I2/3, C1/1, P3/3, M3/3. The anterior incisors are enlarged and procumbent. The upper and lower P4 is molariform, and the upper molars are zalambdodont with metacones absent. The coronoid process extends anteriorly (Asher et al., 2002).

Apternodus Matthew, 1903

Systematics and Evolution The type species of *Apternodus* is *A. mediaevus* Matthew, 1903. The type specimen of *Apternodus* is AMNH 9601, which was found in Chadronian rocks near Pipestone Springs, Jefferson County, Montana.

Distinctive Characters Same as Family. *Apternodus* is the only genus in the family Apternodontidae.

Stratigraphic and Geographic Distribution Fossilized remains from *Apternodus* occur from the Duchesnean to the late Orellan. It has a long temporal range of about 8 million years. Fossils are found in Wyoming, South Dakota, Montana, Nebraska, Texas, Colorado, and Saskatchewan. In the White River Badlands, *Apternodus* is found in the Chadron Formation.

Family Oligoryctidae

The posterior braincase is unspecialized, and a prominent entoglenoid process is present. The foramen ovale is enlarged, and the alisphenoid canal is absent. The ethmoid foramen and sinus canal exit anterior to the sphenorbital fissure. The lacrimal foramen is large and laterally oriented (Asher et al., 2002).

The molars are zalambdodont. The upper molars have distinct protocones, anterior cingula, and lack metacones. The lower second and third incisors are tricuspid. The lower molars have reduced talonid basins. The M3 talonid cusp is taller than the paraconid. The coronoid process is pocketed medially (Asher et al., 2002).

Oligoryctes Hough, 1956

Systematics and Evolution The type species of *Oligoryctes* is *O. cameronensis* Hough, 1956. The type specimen of *Oligoryctes* is USNM 19909, which was collected from the White River Formation along Beaver Divide in Wyoming. *Oligoryctes altitalonidus* Clark, 1936, was found in the Chadron Formation in the White River Badlands.

Distinctive Characters Same as Family. *Oligoryctes* is the only genus in the family Oligoryctidae.

Stratigraphic and Geographic Distribution Fossilized remains from *Oligoryctes* can be found in the Uintan to the Whitneyan. The genus has a long temporal range of about 14 million years. Fossils are found in Wyoming, Nebraska, Montana, South Dakota, California, North Dakota, Colorado, and Saskatchewan.

Family Soricidae

This family is commonly referred to as shrews. The skull has a relatively short rostrum in which the bones fuse early. The cranium, although small, is inflated. The lacrimal duct curves upward and forward from a small, ventrally exiting foramen on an infraorbital bridge. The zygomatic arches are thin and the jugal is absent. The tympanic ring is large, lying nearly horizontal and close to the skull. The auditory bullae are not ossified (Gunnell et al., 2008:112).

The dentary tapers anteriorly with a long procumbent symphysis. The coronoid process is oriented vertically to a horizontal ramus. The upper first incisor is hook shaped, and the teeth between the I1 and P4 are unicuspid and single rooted. The P4 is semimolariform with a well-developed and basined lingual shelf and a large paracone extended posteriorly into a prominent blade. The M1–2 is dilophodont and roughly rectangular with a well-developed cingular hypocone and crested protocone. The upper dentition is reduced by one or more antemolars, and the lower dentition is reduced by two or more antemolars (Gunnell et al., 2008:113). The incisors of shrews are red from hematite in the enamel. Little is known of the postcranial skeleton of most fossil shrews. The skeleton of shrews in general is unspecialized (Reed, 1951).

Domnina Cope, 1873

Systematics and Evolution The type species for *Domnina* is *D. gradata* Cope, 1873. The type specimen for *Domnina* is AMNH 5353, collected from the White River Formation, Logan County, Colorado. *Domnina gradata* is also found in the White River Badlands of South Dakota.

Distinctive Characters *Domnina* is the least specialized of the shrews, with only a weakly differentiated condyle on the jaw and five or six lower antemolars. The m1–3 entoconid crest is high and joined to the metaconid but distinctly separated from the postcristid (Gunnell et al., 2008:113).

Stratigraphic and Geographic Distribution Fossilized remains from *Domnina* can be found in the Uintan to the Hemingfordian. It has a broad temporal range, lasting for 25 million years. Fossils are found in Wyoming, Colorado, South Dakota, Nebraska, Saskatchewan, Montana, California, North Dakota, and Idaho. At Badlands National Park, *Domnina* has been found in the Scenic Member of the Brule Formation in the North Unit of the park.

Natural History and Paleoecology *Domnina* is considered the first soricid, or true shrew. A large radiation of the soricids begins in the late Arikareean with a broad diversity through the Miocene and Pliocene. In North America today, the two most common genera are *Sorex* and *Blarina*.

Order Lagomorpha

The Order Lagomorpha includes pikas, rabbits, and hares. The earliest record of the order is from the Eocene of Asia, and the group first appears in the late Uintan of North America (Dawson, 2008). Early rabbits are placed in the subfamily Palaeolaginae, and this includes *Palaeolagus* and *Megalagus*, the two rabbit taxa found in the White River Badlands. The various subfamilies of rabbits are distinguished by differences in the folding of the enamel patterns of the lower third premolar and upper second premolar (Dawson, 2008).

Members of the order are characterized by two pairs of upper incisors in which a smaller peglike incisor is positioned behind the larger anterior grooved tooth of the pair. The cheek teeth, three upper and two lower premolars, and two to three upper and lower molars are hypsodont and ever-growing. The facial portion of the maxillae may have a single or multiple fenestrae. The bony palate is short and often a narrow bar. The auditory bullae is enlarged and formed from the ectotympanic. The distal end of the humerus is lacking an epicondylar foramen. The tibia and fibula are fused distally with the fibula contacting the calcaneum (Dawson, 2008).

Family Leporidae

The Leporidae includes both rabbits and hares and today is represented by four genera in North America: *Lepus*, *Sylvilagus*, *Brachylagus*, and *Romerolagus*. Characteristic locomotion in the family includes short bursts of speed, as in rabbits, or sustained bounding locomotion, as in hares (Dawson, 2008).

Palaeolagus Leidy, 1856

Systematics and Evolution The type species of *Palaeolagus* is *P. haydeni* Leidy, 1956. The type specimen of *Palaeolagus*, ANSP 11031, was collected in at the head of Bear Creek in the Badlands of South Dakota (Leidy, 1869). *Palaeolagus* is a common member of many Oligocene mammalian faunas, and many species have been described. The number of species also reflects the long chronology of the genus, which extended into the early Miocene.

Distinctive Characters The teeth in *Palaeolagus* are more hypsodont and have better-developed cement tracts than *Megalagus* and earlier rabbits. The overall size is smaller than that of *Megalagus* (Fig. 5.12) (Dawson, 2008).

Stratigraphic and Geographic Distribution *Palaeolagus* is found in the Chadronian through the Arikareean. In the White River Badlands, *Palaeolagus* is found in the Brule Formation. Fossils are found in South Dakota, Nebraska,

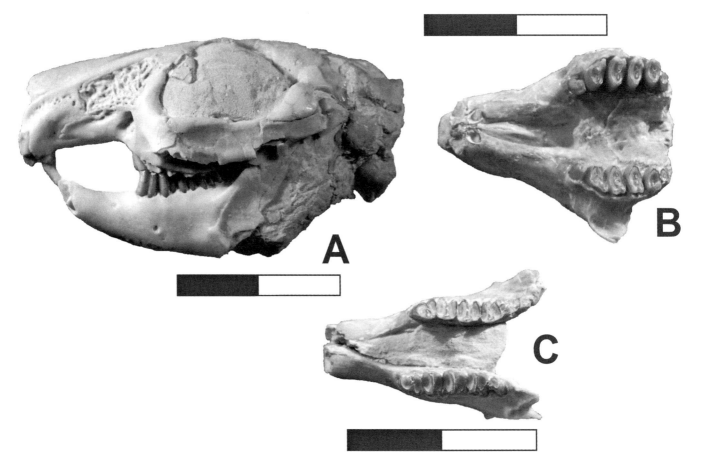

5.12. *Palaeolagus haydeni.* (A) Skull and jaws. (B) Partial skull, occlusal view. (C) Right and left mandibles, occlusal view. Scales in centimeters. Photos by the authors of specimens from the Department of Earth Sciences teaching collection, University of Northern Colorado.

Wyoming, Colorado, Saskatchewan, Montana, North Dakota, and Oregon (Dawson, 2008).

Natural History and Paleoecology In lagomorphs, one of the evolutionary changes seen is in the increase in hypsodonty, which coincides with the cooling and drying that occurred at the Eocene–Oligocene transition and the related changes in vegetation. Change in dentition was also accompanied by changes in posture that affected locomotion (Dawson, 2008).

Megalagus Walker, 1931

Systematics and Evolution The type species of *Megalagus* is *M. turgidus* (Cope, 1873), which was originally described as *Palaeolagus turgidus*. The type specimen of *Megalagus*, AMNH 5635, was collected from the White River Formation, Logan County, Colorado. Evolutionary trends in *Megalagus* parallel similar trends in *Palaeolagus*, but at any given time, *Megalagus* tends to lag behind *Palaeolagus* in most dental characters (Dawson, 2008).

Distinctive Characters *Megalagus* is characterized by upper molariform teeth with buccal roots. The overall size is

larger than *Palaeolagus* except for one species, *P. intermedius* (Dawson, 1958).

Stratigraphic and Geographic Distribution *Megalagus* is found in the Chadronian through the Hemingfordian. In the White River Badlands, *Megalagus* is found in the Chadron and Brule formations. Fossils are found in South Dakota, Nebraska, Saskatchewan, Montana, North Dakota, Colorado, and Florida.

Order Rodentia

Rodents are considered to be a successful group, a result of their taxonomic diversity and wide geographic distribution. Over 1800 living species are contained within this group, making up 50 percent of all living mammals. The rodent fossil record can be found on all continents except Antarctica. One of the defining features of rodents is their unique teeth and jaws, which played a significant role in their adaptations and evolutionary radiation. Rodents are characterized by a single pair of deep-rooted incisors that grow throughout their lives. The incisors are used to gnaw wood and nuts and to clip vegetation and husks of fruit. The incisors bear enamel

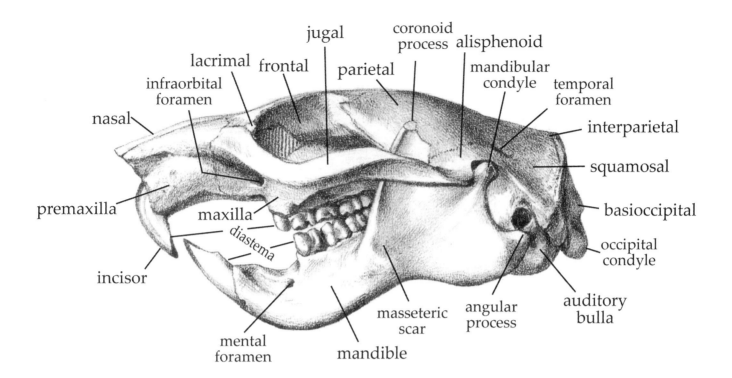

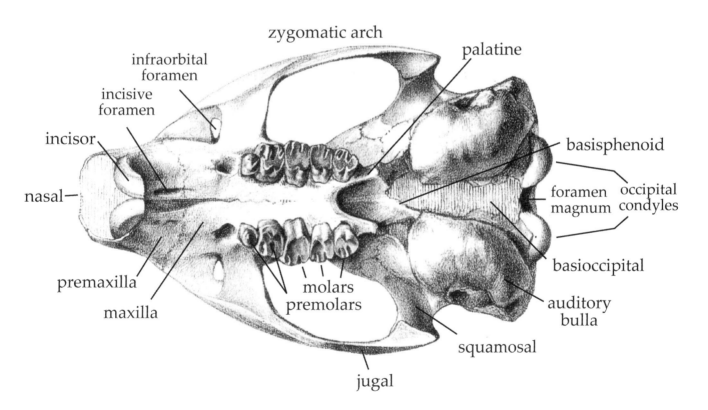

5.13. Diagram showing individual features of a rodent skull and jaws (*Ischyromys* sp.). Images modified from Wood (1937:plate 27, figs. 1, 1a).

only on the anterior surface so that the softer dentine behind wears faster and provides a sharp edge that is self-sharpening. Rodent jaws are also characterized by a diastema between the single incisor and the first cheek tooth, which is either the fourth premolar or first molar, a gap that in other mammals would be located at the second and third incisors, the canine, and the first three premolars (Fig. 5.13) (Benton, 2000).

Rodents first appear in North America during the late Paleocene and early Eocene. By the late Eocene and Oligocene, they are quite diverse and well adapted to a variety of environments.

Family Ischyromyidae

The family Ischyromyidae is one of the earliest rodent families to appear in the fossil record, first appearing in the late Paleocene to early Eocene in North America and also in Europe. They were the most diverse taxonomically during the Uintan (Korth, 1994b). Ischyromids are the most common rodents in the Orellan and have been used for biostratigraphic correlation (Prothero and Whittlesey, 1998). During the early Oligocene, their diversity decreases and they become extinct in the Whitneyan (Korth, 1984, 1994b; Heaton, 1993; Howe, 1996). The family Ischyromyidae may have been the ancestor to later Tertiary rodent families, such as Sciuridae and Aplodontidae (Emry and Korth, 1996).

Ischyromys Leidy, 1856

Systematics and Evolution The type species of *Ischyromys* is *I. typus* from the Badlands of South Dakota. Currently seven species are recognized. The genus first appears in the middle Wasatchian and became extinct at the end of the Whitneyan. *Ischyromys typus* is known from the Scenic and Poleslide members of the Brule Formation. *Ischyromys parvidens* is from the Scenic Member of the Brule Formation. *Ischyromys* spp. from the Chadron Formation have not been identified to species (Anderson, 2008).

Distinctive Characters Members of the family Ischyromyidae are defined by a robust skull, mandible, and skeleton. The skull has a heavy rostrum, a small incisive foramina, and a sagittal crest on the cranium (Fig. 5.14A, D). These rodents have a protrogomorphous morphology where the masseter muscles attach to the zygomatic arch (primitive form). They also have the complete rodent dental formula of two upper premolars and one lower. The upper third premolar is a peg. In the subfamily Ischyrominae, the auditory bullae were co-ossified with the skull, and the cheek teeth were lophodont (Fig. 6.14B, C, E) (Black, 1968, 1971; Wahlert, 1974; Anderson, 2008). *Ischyromys* was a medium-size rodent with incisors that were oval in cross section; the lower molar

crown pattern was crestlike; and the length of the molars was slightly greater than the width. Also the scaphoid and lunate bones (wrist) are fused in *Ischyromys* (Anderson, 2008).

Stratigraphic and Geographic Distribution Fossilized remains of *Ischyromys* occur in the Wasatchian through the early Whitneyan. Fossils are found in Wyoming, Colorado, South Dakota, California, Nebraska, Texas, Montana, Saskatchewan, North Dakota, and Utah.

Natural History and Paleoecology On the basis of its skeletal anatomy, *Ischyromys* was thought to have adaptations for terrestrial quadripedal locomotion ranging from scampering to subfossorial (Korth, 1994b; Anderson, 2008). It was also considered to be the most common rodent in the Orellan of the White River Group and the last genus of this family in the fossil record (Korth, 1994b).

Family Aplodontidae

The family consists of a holarctic radiation of small to medium and sometimes large rodents. There is a single surviving species, *Aplodontia rufa*, the mountain beaver from the Pacific Northwest of North America. Fossils from the family Aplodontidae are found in North America, Europe, and Asia. The center of diversity was during the mid-Cenozoic in North America, with repeated dispersals. This family was considered diverse in the western part of North America during the Oligocene and Miocene epochs. There were two distinctive groups that evolved during the mid-Tertiary. The Allomyinae–Aplodontinae subfamilies appeared in the later early Oligocene, diversified, and survives today as *Aplodontia*. The Aplodontids are small- to medium-size rodents that show various dental specializations including extra cusps on cheek teeth, development of an ectoloph on the upper cheek teeth, and the presence of molar styles. The skull and skeleton show a terrestrial and fossorial skull and postcranial adaptations. Aplodontids possibly evolved from the early Tertiary ischyromyoid rodents from North America. A recent molecular study shows a close link to squirrels (Adkins, Walton, and Honeycutt, 2002). Early morphologists noted that aplodontoid molars resemble squirrel molars in the arrangement of cusps and the lack of a hypocone (Flynn and Jacobs, 2008).

Prosciurus Matthew, 1903

Systematics and Evolution The type species of *Prosciurus* is *P. vetustus* Matthew, 1903. The type specimen is AMNH 9626, collected from an unknown location. *Prosciurus relictus* Cope, 1873, has been collected from the Brule Formation, in the White River Badlands of South Dakota.

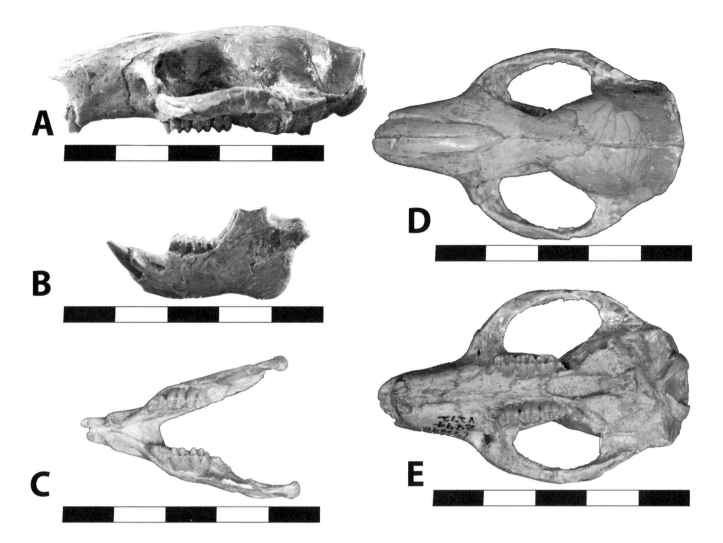

5.14. *Ischyromys plaicus,* BADL 53890/SDSM 2994. (A) Skull, left lateral view. (B) Mandibles, left lateral view. (C) Mandibles, occlusal view. (D) Skull, dorsal view. (E) Skull, occlusal view. Scales in centimeters. Photos by the authors of specimen from the Museum of Geology, South Dakota School of Mines and Technology, Rapid City, South Dakota, U.S.A., and is the property of the U.S. government.

Distinctive Characters *Prosciurus* had brachydont and cuspate upper molars with a continuous anteroloph–protocone–posteriorloph crest. The hypocone is absent or minute, and the metaloph is incomplete. The lower molars are rhombic, with broad, shallow basins (Flynn and Jacobs, 2008).

Stratigraphic and Geographic Distribution Fossilized remains of *Prosciurus* can be found in the Uintan to Whitneyan in North America. Fossils are found in Wyoming, Nebraska, Saskatchewan, North Dakota, Montana, Colorado, and South Dakota.

Natural History and Paleoecology *Prosciurus* was originally proposed as a subgenus for *Sciurus* (Matthew, 1903). It is one of the earliest aplodontines and possibly was the ancestor to the later Mylagualidae (Flynn and Jacobs, 2008).

Pelycomys Galbreath, 1953

Systematics and Evolution The type species of *Pelycomys* is *P. rugosus* Galbreath, 1953. The type specimen is KU 8343,

which was collected from the White River Formation, Logan County, Colorado. *Pelycomys* is only identified to genus in the White River Badlands of South Dakota.

Distinctive Characters *Pelycomys* was a large rodent with brachydont cheek teeth that were subtriangular to subrhombic. The cusps were round and inflated (Flynn and Jacobs, 2008).

Stratigraphic and Geographic Distribution *Pelycomys* occurs in the Chadronian to Arikareean in North America. Fossils are found in Colorado, Wyoming, Nebraska, Montana, and South Dakota. In the White River Badlands, *Pelycomys* can be found in the Sharps Formation.

Family Sciuridae

The family Sciuridae, also known as squirrels, chipmunks, prairie dogs, and marmots, comprise about 275 living and numerous extinct species. Squirrels are found worldwide except for Australia, southern South America, and Antarctica (Jones

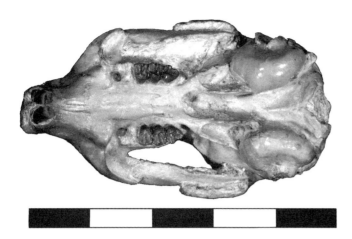

5.15. *Cedromus wilsoni,* USNM 256584. Skull, occlusal view. Scale in centimeters. Photo by the authors. Courtesy of Smithsonian Institution, Washington, D.C., U.S.A.

et al., 1983). The first sciurids appear in the Chadronian of North America and in similar age rocks in Europe (Goodwin, 2008). The oldest squirrel, *Douglassciurus jeffersoni,* from Flagstaff Rim, Wyoming, is one of the most complete and best described of the fossil squirrel specimens from North America (Emry and Thorington, 1982). *Douglassciurus jeffersoni* has a similar skeleton to a modern tree squirrel. All members of Family Sciuridae have sciuromorphic zygomasseteric structures except for *D. jeffersoni.* Within Badlands National Park, only two sciurid genera are found, *Cedromus* and *Protosciurus* (Goodwin, 2008).

Cedromus Wilson, 1949

Systematics and Evolution The type species for *Cedromus* is *C. wardi* Wilson, 1949. The type specimen is UCM 19808 from the Orellan rocks of Colorado.

Distinctive Characters *Cedromus* is characterized by a low, broad skull, short rostrum, and a primitive zygomasseteric structure. The zygoma is anteriorly broadened and tilted lateral to the infraorbital foramen. The upper molars had a small hypocone and large metaconules (Fig. 5.15). The lower molars are characterized by a posterior cingulum confluent with the entoconid and by the mesostylids having a greater buccal expansion (Korth and Emry, 1991).

Wilson (1949) interpreted *Cedromus* to be an ischyromyid but suggested the possibility that it could have been a primitive squirrel. Korth and Emry (1991) placed *Cedromus* in with the sciurids on the basis of well-preserved skull material.

Stratigraphic and Geographic Distribution Fossilized remains from *Cedromus* can be found in the Orellan through the Whitneyan. Fossils are found in South Dakota, Wyoming, Nebraska, Colorado, and Saskatchewan. In the White River Badlands, *Cedromus* is found in the Brule Formation.

Natural History and Paleoecology There are four major groups of squirrels: chipmunklike squirrels, tree squirrels, flying squirrels, and terrestrial squirrels, which include ground squirrels, prairie dogs, and marmots. The fossil squirrels found in the White River Badlands were believed to be arboreal in nature because terrestrial squirrels had not yet evolved. They are believed to have appeared and radiated in the late Arikareean to Barstovian (early to middle Miocene) (Goodwin, 2008).

Protosciurus Black, 1963

Systematics and Evolution The type species of *Protosciurus* is *P. condoni* Black, 1963. The type specimen is UO F-5171, from the early Arikareean of Oregon. *Protosciurus* is only identified to genus in the White River Badlands of South Dakota. The skull of *Protosciurus* is not fully sciuromorphous but is more advanced than *Douglassciurus* or *Cedromus.* Overall, *Protosciurus* is still considered to be primitive compared to other squirrel groups (Goodwin, 2008).

Distinctive Characters The teeth are low crowned, subquadrate, with low lophs and a broadened protocone. The skeleton is gracile and is similar to those of nut- and seed-eating squirrels today (Black 1963, 1972).

Stratigraphic and Geographic Distribution Fossilized remains of *Protosciurus* occur in the Orellan to Hemingfordian in North America. Fossils are found in South Dakota, Florida, Oregon, Nebraska, Texas, California, and Saskatchewan. In the White River Badlands, *Protosciurus* is found in the Sharps Formation.

Natural History and Paleoecology On the basis of a single tibia assigned to this genus, it is thought that *Protosciurus* was more arboreal than terrestrial (Frailey, 1978).

Family Castoridae

The family Castoridae includes fossil and modern beavers. Living members of this family can weigh as much as 40 kg, with fossilized forms weighing as little as 1 kg. Beavers reached their maximum size in the Pleistocene with *Castoroides,* which may have weighed as much as 100 kg. Modern beavers are semiaquatic, but ancient forms were terrestrial, and some showed burrowing adaptations. Early beavers had relatively low crowned teeth but achieved full hypsodonty in the late Tertiary. Beavers originated in North America and migrated to Europe (Flynn and Jacobs, 2008). The greatest diversity of castorids occurred in the Arikareean with the radiation of palaeocastorine beavers. A third radiation involved the castorine beavers, which occurred in the Blancan.

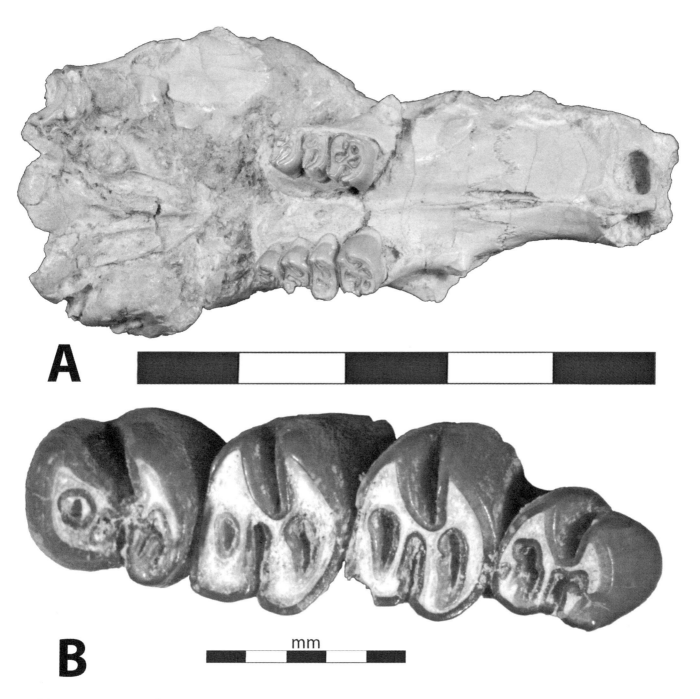

5.16. (A) *Agnotocastor praetereadens,* AMNH 1428, skull, occlusal view. (B) *Agnotocastor* sp., BADL 22370/SDSM 56624, partial right mandible with lower p4–m3, occlusal view. Scales in centimeters unless otherwise noted. Photo A by the authors; and photo B by Edward Welsh. AMNH 1428 is from the American Museum of Natural History, New York, New York, U.S.A. BADL 22370/SDSM 56624 is from Badlands National Park, Interior, South Dakota, U.S.A., and the Museum of Geology, South Dakota School of Mines and Technology, Rapid City, South Dakota, U.S.A., and is the property of the U.S. government.

Agnotocastor Stirton, 1935

Systematics and Evolution The type species of *Agnotocastor* is *A. praetereadens* Stirton, 1935. The type specimen of *Agnotocastor* is AMNH 1428, collected near the Cheyenne River, South Dakota, in the Poleslide Member of the Brule Formation (Whitneyan). *Agnotocastor* is considered the earliest and most primitive castorid (Flynn and Jacobs, 2008). Because it appeared in both Asia and North America at the same time, it is unclear what direction the migration occurred (Korth, 1994b).

Distinctive Characters *Agnotocastor* is characterized by relatively low crowned teeth, and a narrow rostrum and skull (Fig. 5.16). The auditory bulla is kidney shaped. The lower incisors are rounded. The P3 is present and the P4 is square. All of the molars are rectangular (wider than long) (Fig. 5.16) (Flynn and Jacobs, 2008).

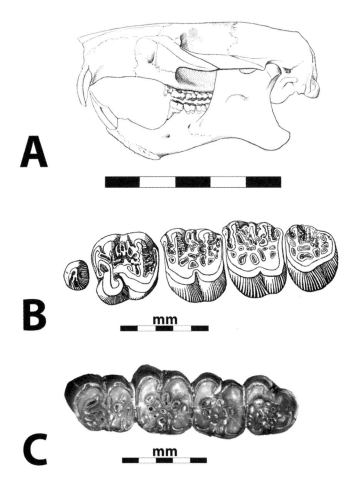

5.17. *Eutypomys* sp. (A) Skull and jaws, left lateral view (Wood, 1937:224). (B) *Eutypomys thomsoni*, AMNH 12254, left P3–M3 (Wood, 1937:225). (C) *Eutypomys* sp. left p3–m3, BADL 63385. Scales in centimeters unless otherwise noted. Photo C by Levi Moxness. BADL 63385 is from Badlands National Park, Interior, South Dakota, U.S.A., and is the property of the U.S. government.

Stratigraphic and Geographic Distribution Fossilized remains from *Agnotocastor* occur in the Chadronian to the early Arikareean. Fossils are found in South Dakota, Colorado, Wyoming, Nebraska, Florida, and Montana. In the White River Badlands, *Agnotocastor* is found in the Chadron and Brule formations.

Natural History and Paleoecology *Agnotocastor* was a terrestrial beaver, but there is no evidence that it burrowed. The radiation of *Palaeocastor*, with its famous *Daemonelix* burrows (devil's corkscrews), is separate from the *Agnotocastor* lineage. *Palaeocastor* is only contemporaneous with *Agnotocastor* during the Whitneyan and continues into the Miocene (Korth, 1994b).

Palaeocastor Leidy, 1869

Systematics and Evolution The type species of *Palaeocastor* is *P. nebrascensis* (Leidy, 1856), which originally was described as *Steneofiber nebrascensis*. The exact location where the type was found is unknown.

Distinctive Characters *Palaeocastor* is characterized by high-crowned cheek teeth, which readily distinguishes it from *Agnotocastor*; large, rounded bullae; convergent suborbital ridges; moderate lambdoidal crest; incisors that are somewhat rounded; and premolars that are close to molars in size (Flynn and Jacobs, 2008). Reynolds (2002) estimated the body mass of *Palaeocastor* to be about 1 kg.

Stratigraphic and Geographic Distribution *Palaeocastor* occurs in the Whitneyan to the Hemingfordian. Fossils are found in Wyoming, South Dakota, Nebraska, North Dakota, Oregon, Texas, Saskatchewan, and Montana. The fossil record for *Palaeocastor* at Badlands National Park is limited to a few localities within the Sharps Formation (Parris and Green, 1969; Flynn and Jacobs, 2008).

Natural History and Paleoecology Martin (1987b) noted that skeletal remains of both *P. fossor* and *Capacikala magnus* were recovered from *Daemonelix* burrows, showing that *Palaeocastor* and *Capacikala* dug these amazing spiral features. No *Daemonelix* burrows have been found in Badlands National Park.

Capacikala Macdonald, 1963

Systematics and Evolution The type species of *Capacikala* is *Steneofiber gradatus* Cope, 1878, and the species was originally placed in the genus *Steneofiber*. The type specimen of *Capacikala*, AMNH 7008, was collected from the John Day Formation, Grant County, Oregon. Martin (1987b) and Korth (1994b, 2001) proposed the subfamily Palaeocastorinae, which included both *Palaeocastor* and *Capacikala*.

Distinctive Characters *Capacikala* is considered a small beaver with a short, deep skull and orthodont incisors that have flattened anterior faces. There are double temporal muscle scars that do not meet in a single sagittal crest. The bullae are large and expand ventrally. The cheek teeth decrease in size posteriorly (Flynn and Jacobs, 2008).

Stratigraphic and Geographic Distribution Fossilized remains from *Capacikala* can be found in the early to middle Arikareean in North America. Fossils are found in Wyoming, South Dakota, Montana, and Oregon. Both *Palaeocastor* and *Capacikala* were found in channel deposits within the Sharps Formation in Cedar Pass at Badlands National Park (Parris and Green, 1969).

Natural History and Paleoecology The subfamily has adaptations for a subterranean habitat, including a flattening of the enamel surface of the incisors to form a chisellike digging tools and a shortened rostrum (Flynn and Jacobs, 2008).

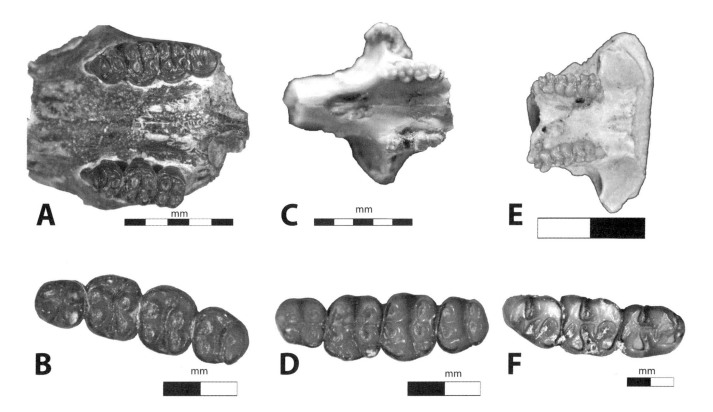

5.18. Rodents from the White River Badlands (A) *Proheteromys nebraskensis,* palate with right P4–M2 and left P4–M3, BADL 18589/SDSM 10001. (B) *Proheteromys nebraskensis* right p4–m3, BADL 18588/SDSM 10000. (C) *Heliscomys* sp., rostrum with no incisors, UCM 45468. (D) *Heliscomys* sp., left p4–m3, BADL 22286/ SDSM 56540. (E) *Eumys elegans,* palate with right and left M1–M3, FM 38871. (F) *Eumys* sp., left m1–m3, BADL 22282/SDSM 56536. Scales in centimeters unless otherwise noted. Photos (A), (B), (D), and (F) by Edward Welsh; photos (C) and (E) by the authors. BADL 18589/SDSM 10001, BADL 18588/SDSM 1000, BADL 22286/ SDSM 56540 are from Badlands National Park, Interior, South Dakota, U.S.A., and the Museum of Geology, South Dakota School of Mines and Technology, Rapid City, South Dakota, U.S.A. UCM 45468 is from the University of Colorado Museum of Natural History, Boulder, Colorado, U.S.A. FM 38771 is from the American Museum of Natural History, New York, New York, U.S.A.

Family Eutypomyidae

Eutypomyidae is a sister group with Castoridae, on the basis of a mix of primitive and advanced tooth patterns. The teeth are high crowned and based on a four-cusp pattern, which is an advanced feature in the family Castoridae. The muscle and jaw morphology demonstrates a fully sciuromorphous condition (Flynn and Jacobs, 2008). The rostrum is elongate.

Eutypomys Matthew, 1905

Systematics and Evolution The type species of *Eutypomys* is *E. thomsoni* Matthew, 1905. The type specimen of *Eutypomys*, AMNH 12254, is from the Scenic Member of the Brule Formation on Quinn Draw, Cheyenne River, South Dakota.

Distinctive Characters *Eutypomys* has a complicated crown pattern consisting of a complex of secondary loops that result in separate enamel islets. The complex eutypomyid tooth morphology is built upon a fundamentally primitive dentition including a 5/4 cheek tooth formula (Fig. 5.17) (Flynn, Lindsay, and Martin, 2008).

Stratigraphic and Geographic Distribution Fossilized remains of *Eutypomys* occur in the Duchesnean to Arikareean in North America. Fossils are found in South Dakota, Nebraska, Saskatchewan, Montana, North Dakota, Wyoming, and Texas. *Eutypomys* is found through a long section of the fossil record, and individual species are valuable stratigraphic indicators. *Eutypomys thomsoni* is an indicator species of the Early Late Orellan *Miniochoerus gracilis* Interval Zone (Prothero and Whittlesey, 1998), and the genus *Eutypomys* helps typify the Orellan land mammal age.

Family Cricetidae

The cricetids are a successful group of rodents. Their record extends from the late Eocene to the Recent. They originated in Asia, migrated into North America, and radiated into the Neotropics during the Pliocene. They are small rodents and are adapted to a large variety of environments, including xeric, boreal, aquatic, and semifossorial. Most species are herbivorous. Both fossil and modern cricetids are successful because of their high fecundity and broad ecological tolerance (Lindsay, 2008). In terms of jaw musculature, the family

Cricetidae represents a transition from subhystricomorphous to a myomorphous zygoma. The mandible is slender, and the cheek teeth are restricted to three molars, upper and lower and on each side. The incisors are narrow anteriorly and gently convex. The enamel surface is ornamented by numerous ridges. The braincase is inflated and the rostrum is narrow. The interorbital area has a smooth hourglass shape, and the auditory bullae are inflated and usually small. The hind limb is longer than the forelimb, and the fibula is fused to the tibia (Korth, 1994b).

Eumys Leidy, 1856

Systematics and Evolution The type species of *Eumys* is *E. elegans* Leidy, 1856. The type specimen of *Eumys* is ANSP 11027, from the Scenic Member, Brule Formation, South Dakota. Besides the type species, eight other species are considered valid (Lindsay, 2008:465). *Eumys* represents the first known cricetid in North America. The eumyines represented the first radiation of cricetids, which achieved its greatest diversity in the Orellan with seven species (Lindsay, 2008:465).

Distinctive Characters *Eumys* was a small- to medium-size rodent with brachydont dentition. The skull has a broad rostrum and a median sagittal crest. The dentition is slightly lophodont with robust cusps. Both upper and lower third molars are quite large (Fig. 5.18E, F) (Lindsay, 2008).

Stratigraphic and Geographic Distribution Fossilized remains of *Eumys* can be found in the Duchesnean to Arikareean in North America. Fossils are found in Wyoming, Colorado, South Dakota, Nebraska, Saskatchewan, Montana, and North Dakota. *Eumys brachyodus* is reported from the Poleslide member of the Brule Formation, and *E. obliquidens* and *E. parvidens* from the Scenic Member of the Brule Formation (Lindsay, 2008).

Natural History and Paleoecology *Eumys* is generally considered the most common and best-known Oligocene cricetid (Lindsay, 2008).

Scottimus Wood, 1937

Systematics and Evolution The type species of *Scottimus* is *S. lophatus* Wood, 1937. The type specimen of *Scottimus*, MCZ 5064, is from the Poleslide Member of the Brule Formation from South Dakota. Besides the type species, currently five other species in the genus are considered valid.

Distinctive Characters *Scottimus* was a small brachydont rodent with a parallel-sided snout, parasagittal crests, and cheek teeth developing lophs directed anteroposteriorly between the labial cusps and on the lingual side. The upper molars decrease in size posteriorly (Lindsay, 2008:462).

Stratigraphic and Geographic Distribution Fossilized remains of *Scottimus* occur in the Orellan to Arikareean. Fossils are found in South Dakota, Wyoming, Nebraska, Montana, Colorado, and Saskatchewan.

Natural History and Paleoecology *Scottimus* was part of the first radiation of cricetids during the late Eocene to Oligocene pulse that began in the Chadronian with *Eumys* and climaxed in the early Arikareean. Cricetids declined during the latter part of the Arikareean. A second radiation began in the Barstovian (Lindsay, 2008).

Family Dipodidae

The family Dipodidae includes the modern jumping mice and the birch mice of North America and Asia, which are often referred to as zapodids. They are small- to medium-size rodents and have saltatorial adaptations, especially in the hind limb. There are not many species within this family. Because of their preference for moist conditions, dipodoids can be useful climate indicators. The skull has a hystricomorphous zygomasseteric structure. The molars show a "cricetid" plan, and opposite cusps are joined by transverse crests. Evolutionary trends are toward increasing lophodonty and an emphasis on shearing crests. The upper incisors contain a longitudinal groove (Flynn, 2008a). The postcrania are rarely preserved in the fossil record. Living forms are gracile with limbs modified for saltatorial locomotion.

Diplolophus Troxell, 1923

Systematics and Evolution The type species of *Diplolophus* is *D. insolens* Troxell, 1923. The type specimen of *Diplolophus*, YPM 10368, was collected from the Brule Formation near Scottsbluff, Nebraska. *Diplolophus* is an early rodent characterized by three cheek teeth. Close affinities with the geomyids (Korth, 1994b; Wood, 1980) suggest a relationship with Geomorpha (extant pocket gophers, pocket mice, and kangaroo rats).

Distinctive Characters *Diplolophus* is characterized by its large body size, mesodont, and strongly lophodont dentition. The cheek teeth have weak longitudinal connections in the upper molars and are absent in the lowers. The second molars are wider than long (2008a).

Stratigraphic and Geographic Distribution *Diplolophus* occurs in the Duchesnean through Orellan. Fossils are found in Colorado, South Dakota, Nebraska, and California. In the White River Badlands, *Diplolophus* is found in the Brule Formation.

Plesiosminthus Viret, 1926

Systematics and Evolution The type species of *Plesiosminthus* is *P. schaubi* from Saint-Gérand-le-Puy, France (early Miocene), and five species are recognized in North America. *Plesiosminthus clivosus* Galbreath, 1953, the type specimen of *Plesiosminthus*, is KU 9279, collected from the Pawnee Creek Formation in Logan County, Colorado. In Badlands National Park, *Plesiosminthus* is only identified to genus (Parris and Green, 1969).

Distinctive Characters *Plesiosminthus* is described as a small rodent with grooved upper incisors. The cheek teeth are brachydont, and the M1 has a strong anterocone. The upper molars have a long mesoloph and a reduced M3. The m1 has a prominent mesoconid, variable mesolophid, and a low and tiny anteroconid (2008a). *Plesiosminthus* is in the same subfamily (Sicistinae) as the modern birch mice.

Stratigraphic and Geographic Distribution Fossilized remains of *Plesiosminthus* can be found in Arikareean through Clarendonian rocks. Fossils are found in South Dakota, Nebraska, Saskatchewan, Wyoming, Montana, Colorado, and Delaware. In the White River Badlands, *Plesiosminthus* is found in the Sharps Formation (Parris and Green, 1969).

Natural History and Paleoecology The extant Sicistinae excavate shallow burrows and are found in both forest and steppe environments.

Family Heteromyidae

The Heteromyids are a diverse family, with six genera and 300 species. Extant genera include the pocket mice and kangaroo rats. Their locomotor adaptations range from bipedal hoppers to burrowers. These small to medium sciuromorphous rodents have a rich fossil record, especially in the Oligocene and Miocene of the Great Plains (Flynn, Lindsay, and Martin, 2008). The rodents within this family have thin skull bones, and the side of the snout is perforated. Their molars have six cusps with large styles and stylids. The p4 is small and narrow anteriorly. The incisors are laterally compressed, and the postcrania are gracile. The limbs are elongated, and the tibia and fibula are fused (Korth, Wahlert, and Emry, 1991).

Proheteromys Wood, 1932

Systematics and Evolution The type species of *Proheteromys* is *P. floridanus* Wood, 1932. The type specimen of *Proheteromys*, FSGS V5329, was collected from the Torreya Formation (Miocene) in Gadsden County, Florida. *Proheteromys* is only known at the genus level in Badlands National Park.

Distinctive Characters *Proheteromys* is characterized by smooth asulcate upper incisor enamel, mesodont dentition, and four-cusped premolars. The crown height varies with individual species. The anterior cusps on the p4 are close and joining early in wear to isolate a central basin. The lower molars have weak anterior and posterior cingula (Fig. 5.18A, B) (Flynn, Lindsay, and Martin, 2008).

Stratigraphic and Geographic Distribution *Proheteromys* has a broad range stratigraphically and geographically. However, its occurrence in Badlands National Park is rare. Fossilized remains of *Proheteromys* occur in the Whitneyan to Barstovian. Fossils are found in Texas, Florida, Colorado, South Dakota, Nebraska, Saskatchewan, California, Delaware, Wyoming, New Mexico, Oregon, and Montana. In the White River Badlands, *Proheteromys* is found in the Brule and Sharps formations.

Family Heliscomyidae

The heliscomyids are some of the smallest rodents in North America, both fossil and recent (Korth, 1994b). The skull has a sciuromorphous zygomasseteric structure. The family Heliscomyidae was created by Korth, Wahlert, and Emry (1991). Taxa within this group used to be included within the primitive heteromyids. Green and Bjork (1980) noted indicators of a more complex evolution, and many species were grouped together that had divergent characteristics. This family is part of a larger group called Geomorpha (pocket gophers, pocket mice, and kangaroo rats) and is believed to be an early distinct radiation of the Geomyoidea (Korth, Wahlert, and Emry, 1991). The family was considered primitive because it lacked many of the features that characterize the modern pocket gophers, pocket mice, and kangaroo rats.

Heliscomys Cope, 1873

Systematics and Evolution The type species of *Heliscomys* is *H. vetus* Cope, 1873. The type specimen of *Heliscomys* is AMNH 5461, collected from the White River Formation in Logan County, Colorado. Besides the type species, three other species are currently recognized as valid.

Two distinct lineages are found within the genus *Heliscomys* (Korth, Wahlert, and Emry, 1991). One lineage, *H. vetus–H. mcgrewi–H. woodi*, has a reduction of the premolars and the simplification of molars (Orellan and Arikareean). A separate lineage, *H. subtilis*, has enlarged and an increased complexity of the premolars (Chadronian–Barstovian). The greatest diversity is in the Orellan, with three genera and five species.

Distinctive Characters *Heliscomys* is a small geomyine with brachydont teeth showing isolated cusps and strong cingulae-bearing accessory cusps. The upper premolar is triangular with one main anterior cusp and two posterior cusps. The lower premolar has four main cusps and narrows anteriorly. The mandible was dorsoventrally slender, and the diastema is shallow and short. The incisors are small, delicate, and laterally compressed with a narrow, gently convex anterior enamel surface (Fig. 5.18C, D) (Flynn, Lindsay, and Martin, 2008).

Stratigraphic and Geographic Distribution *Heliscomys* occurs in the Duchesnean to Hemingfordian. Fossils are found in Saskatchewan, California, Wyoming, South Dakota, Nebraska, Montana, Colorado, and Florida. Only *H. vetus* has been reported from the Scenic Member of the Brule Formation in the Badlands. There are no species known in the Whitneyan, but this absence of *Heliscomys* spp. could be due to a collecting bias (Korth, 1994b).

Family Florentiamyidae

The family is characterized by a large optic foramen and reduced incisive foramina. The entostyle is elongated anteroposteriorly. The lower molars have lingual styles, and the cheek teeth are brachydont (Wahlert, 1983).

Hitonkala Macdonald, 1963

Systematics and Evolution The type species of *Hitonkala* is *H. andersontau* Macdonald, 1963. The type specimen of *Hitonkala*, SDSM 56120, was collected from Shannon County, South Dakota.

Distinctive Characters *Hitonkala* is a small florentiamyid with a single anterior cusp on the P4. The p4 is highly molariform. The M1 has a separated protostyle and hypostyle (Flynn, Lindsay, and Martin, 2008). Korth (1993) completed a study of the *Hitonkala* cranial anatomy that helps clarify the unique features of the family Florentiamyidae.

Stratigraphic and Geographic Distribution *Hitonkala* occurs in the early to late Arikareean. Fossils are found in Wyoming, South Dakota, Nebraska, and Saskatchewan. Parris and Green (1969) noted the presence of *Hitonkala* in the Sharps Formation in Cedar Pass at Badlands National Park.

Family Eomyidae

Eomyidae is a complex and rich group that dominated the small-mammal faunas of North America during certain times. They are considered a sister taxa to Geomorpha (living gophers and mice). The family is now extinct. Because of increased screen washing, the group has become better documented in the fossil record. Superficially, eomyids are squirrellike in their body proportions and ground dwelling

to arboreal in habits. They are small- to medium-size rodents and are found in North America, Europe, and Asia. Primitive genera occur in the late to middle Eocene of North America and appear during the Oligocene in Europe and Asia. Eomyids persisted to the end of the Miocene in North America (Flynn, 2008b).

Eomyids have a sciuromorphous jaw musculature, but their molars follow a cricetid morphology and are composed of four major cusps joined by thin strong lophs. The mandible is slender with a long diastema. There are no fossorial adaptations to the skull – for example, heavy bone, strong contacts, or a broad posterior skull. Skulls are generally gracile, and there is no sagittal crest. The premolars are submolariform and nearly rectangular. The molars are often wider than long. The cheek teeth are bunodont to pentalophodont. The eomyid postcrania are slender, which is similar to squirrels, and some taxa are inferred to have had gliding membranes, such as *Eomys quercyi* (Storch, Engresser, and Wuttke, 1996).

Most eomyids were associated with a mesic and forest habitat and with arboreality. Ecomorphological counterparts to Eomyidae might be living small to medium squirrels. However, their dentition follows a more conservative muroid pattern.

Adjidaumo Hay, 1899

Systematics and Evolution The type species of *Adjidaumo* is *A. minutus* (Cope, 1873). The type specimen of *Adjidaumo*, AMNH 5362, was collected from the White River Formation, Logan County, Colorado.

Distinctive Characters *Adjidaumo* is characterized as a small eomyid and has brachydont cheek teeth, cuspate molars with low crests, and a strong anterior cingulum joined centrally. The incisor enamel is smooth and rounded (Flynn, 2008b). *Adjidaumo* is often confused with the European taxon *Eomys*, which migrated from North America.

Stratigraphic and Geographic Distribution *Adjidaumo* occurs in the Bridgerian to Whitneyan. Fossils are found in Texas, Wyoming, Colorado, South Dakota, Nebraska, Saskatchewan, British Columbia, and Montana. In the White River Badlands, *Adjidaumo* is found in the Chadron and Brule formations.

Natural History and Paleoecology Eomyids experienced a range expansion across several continents. The confusion between *Adjidaumo* and *Eomys* is a classic example of taxonomy driven by continent of occurrence. It is difficult to differentiate between the two taxa on the basis of morphology of individual teeth. McKenna and Bell (1997) placed *Adjidaumo* in *Eomys*. Flynn (2008b) concluded that there are enough features to distinguish the two genera.

Paradjidaumo Burke, 1934

Systematics and Evolution The type species of *Paradjidaumo* is *P. trilophus* (Cope, 1873), which was originally described as *Gymnoptychus trilophus*. The type specimen of *Paradjidaumo*, AMNH 5401, was collected from the White River Formation in Logan County, Colorado.

Distinctive Characters *Paradjidaumo* is a small eomyid. Its molars have developed lophs and moderately high crowns. The valleys, especially the central ones, are narrow and deep. The lower incisor is flattened anteriorly and bears a longitudinal ridge (Black, 1965).

Stratigraphic and Geographic Distribution *Paradjidaumo* occurs in the Bridgerian to Hemingfordian. Fossils are found in Utah, Wyoming, Colorado, South Dakota, Nebraska, Montana, North Dakota, California, British Columbia, and Saskatchewan. In the White River Badlands *Paradjidaumo* is found in the Chadron and Brule formations.

Natural History and Paleoecology *Paradjidaumo* has a broad range stratigraphically and geographically but is still fairly rare at Badlands National Park.

Order Creodonta

One of the earliest group of mammals to fill the ecological niche as carnivores are members of the Order Creodonta. A primitive group of mammals, they first appeared in the late Paleocene and survived until the end of the Oligocene in North America. Their extinction probably was the result of increased competition from the more advanced members of the Order Carnivora. There are two families within the Creodonta, Oxyaenidae and Hyaenodontidae, but only one genus of the latter is known from the White River Group. Members of the order range in body size from around a small cat to a hyena or lion. In addition to being active predators, like modern carnivores, they probably also scavenged and ate carrion.

Compared to modern carnivores with a similar body size, the size of the brain in creodonts was smaller and occupies a lower position on the skull. Another distinctive feature is a large, prominent sagittal crest and a large temporal fossa, suggesting they possessed powerful temporalis muscles for closing the jaw. Unlike modern carnivores, the auditory bulla is unossified in most creodonts. There is a well-defined postorbital constriction in the skull that distinctly separates the rostrum from the braincase. As in modern carnivores, the canines are large and pointed for holding and killing prey. Modern carnivores have one set of slicing teeth (carnassials), which are always formed by the same upper and lower tooth. In contrast, creodonts could have two or three pairs of carnassial teeth that could be formed by either the upper first

molar and lower second molar or the upper second molar and lower third molar. These molars were located farther back on the skull than in modern carnivores. The body skeleton is generalized, but some forms were plantigrade and some digitigrade, like modern carnivores. Unlike modern carnivores with a fused scaphoid and lunar, all of the carpals in creodonts are separate. Their claws had a prominent groove down their midline (fissured), a feature not present in modern carnivores. The femur has a distinct third trochanter, a feature not present in modern carnivores.

Family Hyaenodontidae

The Hyaenodontidae first appear in the Wasatchian and became extinct at the end of the early early Arikareean. They are characterized by having sectorial molars (modified for slicing) and highly specialized carnassials. The bones of the limbs and feet of members of the family indicate they were often cursorial.

Hyaenodon (=*Neohyaenodon* Mellett, 1977) Laizer and Parieu, 1838

Systematics and Evolution The type species for *Hyaenodon* is *Didelphis* (*H.*) *leptorhynchus* from the Oligocene of France. Currently 10 species are recognized in North America and range in age from the Duchesnean to earliest Arikareean (Gunnell, 1998).

Distinctive Characters There is a great size range for species of *Hyaenodon*; an estimated body mass of an adult or subadult *H. horridus*, the largest North American species, is about 40 kg (Egi, 2001). In *Hyaenodon* both the upper first and second molars and lower second and third molars are modified as carnassials and were well developed for shearing meat. The feet are digitigrade, indicating it was cursorial and probably ran down its prey (Fig. 5.19A, B).

Stratigraphic and Geographic Distribution The earliest record of the genus is the early Duchesnean, but there are some indications that it may have appeared earlier in the late Uintan. It became extinct at the end of the early early Arikareean. *Hyaenodon* from the White River is represented by *H. horridus*, *H. montanus*, and *H. crucians*. *Hyaenodon horridus* is present in all members of the Chadron Formation, and a second species, *H. crucians*, is reported from the Peanut Peak Member. *Hyaenodon brevirostris* is reported from the Poleslide Member of the Brule Formation. *Hyaenodon* was the last surviving creodont in North America. The genus was widely distributed in North America and is known from South Dakota, California, Utah, Wyoming, Texas, Nebraska, Montana, New Mexico, Colorado, North Dakota, and Saskatchewan.

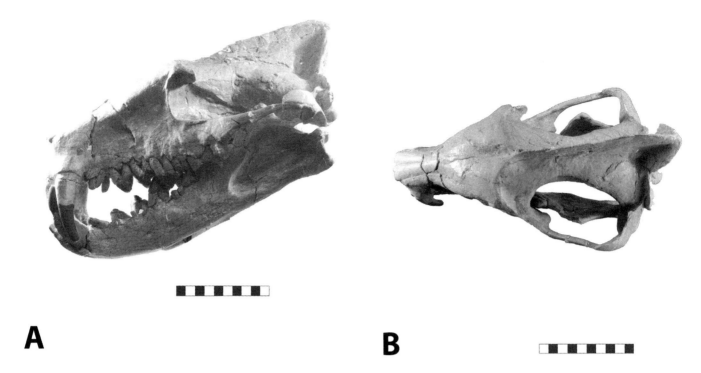

5.19. *Hyaenodon horridus.* (A) Skull and jaws, left lateral view, SDSM 242. (B) SDSM 242, skull, dorsal view. Scales in centimeters. Photos by the authors of specimens from the Museum of Geology, South Dakota School of Mines and Technology, Rapid City, South Dakota, U.S.A.

Natural History and Paleoecology The body size of *Hyaenodon horridus* is small compared to other species of the genus, and it may have preyed on any of the small artiodactyls present in the White River fauna, like *Leptomeryx*, *Hypertragulus*, or *Hypisodus*. A mount at the Denver Museum of Nature and Science has *Hyaenodon* killing a small individual of *Merycoidodon* (Fig. 5.20). The extinction of the creodonts is attributed to competition with members of the modern order Carnivora.

Order Carnivora

While the name of the order is indicative of the primary habits of its members as carnivores, this is not true of every member of the order, and various groups have adapted to quite different dietary preferences. Members of the Ursidae and Procyonidae tend to be omnivorous, and some bears, such as the panda and South American spectacled bear, are strictly herbivores. Likewise, members of the Canidae and Mustelidae will include varying amounts of both animal (including fish) and vegetable material in their diets. The Felidae are the most carnivorous members of the order and are often referred to as hypercarnivores. Carnivores may be both ambush or pursuit predators, and this is reflected in modifications of the skeleton that aided in running down prey, such as the cheetah or wolves. Some mustelids like martens and fishers are fully arboreal. Other carnivores may obtain food from aquatic environments and may be semiaquatic, like otters, or fully aquatic, like seals and sea lions.

All members of the order have large, well-developed canines that aid in catching and holding prey during hunting. Although reduced in the more omnivorous or herbivorous forms, members of the modern order Carnivora are characterized by having the shearing teeth, or carnassials, always being formed by the upper fourth premolar and lower first molar, unlike the creodonts, in which the position of the carnassials were further back in the skull and jaw, and could be formed by either first upper and second lower molars, or the second upper and third lower molars.

Another distinctive feature of the Order Carnivora is the fusion of two carpal bones, scaphoid and lunar, into a composite structure, the scapholunar. Claws are present in all forms and in some groups can be retracted, a common character of the Felidae, but also present in some other families as well. The manus and pes may be plantigrade or digitigrade, and while the first digit on either the manus or pes may be reduced or lost in some forms, the number of digits is never less than four. While the amount may vary, the radius remains mobile and able to rotate around the ulna to aid in the maneuverability of the forelimb while catching prey.

Family Nimravidae

While members of the carnivore family Nimravidae superficially resemble modern cats (family Felidae), they represent

5.20. Reconstruction of the creodont, *Hyaenodon*, attacking the oreodont, *Merycoidodon*. *Hyaenodon crucians* (DMNS 1604) and *Merycoidodon culbertsoni* composite skeleton (DMNS 84 and DMNS 54). Photo by the authors. Denver Museum of Nature and Science, Denver, Colorado, U.S.A. All rights reserved.

a distinct family of carnivores. They first appear in Europe and North America in the late Eocene (early Chadronian in North America), but the family became extinct in the early Miocene (Clarendonian) and did not have any descendants. The general resemblance to true cats is based on the short face, but they retain a short braincase unlike the elongated braincase of members of the Felidae. Like true cats, the carnassials are highly modified for cutting meat, the upper first molar is reduced in size, and they have lost the upper second and third molars. Most of the major characters that distinguish members of the Nimravidae from the Felidae are in the anatomy of the braincase and the ear region, such as an incomplete auditory bulla. Some but not all nimravids developed a prominent elongation of the mental process of the mandible to form a "sheath" for the elongated upper canine.

Primitive members of the family were plantigrade, although there is some evidence that advanced members of the family had become digitigrade, like modern felids. Another feature they shared with modern cats was retractable claws with a well-developed sheath housing the base of the nail. Their postcranial skeleton suggests they were capable of climbing trees, although *Dinictis* and *Nimravus* appear to have more cursorial habits than other nimravids. The scapula has a distinct process for the teres major muscle, suggesting the forearms could be retracted with force.

Except for a later genus, *Dinaelurus*, which is found only at John Day, Oregon, which has conical canines, all other nimravids have enlarged bladelike upper canines and were functional saber-tooths. The elongated (saber-tooth) upper canine morphology can be divided into the dirk-toothed forms with fine serrations (*Hoplophoneus* and *Eusmilus*) and a scimitar-toothed group (*Dinictis* and *Nimravus*) in which the serrations are coarse. The canines had serrations on both their mesial and distal edges. The coarseness of the serrations seems to be correlated with locomotor habits, with the dirk-toothed forms being short legged and plantigrade, so were more similar to bears in their body proportions, while the scimitar-toothed forms were more cursorial, like modern felids, in their postcranial anatomy. The dirk-toothed forms are thought to have been ambush predators, while the scimitar-toothed forms were probably capable of chasing their prey over short distances (Martin, 1998). It is interesting that the White River fauna contained five genera of nimravids with saber-toothed adaptations, although it is only during the Whitneyan that all five genera coexisted. It is possible that *Eusmilus* replaced *Hoplophoneus* as a dirk-toothed form and *Nimravus*, a scimitar-toothed form, replaced *Dinictis*. Both *Eusmilus* and *Nimravus* appear at the beginning of the Whitneyan, and *Hoplophoneus* and *Dinictis* had become extinct by the end of the Whitneyan. It is possible that these new genera may have competed with and ecologically replaced the older genera.

Dinictis Leidy, 1854

Systematics and Evolution The type species of *Dinictis* is *D. felina* and is based on a poorly preserved skull and jaw from the White River of the Dakota Territory. The genus is monotypic, and only one species is recognized as valid. Disappearing at the end of the Whitneyan, *Dinictis* may have been the ancestor to *Pogonodon*, which appeared in the Orellan.

Distinctive Characters *Dinictis* is a dirk-toothed nimravid, while *Hoplophoneus* is a scimitar-toothed form, suggesting they occupied different ecological niches as predators and probably focused on different prey species. The upper canine is shorter and broader than in *Hoplophoneus*. The occiput of *Dinictis* is inclined while in *Hoplophoneus* it is vertical (Martin, 1998).

Stratigraphic and Geographic Distribution *Dinictis*, along with *Hoplophoneus*, are the two earliest genera of nimravids in North America, and both first appear in the early Chadronian. In the Big Badlands of South Dakota, *Dinictis* is present in the Peanut Peak Member of the Chadron Formation and the Poleslide Member of the Brule Formation. It is also found in faunas in Wyoming, Colorado, Nebraska, and Saskatchewan.

Natural History and Paleoecology Nimravids appear in North America in the late Eocene, at about the same time or perhaps a little later than their appearance in Eurasia. While the first North American forms are already represented by cursorial forms, some of the early nimravids have features of the skeleton that suggest arboreal adaptations. *Dinictis* is more cursorially adapted than its contemporary, *Hoplophoneus*, a shorter-limbed form, which was probably, in contrast, a solitary ambush predator. Except for the brontotheres, most of the available prey species were small, and because the late Eocene–early Oligocene fauna was dominated by oreodonts, small horses, and small rhinocerotids, these were the most likely prey for *Dinictis* and other nimravids (Martin, 1998).

Pogonodon Cope, 1880

Systematics and Evolution The type species of *Pogonodon* is *P. platycopis* from the John Day Formation, Oregon, and the genus includes two other species, *P. eileenae* and *P. paucidens*. Species now referred to *Pogonodon* have at times been placed in *Dinictis*, but the two genera seem to have been distinct lineages since the early Oligocene (Orellan) (Martin, 1998).

Distinctive Characters *Pogonodon* is a large nimravid in the size range of a mountain lion to jaguar. Like *Dinictis*, it is a scimitar-toothed form, but the upper canine is short, broad, and coarsely serrated. The skull has a high sagittal crest, and there is a distinct flange on the mandible. *Pogonodon* retains a lower second premolar, which is absent in *Hoplophoneus* but has lost the lower second molar, which is retained in most specimens of *Dinictis* (Bryant, 1996). A distinctive feature of the pes is the presence of an articulation between the calcaneum and the navicular.

Stratigraphic and Geographic Distribution The biostratigraphic range of the genus is from the Orellan to the late early Arikareean. *Pogonodon platycopis* is the only species present in the White River Group in South Dakota and is found in the Brule Formation (Orellan to Whitneyan) (Martin, 1998). The species is also known from Wyoming, Montana, and Oregon.

Natural History and Paleoecology The extinction of *Pogonodon* coincides with that of *Eusmilus* and *Nimravus* at the end of the late early Arikareean. Although *Pogonodon* appears first in the Orellan and the other two genera first appear in the Whitneyan, their biostratigraphic ranges essentially overlap. This would suggest that they may have been ecological competitors for most of their existence because they overlapped not only in time but also in geography and have been found together in the same faunas. Martin (1998) noted that the late Eocene–Oligocene fauna was dominated by oreodonts and small rhinocerotoids, and this prey diversity may have permitted the coexistence of these three genera, perhaps ecologically equivalent to that of lions, leopards, and cheetahs in Africa today, although none of these genera have the advanced cursorial adaptations seen in cheetahs.

Hoplophoneus Cope, 1874

Systematics and Evolution The type species of *Hoplophoneus* is *Machairodus primaevus* from the *Oreodon* Zone of the White River Group in Colorado. Eleven species have been described for *Hoplophoneus*—more than for any other nimravid. Currently three other valid species besides the type are recognized.

Distinctive Characters It is a dirk-toothed form with fine serrations on the upper canine, and the mandibular flange is larger and more prominent than in *Dinictis*. *Hoplophoneus* also has a prominent sagittal crest, and the occiput is vertical (Fig. 5.21A). The mastoid process is enlarged, suggesting powerful neck musculature. Like other saber-toothed cats, the deciduous canine was retained until the permanent canine was fully erupted. The permanent canine erupted along the inner edge of the deciduous canine (Fig. 5.21D, E). In the lower jaw it has only two premolars and one molar, while in

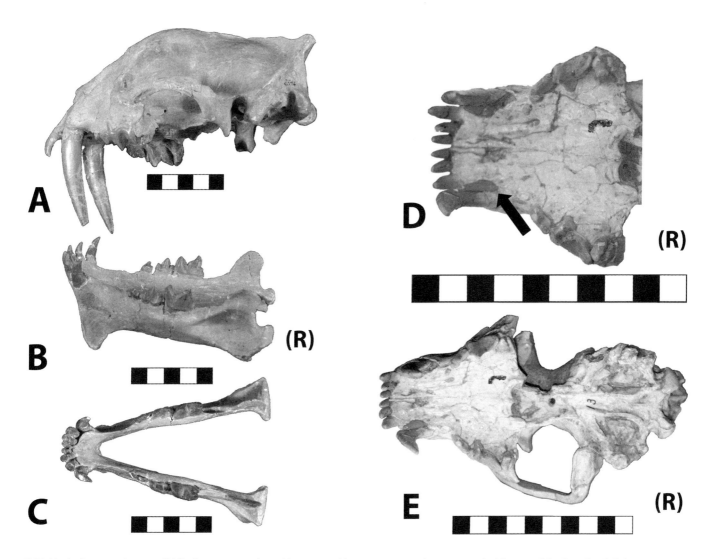

5.21. *Hoplophoneus primaevus.* (A) Skull, left lateral view, SDSM 2544. (B) Mandibles, right lateral view, reversed, SDSM 2544. (C) Mandibles, occlusal view, SDSM 2544. (D) *Hoplophoneus* sp. skull, anterior end, occlusal view with arrow marking permanent canines beginning to erupt, reversed, UCM 19160. (E) *Hoplophoneus* sp., skull, occlusal view, reversed, UCM 19160. Scales in centimeters. Photos by the authors. SDSM 2544 is from the Museum of Geology, South Dakota School of Mines and Technology, Rapid City, South Dakota, U.S.A. UCM 19160 is from the University of Colorado Museum of Natural History, Boulder, Colorado, U.S.A.

Dinictis the lower dentition consists of three premolars and two molars (Fig. 5.21C). The mandible has a large flange (Fig. 5.21B) (Martin, 1998).

Stratigraphic and Geographic Distribution The stratigraphic range for the genus is from the Chadronian through Whitneyan. *Hoplophoneus mentalis* is present in the Crazy Johnson and Peanut Peak members of the Chadron Formation. *Hoplophoneus primaevus* and *H. occidentalis* are present in the Brule Formation. The genus is also found in Wyoming, Nebraska, and Montana.

Natural History and Paleoecology *Hoplophoneus* was a contemporary of *Dinictis*, and both genera first appear in the early Chadronian and became extinct at the end of the Whitneyan. *Hoplophoneus* was larger than *Dinictis*, with a size ranging from similar to a mountain lion (*H. primaevus*) to that of a jaguar (*H. occidentalis*) (Fig. 5.22).

On May 30, 2010, a *Hoplophoneus* skull was discovered in Badlands National Park by a 7-year-old girl while participating in a park-led Junior Ranger program. The skull was preserved in a carbonate-cemented mudstone bed in the middle to upper Scenic Member of the Brule Formation. The cranium is exceptionally well preserved and shows unique punctures, possibly bite marks from another nimravid. Badlands National Park, in cooperation with Rapid City Regional Hospital, scanned the skull by computed tomography (CT). The CT scan provided evidence that another *Hoplophoneus* attacked this individual (Fig. 8.1).

5.22. *Hoplophoneus primaevus,* SDSM 2528, skeleton (reversed), right lateral view. Photo by the authors of specimen from the Museum of Geology, South Dakota School of Mines and Technology, Rapid City, South Dakota, U.S.A.

Eusmilus (=Egmoiteptecela Macdonald, 1963) Gervais, 1876

Systematics and Evolution The type species of *Eusmilus* is *E. bidentatus* from the fissure deposits of Quercy, France, are considered to be Oligocene in age. Another genus from South Dakota, *Egmoiteptecela*, is considered to be a junior synonym of *Eusmilus* (Martin, 1998). *Egmoiteptecela* is from the Lakota *Egmo*, "cat"; *ite*, "face"; and *ptecela*, "short." The name thus literally means "short-faced cat."

Distinctive Characters *Eusmilus* is the smallest of the nimravids in North America, and unlike other nimravids, there does not appear to have been an increase in size through time. It is a dirk-toothed form with its cheek teeth greatly reduced in number, with two upper premolars and one to two lower premolars and a single upper and lower molar. Of all the nimravids found in the White River, its mandible has the greatest development of the flange in proportion to the animal's size (Martin, 1998).

Stratigraphic and Geographic Distribution *Eusmilus* first appears at the beginning of the Whitneyan and became extinct at the end of the late early Arikareean at the Oligocene–Miocene transition. *Eusmilus* sp. has been identified from the Poleslide Member of the Brule Formation in the Big Badlands. The genus is also known from Nebraska and Oregon.

Natural History and Paleoecology All members of the tribe Eusmilini are small (Bryant, 1984), including *Eusmilus*, which is about the size of a modern bobcat, making it the smallest of all the saber-toothed predators present in the White River faunas (Bryant, 1984; Martin, 1992). Emerson and Radinsky (1980) estimated the body mass of *Eusmilus* at about 33.8 kg. Presumably this size difference allowed *Eusmilus* to avoid competition with other contemporary saber-tooths by preying on smaller species. Despite its smaller body size, the cranial morphology of *Eusmilus* was in many

respects as derived as that of the later *Barbourofelis*, including the enlargement of the carnassials and reduction of the premolars (Martin, 1992). The presence of *Eusmilus* in contemporary faunas in both North America and Europe suggest a widespread Holarctic distribution for the genus during the Oligocene. Martin (1992) proposed that although the smaller species of oreodonts may have been the principal prey for *Eusmilus*, given its cranial specializations, it may have been able to have taken larger prey–up to the size of modern pigs or deer. In contrast, Emerson and Radinsky (1980) pointed out that the relationship between body size of predator and prey might be modified if a smaller predator such as *Eusmilus* preyed on different age classes, i.e., younger and hence smaller individuals of a given prey species, or if it was hunting socially and was thus able to take down larger individuals. We do not yet have any information regarding *Eusmilus* as to whether either or both of these alternatives are applicable. Van Valkenburgh (1985) concluded, on the basis of limb morphology, that although all members of the predatory guild in the early Oligocene were slower and more robust than their modern counterparts, the spacing of predator types was similar to that seen in modern faunas, and that the ecological processes that structured the guild were not appreciably different than those that structure the guild of predators in modern ecosystems.

Nimravus Cope, 1879

Systematics and Evolution The type species of *Nimravus* is *N. brachyops* from the John Day Formation, Arikareean of John Day, Oregon. Currently there are five species considered to be valid.

Distinctive Characters This was a scimitar-toothed form comparable in size to a leopard. The upper canine is coarsely serrated. The protocone is reduced on the upper fourth premolar, and the lower first molar has a distinct talonid. The second and third lower molars may be present in some individuals. Unlike *Hoplophoneus*, it did not develop an enlarged flange on the mandible. Its jaw is also distinguished from the other nimravids by having an enlarged masseteric fossa that extends anteriorly to under the lower carnassial (Martin, 1998).

Stratigraphic and Geographic Distribution The genus first appears in the Whitneyan and becomes extinct at the end of the late early Arikareean, at the Oligocene–Miocene boundary. *Nimravus brachyops* is present in the Poleslide Member of the Brule Formation in the Big Badlands. The genus is also known from California, Nebraska, North Dakota, and Oregon.

Like *Eusmilus*, *Nimravus* first appears in the Whitneyan and becomes extinct at the Oligocene–Miocene boundary, although there are some fragmentary specimens that suggest this genus may have persisted into the early late Arikareean, so *Nimravus* may have survived into the earliest Miocene.

Natural History and Paleoecology Intraspecific fights between the different saber-toothed taxa in the White River seem to have taken place. A skull of *Nimravus* in the Museum of Geology at the South Dakota School of Mines and Technology has an elongate puncture in the left frontal sinus that had partially healed. The size of the puncture is similar in size to the canine of *Eusmilus* (Scott and Jepsen, 1936).

Family Amphicyonidae

The Amphicyonidae first appear in North America in the late Eocene (Duchesnean) and became extinct in the early late Miocene (Clarendonian), although they survived later in Europe and India. In North America, their greatest diversity was in the Miocene, and there are three genera in the White River: *Daphoenus*, *Paradaphoenus*, and *Brachyrhynchocyon*.

The general skull form is a short rostrum with an elongated postorbital region. As some members of the family increased in size, the skull became disproportionately enlarged, including a greatly enlarged area for the temporalis muscle and an extreme development of the sagittal crest. Like the nimravids, a major feature that distinguishes amphicyonids from other carnivores is the construction of the ear region, with the ectotympanic bone as the primary bone contributing to the formation to the bulla, which was not inflated and thus is somewhat similar to what is seen in bears. Most amphicyonids retain the basic eutherian tooth formula of I3/3-C1/1-P4/4-M3/3, and only in a few derived forms are premolars lost or the upper third molar, which happens independently in some genera. The incisors and canines tend to be large. In some lineages, the molars tend to decrease in size from the first to third, and in many the third molar is eventually lost; in other lineages, the posterior molars becomes the largest and with their enlargement become broad, flat, crushing surfaces, especially in the second molar, so they look like the molars of bears, giving rise to the popular name for the group, bear-dogs. Larger amphicyonids are plantigrade; smaller forms like those from the White River are digitigrade, but not as advanced in cursorial abilities as dogs. Digitigrade amphicyonids like *Daphoenus* retain slicing carnassials, while the larger plantigrade forms tend to lose the shearing capacity of the carnassials and have broader molars, thus more closely resembling bears in two ways. There is pronounced sexual dimorphism in amphicyonids. The males had a baculum, and a significant difference is size has been documented in *Brachyrhynchocyon dodgei* from the late Eocene (Chadronian).

5.23. *Daphoenus vetus.* (A) Skull and jaws, right lateral view, YPM PU 12651. (B) Skull, occlusal view, YPM PU 12648. Scales in centimeters. Photos by the authors. Courtesy of the Division of Vertebrate Paleontology, YPM PU 12651 and YPM PU 12648, Peabody Museum of Natural History, Yale University, New Haven, Connecticut, U.S.A.

Daphoenus Leidy, 1853

Systematics and Evolution The type species for *Daphoenus* is *D. vetus* from South Dakota, but the exact locality is unknown. The type was collected in the White River beds of the Dakota Badlands and was described by Leidy in 1869 as *Amphicyon vetus*. It is one of the earliest finds of a White River mammal. Currently there are five described species.

Distinctive Characters Species of *Daphoenus* range in size from small- to midsize within the amphicyonids. The skull form and the canine teeth are dimorphic, with specimens interpreted as males larger with a more robust rostrum and canines. The skull is more elongate then the short, broad skull of *Brachyrhynchocyon*, but in both genera, the premolars increase in size from the front to the back. It retains the primitive eutherian dental formula with mesiodistally elongate and narrow premolars, especially the upper and lower second and third premolars (Fig. 5.23).

Stratigraphic and Geographic Distribution The genus first appears in the late middle Eocene (Duchesnean) and becomes extinct in the late Oligocene (late early Arikareean).

It is the most common genus within the family, and for the Great Plains, essentially all known specimens are from the White River Group of South Dakota, Nebraska, Wyoming, and Colorado. *Daphoenus vetus* is the only species identified from the Big Badlands and is found in the Crazy Johnson and Peanut Peak members of the Chadron Formation and throughout the Brule Formation (Hunt, 1998b). The genus is also found in Texas, California, Wyoming, Oregon, Nebraska, Colorado, Montana, North Dakota, and Saskatchewan.

Paradaphoenus Wortman and Matthew, 1899

Systematics and Evolution The type species is *P. cuspigerus* from the John Day Formation of Oregon, and a second species, *P. minimus*, is recognized both on the Great Plains and at John Day.

Distinctive Characters This genus is a small amphicyonid, with an estimated body mass between 4 and 6 kg (Figueirido et al., 2011); the general form of the premolars is similar to those of *Daphoenus*, except for their smaller size. While the sample size is small, the lower premolars tend to be laterally compressed and bladelike. *Paradaphoenus cuspigerus* is the oldest and most primitive species and can be distinguished from contemporary amphicyonids such as *Daphoenus* by the closed trigonid with a slightly developed paraconid. The upper first molar has a V-shaped protocone. The talonid of the lower first molar is basined with the hypoconid and entoconid, creating a low ridge that forms the border of the basin. The genus is currently only known from two geographic areas: the Great Plains and the John Day Basin. The two populations are distinguished by the presence of a small upper third molar in the John Day sample that appears to be absent in the Great Plains sample. The genus lacks an enlarged or inflated auditory bulla, so the basioccipital is wide (Hunt, 2001).

Stratigraphic and Geographic Distribution The genus first appears in the Orellan and became extinct in the late early Arikareean. *Paradaphoenus cuspigerus* is present in the Scenic Member of the Brule Formation in South Dakota and it is also known from the Brule Formation in Nebraska (Hunt, 1998b).

Natural History and Paleoecology There is little evolutionary change in body size or skull proportions of the genus from the Orellan to the Arikareean, including the advanced form *P. cuspigerus* from John Day. The genus seems to be strictly North American. The rarity of specimens prevents any real determination of its probable diet.

Brachyrhynchocyon Loomis in Scott and Jepsen, 1936

Systematics and Evolution The type species of *Brachyrhynchocyon* is *B. dodgei* from the Seaman Hills of Wyoming, and a second species, *B. montanus*, is also recognized.

Distinctive Characters It is a small form but highly sexually dimorphic, with males having a broad, massive skull while in females the skull is narrower and more gracile (Gustafson, 1986). The skull is short and broad, thus giving rise to its name, *Brachyrhynchocyon*, "short-rostrum dog." The genus retains the full primitive eutherian dental formula. It can be distinguished from its contemporary in the White River Group, *Daphoenus*, in having mesiodistally short, wide, and robust premolars. In both forms, the premolars increase in size posteriorly and will develop accessory cusps on the posterior margins of the second, third, and fourth premolars. The mandible is thick and massive (Hunt, 1996).

Stratigraphic and Geographic Distribution The genus is known only from the late Eocene (Chadronian). In the Big Badlands, it is found in all members of the Chadron Formation (Hunt, 1998b). Outside of South Dakota, it is found in Texas, Wyoming, Colorado, Nebraska, Montana, and Saskatchewan.

Natural History and Paleoecology A few postcranial skeletons are known for *Brachyrhynchocyon*. They indicate a robust, muscular, wolverinelike build to the body that, coupled with the shortened muzzle, which is also wolverinelike, suggests it may have filled a similar niche.

Family Canidae

The Canidae includes the living dogs such as wolves, coyotes, jackals, and foxes. The distribution of the family today is widespread and includes North and South America, Eurasia, and Africa, with the human introduction of the dingo into Australia. The early fossil record and evolutionary history of the family is in North America, and they did not disperse into other continents until the late Miocene (Hemphillian). The family first appears in the late middle Eocene (Duchesnean). Three subfamilies are recognized: Hesperocyoninae, Borophaginae, and Caninae. The first two are extinct (Wang and Tedford, 2008).

Members of the Canidae are generally considered the most primitive or generalized of all the carnivores. As is the case with most carnivores, one of the primary characters that distinguishes the family is the ear region. Canids have a fully ossified and inflated auditory bulla formed primarily by the entotympanic and a smaller contribution from the ectotympanic. The overall shape of the skull is elongate, with a prominent tapering rostrum and strong zygomatic arches. The temporal fossae are well defined by either a prominent

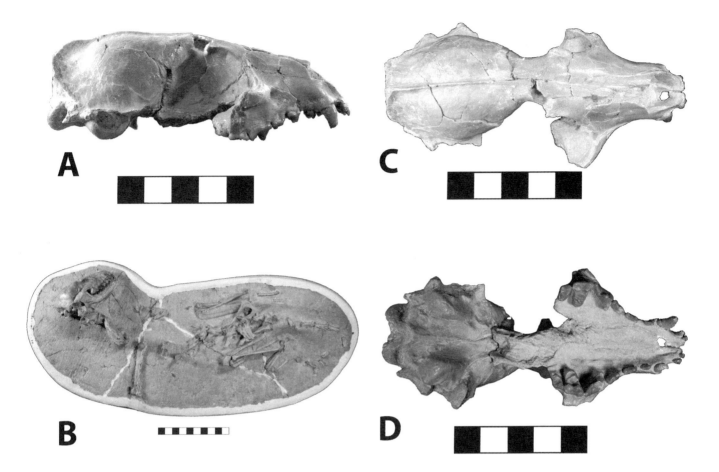

5.24. *Hesperocyon* sp., SDSM 4385. (A) Skull, right lateral view. (B) Skeleton, right view, SDSM 2513 *Hesperocyon gregarious*. (C) Skull, dorsal view, SDSM 4385, *Hesperocyon* sp. (D) Skull, occlusal view, SDSM 4385, *Hesperocyon* sp. Scales in centimeters. Photos by the authors of specimens from the Museum of Geology, South Dakota School of Mines and Technology, Rapid City, South Dakota, U.S.A.

sagittal crest, generally better developed in males, or paired temporal cristae. The dental formula is similar to that of primitive eutherian mammals except that the upper third molar has been lost. Most canids are mesocarnivores with an omnivorous diet, although in the fossil record there are forms that were hypercarnivores, feeding almost exclusively on meat, as well as hypocarnivorous forms where meat is only a minor component of their diet.

Subfamily Hesperocyoninae

The extinct subfamily Hesperocyoninae is considered the most primitive of the canids. Although members of the subfamily have the canid type of auditory bulla, there are some genera that lack a bony tubular external auditory meatus. Recent analysis of this subfamily (Wang, 1994) did not support a monophyly for the subfamily, so even though it is still used, it is most likely that it is a polyphyletic group with multiple clades that only share primitive characters of the family.

Hesperocyon Scott, 1890

Systematics and Evolution The type of *Hesperocyon* is *H. gregarius*. The locality given by Cope (1873a:3) in the type description is "abundant in the *Oreodon* beds of the Miocene formation of Colorado." This has subsequently been interpreted as a reference to the Orellan rocks of the White River Formation of northeastern Colorado. Two other species are presently included in the genus, "*H.*" *coloradensis* and "*H.*" *pavidus* (Wang, 1994).

When Cope (1873) described the type of *Hesperocyon*, he originally placed it in the genus *Canis* as *C. gregarius*. The species was placed in the new genus *Hesperocyon* by Scott (1890). Cope did not designate a type specimen but merely stated that there were abundant specimens of *Canis gregarius* in his collections. Other species traditionally included in the genus were subsequently transferred to other genera by Wang (1994). *Hesperocyon temnodon* is now in *Mesocyon* and *H. wilsoni* in *Prohesperocyon*. The distinctions between the three remaining genera are often made on the basis of size and minor differences in morphology. Given how common sexual dimorphism is in the canids, it may be that this, as well as stage of ontogenetic development, may account for the features used to distinguish the different species.

Distinctive Characters *Hesperocyon* is a small animal with an elongate body and long tail. The manus is subdigitigrade, and it appears to have had retractable claws, a character not present in most canids. The protocone of the upper fourth premolar (carnassial) is reduced in size and medially positioned on the tooth. The lower carnassial (first molar) has an elongated shearing blade. Fossil bacula have been found with some articulated specimens. Fortunately a number of articulated skeletons of *Hesperocyon*, primarily *H. gregarius*, have been found, and mounted skeletons of the species are on display in museums (Fig. 5.24).

Stratigraphic and Geographic Distribution *Hesperocyon gregarius* ranges from the middle Duchesnean to the end of the Whitneyan, "*H.*" *parvidus* ranges from the beginning of the Orellan into the mid-Arikareean, and "*H.*" *coloradensis* is restricted to the Orellan. *Hesperocyon gregarius* is present in the Peanut Peak Member of the Chadron and both members of the Brule formations; a second species, "*H.*" *parvidus*, has been reported from the Poleslide Member of the Brule Formation. The genus is also known from California, Wyoming, Colorado, Nebraska, Montana, North Dakota, Saskatchewan, and Oregon, as well as in the southeastern United States in Georgia.

Natural History and Paleoecology *Hesperocyon gregarius* has the best fossil record of any member of the subfamily because of the large number of specimens, including articulated skeletons, from the White River Group. The co-occurrence of *Hesperocyon* and eomyid rodents has been interpreted by past workers as indicative of a predator–prey relationship (Cope, 1874; Clark, Beerbower, and Kietzke, 1967). Although coprolites are common in the White River, none can be attributed unequivocally to *Hesperocyon* to confirm what comprised its primary prey. In contrast, a premolar of *Hesperocyon* was identified in a coprolite along with snake remains that would have been produced by a larger predator in the ecosystem (Parris and Holman, 1978), and LaGarry (2004) reported remains of *Hesperocyon* in coprolites from the Orella Member of the Brule Formation in northwestern Nebraska, indicating it was prey as well as predator. Clark, Beerbower, and Kietzke (1967) proposed that the preferred habitat for *Hesperocyon* was the river-border forest environment.

Mesocyon Scott, 1890

Systematics and Evolution The type species is *M. coryphaeus* from John Day Formation in Oregon. Traditionally the genus has included 10 species, but Wang (1994) transferred three species to *Parenhydrocyon* and one to *Cynodesmus*, restricting *Mesocyon* to three species: the type, *M. brachyops*, and "*M.*" *temnodon*. The genus as defined by Wang may be paraphyletic because the three species are linked by shared primitive characteristics and not by any apomorphies.

Distinctive Characters *Mesocyon* is a primitive midsize canid. The genus is distinguished from *Hesperocyon* and *Parenhydrocyon* by having a long, robust, ventrally directed paroccipital process. Primitive characters shared by the three species in the genus include a wide upper second molar and the absence of robust premolars. In the derived species there is a round fossa on the supraoccipital (Munthe, 1998).

Stratigraphic and Geographic Distribution The genus is restricted to the Oligocene, appearing in the Orellan and becoming extinct in the late early Arikareean, with some tentative records reported from the early Miocene (late Arikareean). The genus has been found in faunas in South Dakota, California, Oregon, Wyoming, Colorado, Nebraska, and North Dakota.

Natural History and Paleoecology Given the taxonomic uncertainty of this genus, nothing has been published on its paleoecology.

Ectopocynus Wang, 1994

Systematics and Evolution The type species of *Ectopocynus* is *E. simplicidens* from the Runningwater (formerly Marsland) Formation of western Nebraska, Middle Hemingfordian, early Miocene. The genus includes two other species, *E. antiguus* and *E. intermedius*. It should be noted that while three species are recognized, they are represented by a total of six specimens (Munthe, 1998), so a better understanding of the taxonomy of this genus and its relationships to other canids cannot be fully evaluated until more specimens become available. Wang (1994) did note some dental similarities between *Ectopocynus simplicidens* and *Hesperocyon gregarius*.

Distinctive Characters The lower premolars are short, blunt, and robust, and the accessory and cingular cusps are extremely reduced or absent. The lower first premolar is absent. The metaconid on the lower first and second molars is also reduced.

Stratigraphic and Geographic Distribution The genus appears in the Whitneyan and became extinct at the end of the early Hemingfordian. *Ectopocynus antiquus* is known from the Poleslide Member of the Brule Formation (Whitneyan) in the Big Badlands. It is also known from Wyoming.

Natural History and Paleoecology This is a poorly known taxon represented by only six specimens, thus limiting inferences about both phylogenetic relationships and paleoecology

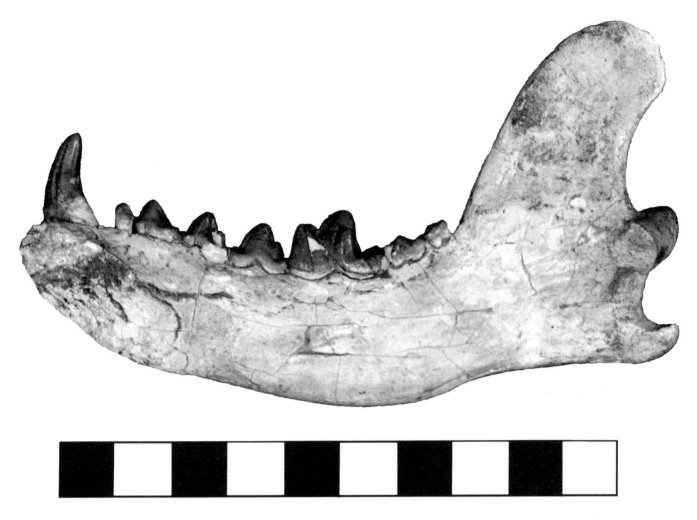

5.25. *Sunkahetanka geringensis,* YPM PU 13602. Right mandible, left lateral view. Scale in centimeters. Photo by the authors. Courtesy of the Division of Vertebrate Paleontology, YPM PU 13602, Peabody Museum of Natural History, Yale University, New Haven, Connecticut, U.S.A.

Cynodesmus Scott, 1893

Systematics and Evolution The type species of *Cynodesmus* is *C. thooides* from the Deep Creek Beds of Montana, considered to be early Arikareean in age. In the past the genus has included species that are now placed in all three subfamilies of the Canidae. Over 14 species have been described for this genus, but currently only the type species and *C. martini* are considered valid (Wang, 1994).

Distinctive Characters As defined by Wang (1994), the genus can be distinguished from other members of the Hesperocyoninae by the strong paroccipital process on the skull with a posterior keel and a broadened upper second molar.

Stratigraphic and Geographic Distribution The stratigraphic range is from the Whitneyan to the end of the early Arikareean of western Nebraska and South Dakota, and the genus is present in the earliest Arikareean of western Montana. Specimens referred to the genotypic species are known from the Big Badlands, so all records of this taxon are Whitneyan in age. It is also known from Nebraska and Montana.

Sunkahetanka Macdonald, 1963

In Lakota, *Sunka* means "dog," *he* means "tooth," and *tanka* means "large." Thus the name means "large-toothed dog."

Systematics and Evolution The type species of *Sunkahetanka* is *S. geringensis* (Barbour and Schultz, 1935) from the Gering Formation in Morrill County, Nebraska. The type specimen is USNM 1092, including the skull, mandible, and skeleton. There is only one species assigned to this genus. *Sunkahetanka* was proposed by Macdonald (1963:214) for *Mesocyon geringensis* as a "derivative of the *Mesocyon* line." Wang (1994) considered *Sunkahetanka* as transitional between the more primitive *Mesocyon* and *Cynodesmus* and the more advanced *Philotrox* and *Enhydrocyon.*

Distinctive Characters *Sunkahetanka* is distinctive from *Cynodesmus* by its more massive imbricated premolars and the reduced metaconid on the lower molars. The skull is also heavier and the dentition more robust than *Mesocyon* or *Cynodesmus* (Fig. 5.25) (Wang, 1994).

Stratigraphic and Geographic Distribution *Sunkahetanka* is restricted to the topmost part of the Brule Formation

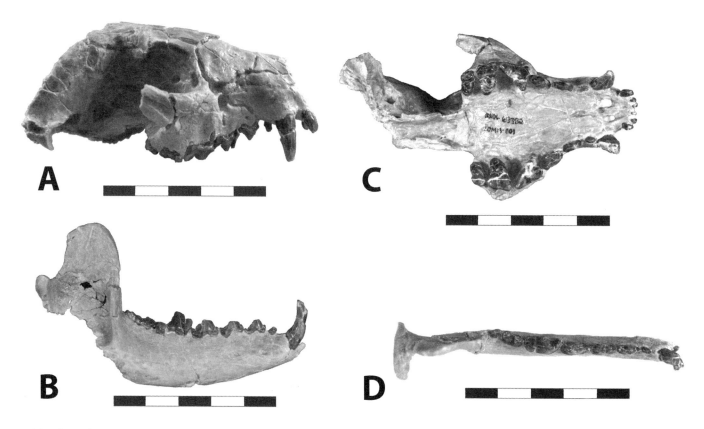

5.26. *Osbornodon* sp., BADL 63382. (A) Skull, right lateral view. (B) Right mandible, right lateral view. (C) Skull, occlusal view. (D) Right mandible, occlusal view. Scales in centimeters. Photos by the authors of specimens from Badlands National Park. Specimens are the property of the U.S. government.

in western Nebraska and to the Sharps Formation of western South Dakota (early Arikareean). At Badlands National Park, *Sunkahetanka* was found in Cedar Pass in the Sharps Formation (Wang, 1994).

Osbornodon Wang, 1994

Systematics and Evolution The type species of *Osbornodon* is *O. fricki* from New Mexico and includes species that originally were placed in *Cynodesmus*, *Mesocyon*, *Tomarctus*, and *Brachyrhynchocyon*. Currently the genus includes five species.

Distinctive Characters *Osbornodon sesnoni* is a large canid. Many features of the skull of *Osbornodon* are similar to those seen in later members of the Caninae, with a long and slender rostrum, slender premolars, quadrate molars, and basined talonids on the lower molars (Fig. 5.26). It is distinguished from the other hesperocyonines by having more quadrate upper molars. The upper and lower second molars are enlarged, and the talonids of the lower first and second molars are basined (Wang, 1994).

Stratigraphic and Geographic Distribution As defined by Wang (1994), this genus has the longest stratigraphic range of the Hesperocyoninae, first appearing in the Orellan and becoming extinct at the end of the early Barstovian. *Osbornodon sesnoni* has been recovered from the Poleslide Member

of the Brule Formation and is Whitneyan in age but is only known from a small number of specimens. *Osbornodon renjiei* occurs in the Poleslide Member of the Brule Formation. The genus is also known from faunas in Florida, Texas, Wyoming, Nebraska, and North Dakota.

Subfamily Borophaginae

The borophagine canids comprise one of the three subfamilies within the Canidae (Wang, Tedford, and Taylor, 1999). The subfamily is closely related to the subfamily Caninae, and the two are distinguished from the Hesperocyoninae by the presence of a basined, bicuspid talonid on the lower first molar. This feature seems to have allowed the borophagines to avoid competition with the hypercarnivorous hesperocyonines by allowing them to move into a hypocarnivorous niche soon after the two lineages diverged. Many of the early primitive borophagines had a dentition reminiscent of the dentition of members of the Procyonidae and were initially considered to be members of that family. Borophagines filled the ecological niches today filled by canids, procyonids, and hyaenids. Many genera of borophagines exploited the hypocarnivorous niche in the Tertiary, but later members became specialized as bone crushers.

The subfamily is known only from North America and first appears in the early Oligocene (Orellan) with the

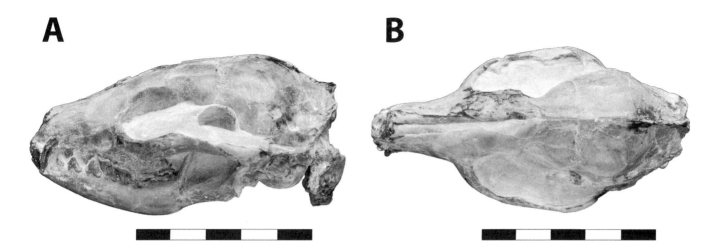

5.27. *Archaeocyon leptodus*, BADL 61910/SDSM 10501. (A) Skull and jaws, left lateral view. (B) Skull, dorsal view. Scales in centimeters. Photos by Xiaoming Wang of specimen housed at the Museum of Geology, South Dakota School of Mines and Technology, Rapid City, South Dakota, U.S.A. Specimen is the property of the U.S. government.

appearance of *Otarocyon macdonaldi*. The borophagines are the dominant canids in the late Tertiary until they became extinct in the Pliocene (Blancan). During this time interval they are among the best represented carnivorans present in a fauna. Among the different lineages of borophagines, the hypercarnivorous forms are better known than the hypocarnivorous, with many species of the latter known from only a handful of specimens. Although many of the genera have long, unbroken records during the Tertiary, there are taxa with significant gaps in their fossil record, such as the lack in the record of *Otarocyon* in the Whitneyan to connect the earlier *O. macdonaldi* (Orellan) and later *O. cooki* (early Arikareean). The eventual extinction of the borophagines is attributed to the appearance of felids, large mustelids, and giant ursids that came from Eurasia, first appearing in North America in the Hemphillian; they may have been ecological competitors. After the appearance of these immigrants, the only surviving borophagines are those adapted to a durophagous (bone-cracking) diet.

Archaeocyon Wang, Tedford, and Taylor, 1999

Systematics and Evolution The type species of *Archaeocyon* is *A. pavidus*, a species originally assigned to *Pseudocynodictis* from the Sespe Formation (late Whitneyan or early Arikareean), Ventura County, California. Two other species, *A. leptodus* and *A. falkenbachi*, are also recognized. *Archaeocyon* is a basal taxon near the separation of the Caninae–Borophaginae lineage from the hesperocyonines, as it has all of the characters that distinguish the two groups. Because of its basal position, *Archaeocyon*, like *Oxetocyon* and *Otarocyon*, occupies an ambiguous position in terms of its relationships to the other borophagines because it lacks

derived characters of its own and in many ways more closely resembles *Hesperocyon*.

Distinctive Characters As a primitive member of the Caninae–Borophaginae lineage, *Archaeocyon* has the distinctive features of the group, including a weak parastyle and a lingual cingulum extending onto the anterior edge of the protocone and surrounding it on the on the upper first molar; the lower first and second molars have basined talonids. However, it lacks the derived characters seen in the Caninae, such as a slender, horizontal ramus of the mandible, narrow and elongate premolars separated by diastema, premolars with the posterior accessory cusps reduced or absent, and a reduced protocone on the upper fourth premolar. Compared to later borophagines, it is primitive in that the auditory bulla is not enlarged, instead having a posteriorly oriented paroccipital process that only rarely contacts the bulla; it has a posteriorly restricted hypocone on the upper first molar; and on the upper second molar, the lingual cingulum does not connect to the metaconule (Fig. 5.27) (Wang, Tedford, and Taylor, 1999).

Stratigraphic and Geographic Distribution *Archaeocyon pavidus* has been reported from the Poleslide Member of the Brule Formation (Whitneyan) in southwestern South Dakota (Jackson and Shannon counties), and *A. leptodus* is reported from the east side of Cedar Pass from the base of the Sharps Formation (early Arikareean), Jackson County, South Dakota (Wang, Tedford, and Taylor, 1999). The genus is known from the Whitneyan of South Dakota, Nebraska, and Wyoming; the late Whitneyan or early Arikareean of California; the early Arikareean (Sharps Formation) of South Dakota, North Dakota, Nebraska, Wyoming, Montana, and Oregon; and the medial Arikareean of Wyoming (Wang, Tedford, and Taylor, 1999).

5.28. *Oxetocyon cuspidatus,* BADL 11010/SDSM 7330. Left maxillary fragment with M1–M2. Photo by Edward Welsh with some modification of image of specimen from Badlands National Park. Specimen is housed at the Museum of Geology, South Dakota School of Mines and Technology, Rapid City, South Dakota, U.S.A. Specimen is the property of the U.S. government.

Natural History and Paleoecology The primitive dentition of *Archaeocyon*, which lacks specializations for slicing meat, suggests a tendency to hypocarnivory in contrast to the dentition seen in *Hesperocyon*, which is more hypercarnivorous. The diet of a hypocarnivore includes less than 30 percent meat. This suggests that *Archaeocyon*, like many early borophagines, was more omnivorous in their diets, perhaps consuming a larger percentage of vegetation in its diet, thus avoiding ecological competition with its contemporary, *Hesperocyon*.

Oxetocyon Green, 1954

Systematics and Evolution *Oxetocyon* is known from a single species, *O. cuspidatus*. The type was collected from the Poleslide Member of the Brule Formation in the Big Badlands. It represents the first of many genera within the subfamily that evolved hypocarnivory.

Distinctive Characters *Oxetocyon* was a small canid with low crowned (bunodont) teeth. The lingual cingulum of the upper first molar has three distinct cusps–thus the species name. The molar is enlarged and squared so that the tooth is symmetrical with a transverse cleavage in the middle of the crown (Fig. 5.28). The modification of the tooth into a more crushing mode for processing food suggests that vegetation may have been predominant in its diet.

Stratigraphic and Geographic Distribution It is only known from the Poleslide Member of the Brule Formation (Whitneyan) in the Big Badlands but has been reported from the Whitneyan and early Arikareean in Nebraska (Tanner, 1973; Wang, Tedford, and Taylor, 1999).

Natural History and Paleoecology *Oxetocyon* is one of the four earliest borophagine canids in North America found in the Big Badlands. Although the borophagines are often popularly characterized as the bone-crushing canids with hyaenalike dentition, this is only true for a few of the later genera like *Borophagus*. Remains of this genus are extremely rare, and little can be deduced regarding its phylogenetic relationships or paleoecology.

Otarocyon Wang, Tedford, and Taylor, 1999

Systematics and Evolution The type species of *Otarocyon* is *O. cooki*, a species originally placed in *Cynodesmus* from the Sharps Formation (early Arikareean), Shannon County, South Dakota. A second species, *O. macdonaldi*, is also recognized. This genus appears abruptly in the Orellan without any apparent predecessor.

Distinctive Characters Compared to *Archaeocyon*, *Otarocyon* is a highly derived taxon that is defined by a number of anatomical features. The auditory bulla is hypertrophied and the rostrum is shortened. The braincase is broadened, with a short temporal fossa defined by a pair of temporal crests. The premolars are single cusped and short with tall crowns. The protocone on the upper fourth premolar is enlarged, and the upper first molar has an elevated cingulum (Wang, Tedford, and Taylor, 1999).

Stratigraphic and Geographic Distribution *Otarocyon macdonaldi* is known from the Scenic Member of the Brule Formation (Orellan), Pennington County, South Dakota, and from the Toston Formation (Orellan?) of Montana (Wang, Tedford, and Taylor, 1999). The genus is known from the Orellan of South Dakota and Montana, the early Arikareean of South Dakota, and the early to medial Arikareean of Wyoming.

Natural History and Paleoecology The closest living analog to *Otarocyon* is the living fennec fox, *Vulpes zerda*, which inhabits the deserts of North Africa and the Arabian Peninsula. Both are small canids and share a number of anatomical similarities, such as the expanded braincase, short nasal processes of the frontal, parasagittal temporal crests, and an enlarged auditory bulla. There is a hiatus of 25 Ma between the two taxa, and the fennec has many derived features that clearly place it with other living foxes, so the characters must have been independently derived. However, this remarkable convergence permits some inferences about the soft anatomy of *Otarocyon*. The enlarged bulla suggests

1. Photographs of (A) the Pierre Shale north of the Sage Creek Campground and (B) Fox Hills Formation north of Wall, South Dakota. Photos by the authors.

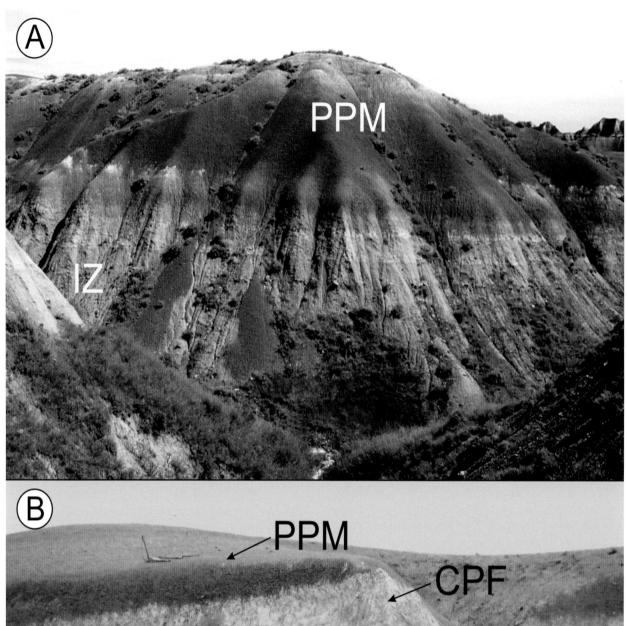

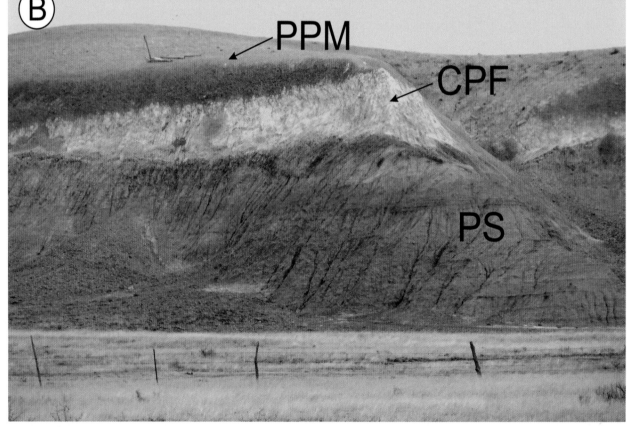

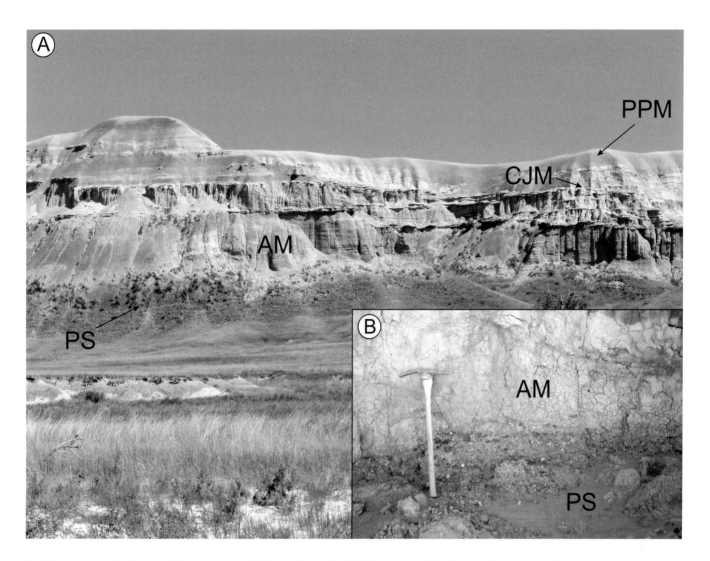

3. (A) Photograph of the Ahearn (AM), Crazy Johnson (CJM), and Peanut Peak (PPM) members of the Chadron Formation in Indian Creek northwest of Sheep Mountain Table. (B) Close-up of the unconformable contact between the Cretaceous Pierre Shale (PS) and the late Eocene Ahearn Member (AM) of the Chadron Formation. Pick for scale is 87 cm long. Photos by the authors.

2. (A) Photograph of the Interior Zone (IZ) and Peanut Peak Member (PPM) of the Chadron Formation in Dillon Pass. (B) Photograph of the channel sandstone facies of the Chamberlain Pass Formation (CPF) underlain by Pierre Shale (PS), which has been modified by the Interior Zone of weathering and overlain by the Peanut Peak Member (PPM) of the Chadron Formation west of Interior, South Dakota. Photos by the authors.

4. Photo of complete White River Sequence, upper Conata Basin. HBm = Hay Butte marker; SPm = Saddle Pass marker; CPwl = Cedar Pass white layer. The rocks here dip 4 degrees to the southwest, away from the viewer. Photo by the authors.

5. Middle and upper Poleslide beds, the Rockyford Ash, and lower Sharps Formation at Cedar Butte. The view is to the west from the overlook at the southwest side of Sheep Mountain Table. Photo by the authors.

6. Photographs of various paleosol features in outcrop and hand sample. (A) Variegated striping is commonly seen with paleosols. (B) Horizons of soil-formed calcium carbonate (C) stand out in relief. Person for scale is 1.8 m tall. (C) Fossil bone locked inside of a calcium carbonate soil nodule. (D) Ped structures. (E) Slickensided surface (S). Coin for scale is 1.8 mm wide. (F) Permineralized roots. Coin for scale is 2 cm wide. Photos by the authors.

7. Additional paleosol features. (A) In situ silicified tree stump along an ancient stream channel. Note breakage suggesting water flow to the left. (B) Clay-filled root traces. Scale in centimeters. (C) Drab-haloed root traces. Scale in centimeters. (D) Burrow structure. (E) Dung balls. Scale in millimeters. (F) Bee larval cell. Coin is 1.8 mm wide. Photos by the authors.

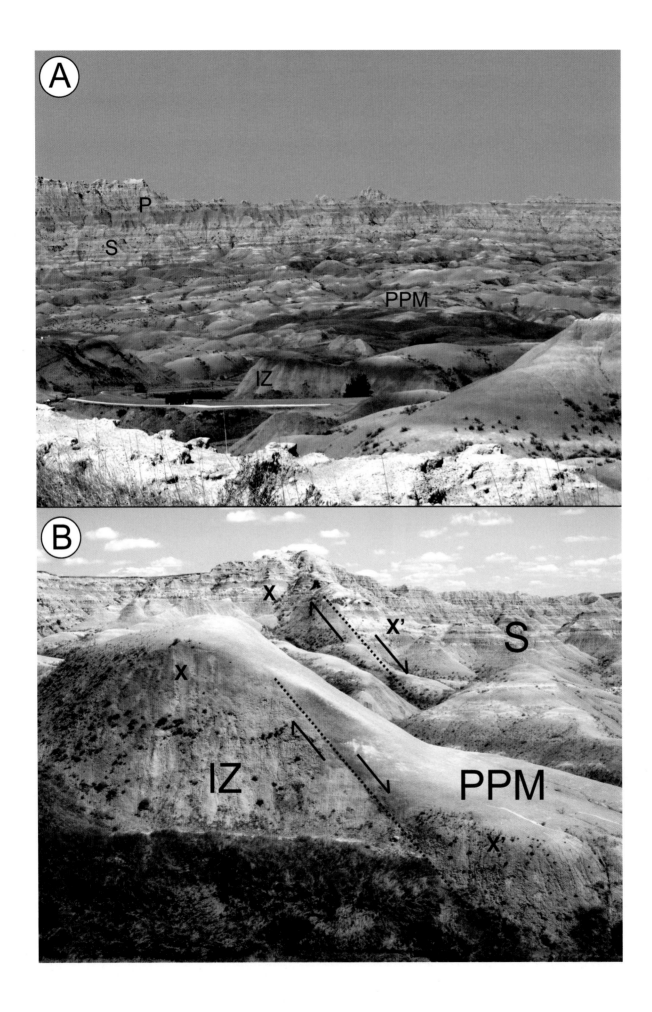

9. Photograph of the original discovery of fossil bones in Conata Basin that started the Big Pig Dig project. Note the greenish-colored layer from which the bones protrude. Photograph courtesy of Jim Carney.

8. Photographs of Dillon Pass looking down the axis of the Sage Arch (A) with strata dipping down to the right and left from the center of the photograph, and the Dillon Pass Fault (B) showing directions of displacement and offset beds (X–X') on either side of the fault. IZ = Interior Zone (Plate 2); PPM = Peanut Peak Member of the Chadron Formation; S, P = Scenic and Poleslide members of the Brule Formation. Photos by the authors.

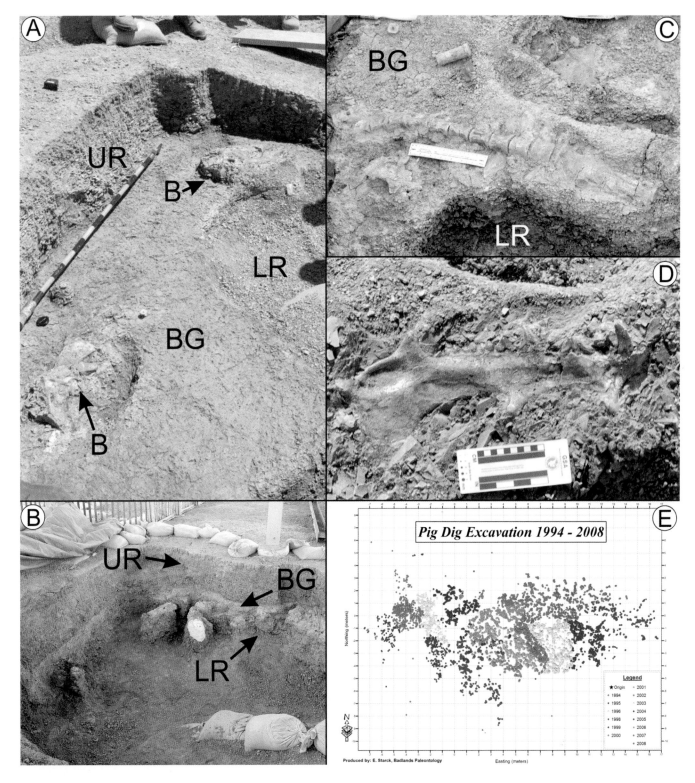

10. Photographs of the Big Pig Dig during excavation. (A) Several bones (B) resting within the bone-bearing green layer (BG). LR and UR = lower and upper red mudstone layers, respectively. Scale rod in decimeters. (B) View of the Big Pig Dig excavation in cross section. Various bones within the bone bearing green layer (BG), some of which have been covered with plaster for removal from the site. LR and UR = lower and upper red mudstone layers, respectively. Photograph courtesy of Diane Hargreaves.com. (C) Articulated backbone and pelvis within the bone-bearing green layer. (D) Lower jaw of an *Archaeotherium*. (E) Map of fossils recovered from the Big Pig Dig over a 14-year period. Photo (E) is courtesy of the National Park Service. Map is the property of the U.S. government. Photos (A), (C), and (D) are by the authors.

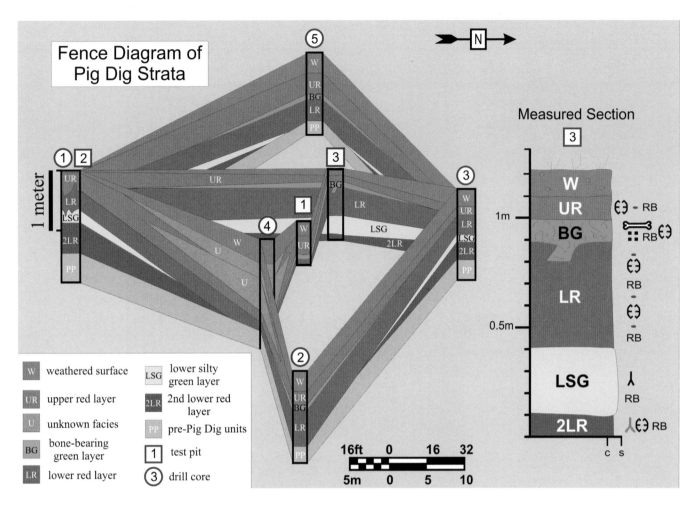

11. Fence diagram of units within the Big Pig Dig and a representative measured section. Modified from Terry (1996b). For a legend of paleosol symbols, see fig. 3.6.

12. Interpretive reconstruction of the Big Pig Dig by Robert Hynes. Property of the U.S. government.

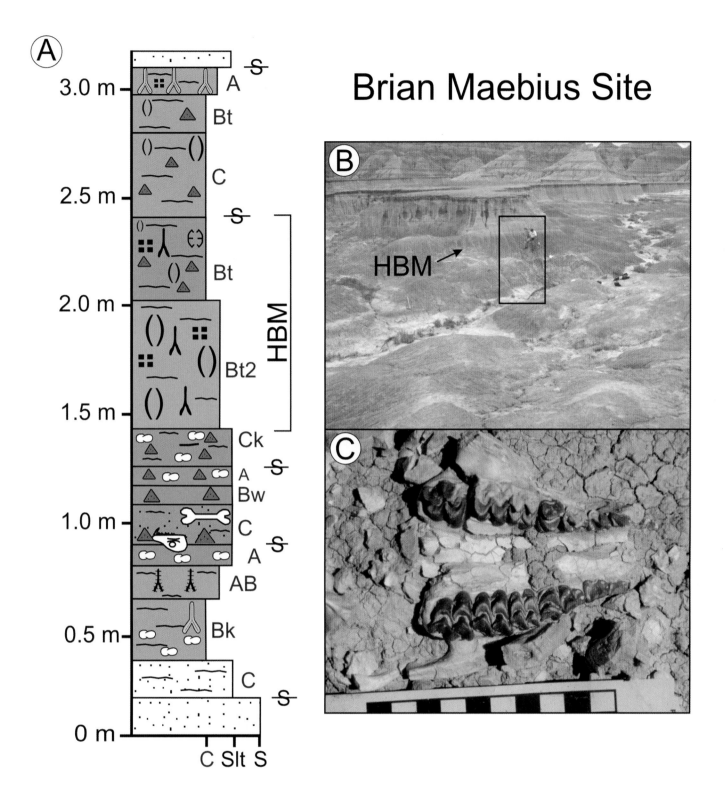

13. Measured section (A) through the Brian Maebius site, with photographs of the site (B) and fossil bone (C). The square in (B) denotes the location of the measured section (A). HBM = Hay Butte marker bed. For a legend of paleosol symbols, see fig. 3.6. Modified from Factor (2002).

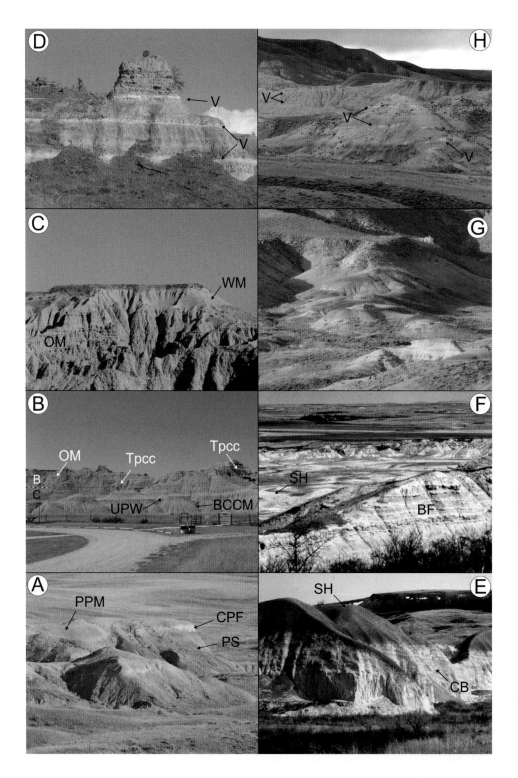

14. Photographs of the White River Sequence across the region. Note that photographs are positioned in relative geologic order from bottom to top in each region. A–D, northwest Nebraska; E–F, southwest North Dakota; G–H, east-central Wyoming. (A) Outcrops near Toadstool Geologic Park showing the Interior Zone, which has modified the Cretaceous black Pierre Shale (PS) into a reddish-purple color, channel sandstone of the Chamberlain Pass Formation (CPF), and the Peanut Peak Member (PPM) of the Chadron Formation. (B) Exposures of the Big Cottonwood Creek Member (BCCM) of the Chadron Formation, the prominent volcanic ash of the Upper Purplish White layer (UPW), the Orella Member of the Brule Formation, and Toadstool Park Channel Complex (Tpcc). The Eocene–Oligocene boundary is several meters above the UPW. (C) The Orella (OM) and Whitney (WM) members of the Brule Formation. (D) The informal Brown Siltstone member of the Brule Formation just west of Fort Robinson State Park in northwest Nebraska. V = volcanic ash layers of the Nonpareil Ash Zone. (E) The Chalky Buttes (CB) and South Heart (SH) members of the Chadron Formation in the Little Badlands. (F) The South Heart Member (SH) of the Chadron Formation overlain by the Brule Formation (BF) in the Little Badlands. (G, H) Exposures of the undifferentiated late Eocene White River Formation at Flagstaff Rim southwest of Casper, Wyoming. V = volcanic ash. Note the rounded haystack morphology in (G), which is similar to the Peanut Peak Member, and the more cliff-forming strata in (H), which is similar to the Big Cottonwood Creek Member. Photos by the authors.

15. Eocene Chadron Formation. Mural by Laura Cunningham illustrating the White River Badlands over 37 Ma, during the deposition of the Chadron Formation. Fossil recreations depicted in the mural include the titanothere *Megacerops* and the oreodont *Merycoidodon*. Ancient forests and a warm subtropical environment, characteristic of the Late Eocene, are represented. Permission granted by Laura Cunningham and the Badlands Natural History Association.

16. Oligocene Brule Formation. Mural by Laura Cunningham illustrating the White River Badlands about 33 Ma. Fossil recreations depicted in the mural include a saber-toothed cat from the family Nimravidae chasing three-toed horses known as *Mesohippus*. Open woodlands developing into an open grassland are represented in the mural. Permission granted by Laura Cunningham and the Badlands Natural History Association.

that like fennecs, *Otarocyon* was sensitive to low-frequency sounds and probably had a large external ear. This implies that it may have lived in an open environment. The presence of at least a patchy network of open environments in the Orellan has been proposed by Retallack (1983b) on the basis of the paleosols.

Cynarctoides McGrew, 1938

Systematics and Evolution The type species of *Cynarctoides* is *C. acridens*, originally placed in the genus *Cynarctus*, from the Upper Harrison Beds (late Arikareean) near Agate, Sioux County, Nebraska. In addition to the type species, six other species are considered valid (Wang, Tedford, and Taylor, 1999). *Cynarctoides*, like many primitive hypocarnivorous borophagines, was originally thought to be a primitive procyonid. The multiple recognized species of the genus form a nearly continuous series of stages that clearly show an evolutionary trend in terms of the peculiar form of hypocarnivory in this genus. This includes retaining a primitive upper fourth premolar that lacks a hypocone and that never has a lingual cingulum. The lower molars tend to have distinctive oblique crests.

Distinctive Characters The many species of *Cynarctoides* are all distinguished by parasagittal crests and a slender, shallow horizontal ramus of the mandible. More derived species have a narrow rostrum, longer jaws with narrow and long premolars, and lower molars with conical and high-crowned cusps.

Stratigraphic and Geographic Distribution *Cynarctoides lemur* has been reported from the Poleslide Member of the Brule Formation in South Dakota and is the only possible Whitneyan record of the genus (Wang, Tedford, and Taylor, 1999). The genus is best known from the Arikareean and is clearly present in the early Arikareean (Sharps Formation) of South Dakota, Nebraska, and Oregon; the medial or late Arikareean of South Dakota and Florida; the late Arikareean of Colorado, Nebraska, Wyoming, and New Mexico; the early Hemingfordian of Nebraska, Idaho, Texas, and New Mexico; the late Hemingfordian of Nebraska, Wyoming, and New Mexico; and the early Barstovian of Nebraska, New Mexico, and California (Wang, Tedford, and Taylor, 1999).

Natural History and Paleoecology As a hypocarnivorous borophagine, the inference is that the genus probably had an omnivorous diet.

Family Ursidae

Bears have a long history in North America, first appearing in the late Eocene (Chadronian), and are represented in the modern fauna by grizzly (*Ursus arctos*) and black (*Ursus americanus*) bears. Three subfamilies are recognized, each representing a separate evolutionary radiation within the family. The fossil history of bears in North America represents repeated dispersal events from Eurasia, followed by extinction and a refilling of the bear niche by the next appearance of bears. Bears tend to have large canines and incisors, but they often reduce in size or lose the anterior premolars. The most distinctive feature of their dentition is the large quadrate molars that permit them to crush their food. Although the shearing ability of the carnassials in modern bears is greatly reduced, many of the fossil species have a more primitive dentition, including carnassials with well-developed shearing capability. Bears have a generalized postcranial skeleton including a greater ability to rotate the forearm than in other carnivores. They retain five digits on the manus and pes and have large, well-developed, nonretractable claws. The scapula has a prominent secondary spine along the posterior margin.

Parictis Scott, 1893

Systematics and Evolution The type of *Parictis* is *P. primaevus*, from the John Day Beds of Oregon, and seven species have been described. In the White River Group, these include *P. dakotensis*, *P. major*, *P. parvus*, and *P. gilpini*. This large number of species from a restricted period of time and geographic area suggests that the taxonomy of the genus needs to be reexamined; many of these species may become junior synonyms.

It is a member of the Amphicynodontinae, the most primitive subfamily of bears in North America, which includes two other genera, *Allocyon* and *Kolponomos* (only found in Alaska, Oregon, and Washington). Only *Parictis* is known from the White River Group and is the earliest representative of the Ursidae in North America

Distinctive Characters *Parictis* is a small animal, comparable in size to the contemporary dog, *Hesperocyon*. The dentitions of the two taxa are quite different, with the premolars of *Parictis* being more robust and transversely widened. Each of the premolars also has a distinctive cingulum (shelf) around the crown. The premolars lack anterior accessory cusps (Fig. 5.29). Post-Chadronian members of the family have more robust premolars that are transversely widened. Bones of the postcranial skeleton have not been described (Hunt, 1998b).

Stratigraphic and Geographic Distribution The genus first appears in the Chadronian and became extinct in the Orellan, with some specimens suggesting possible survival into the Whitneyan. In the White River Group, *Parictis* has been described from the Peanut Peak Member of the Chadron Formation and Scenic Member of the Brule Formation.

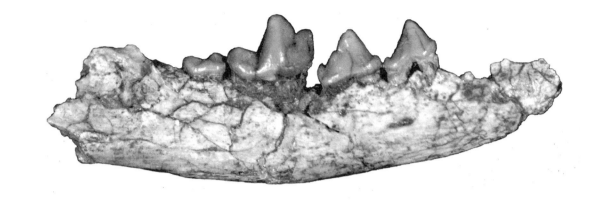

5.29. *Parictis* cf. *parvus.* YPM PU 16265. Right mandible with p3–m2. Scale in centimeters. Photo by the authors. Courtesy of the Division of Vertebrate Paleontology, YPM PU 16265, Peabody Museum of Natural History, Yale University, New Haven, Connecticut, U.S.A.

It is also known from Wyoming, Nebraska, Montana, Colorado, and Oregon.

Family Mustelidae

The Mustelidae includes the living weasels, minks, ferrets, badgers, wolverines, otters, and skunks, although some current taxonomists place skunks in their own family, Mephitidae. The family first appears in North America in the late Eocene (middle Chadronian), approximately 35.5 Ma. The majority of fossil mustelids in North America are the result of multiple dispersal events from the Old World with minimal in situ evolution. Of the 38 recognized fossil mustelid genera in North America, 26 are taxa that originated in the Old World (Baskin, 1998). To understand the fossil history of the North American mustelids, knowledge of the fossil record of the Eurasian members of the family is critical. The first appearance of many of the genera in the Tertiary of North America often provides an important biostratigraphic marker.

Members of the Mustelidae are characterized by having low braincases, wide occiputs, small orbits, anteriorly positioned carnassials, a moderately inflated auditory bulla, and short jaws that permit a powerful bite. The third molars are lost, and the second molars are reduced in size or may be absent as well. They are small- to medium-size carnivores with a tendency to elongated bodies and relatively short legs for their body size.

Mustelavus Clark, in Scott and Jepsen, 1936

Systematics and Evolution The type of *Mustelavus* is *M. priscus*, and the genus is monotypic. This is a primitive mustelid, and the genus has also been placed in the Procyonidae (raccoon family), indicating the similarity of primitive members of the two families. Like other mustelids, the second molars of *Mustelavus* are reduced, whereas in procyonids these teeth tend to be enlarged, so currently it is considered to be a primitive mustelid. It is closely related to *Mustelictis* of the Old World but has a more primitive dentition (Baskin, 1998).

Distinctive Characters The skull of *Mustelavus* has a moderately inflated auditory bulla that contacts the base of the posteriorly projecting paroccipital process and broadly contacts the ventral part of the mastoid process. The dentition is secant, and all four premolars are present. The rounded and conical protocone on the upper carnassial is anteriorly placed, and the tooth has a deep carnassial notch. The upper first molar has a subtriangular outline of the occlusal surface. The upper second molar is small with three roots. The lower carnassial has a tall trigonid and a low-basined talonid (Baskin, 1998).

Stratigraphic and Geographic Distribution *Mustelavus* is known from Peanut Peak Member of the Chadron Formation (late Chadronian) and the Scenic Member of the Brule Formation (Orellan). It is only known from the Big Badlands (Baskin, 1998).

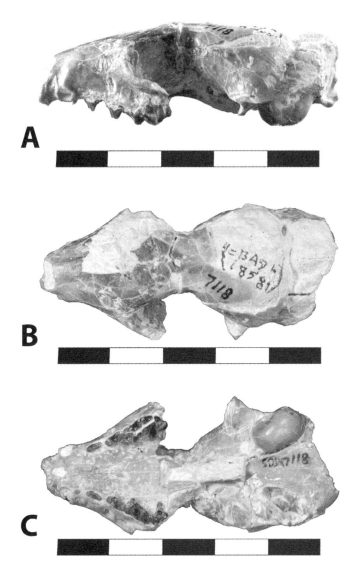

5.30. *Palaeogale* sp., SDSM 7118. (A) Skull, left lateral view. (B) Skull, dorsal view. (C) Skull, occlusal view. Scales in centimeters. Photos by the authors of specimen from the Museum of Geology, South Dakota School of Mines and Technology, Rapid City, South Dakota, U.S.A.

Palaeogale (=*Bunaelurus* Cope, 1873) von Meyer, 1846

Systematics and Evolution The type species is *P. minuta* from Europe, and three species have been described in North America, *P. sectoria*, *P. dorothiae*, and *P. sanguinarius*, from the middle Chadronian to early Hemingfordian. *Palaeogale* is an enigmatic carnivore. Although it is currently placed with the mustelids, its relationships within the carnivora is still under debate. Relationships to an extinct felinelike family, Viverravidae, have been proposed (Hunt, 1974, 1989; Flynn and Galiano, 1982).

Distinctive Characters *Palaeogale* is a small animal. The auditory bulla is complete, inflated, and undivided. The upper fourth premolar has a parastyle and carnassial notch.

5.31. *Archaeotherium* sp., BADL 34255/SDSM 67299. Articulated manus. Specimen was collected from the Big Pig Dig at Badlands National Park. Photo by the authors. Specimen is the property of the U.S. government. Specimen is housed at the Museum of Geology, South Dakota School of Mines and Technology, Rapid City, South Dakota, U.S.A.

The upper second molar is vestigial to absent, and it lacks an upper third molar (Fig. 5.30A, B) (Baskin, 1998).

Stratigraphic and Geographic Distribution The genus has a long stratigraphic history, first appearing in the middle Chadronian and becoming extinct at the end of the early Hemingfordian. There is a possible record of *P. sectoria* from the Scenic Member of the Brule Formation (Baskin, 1998).

It is also known from Wyoming, Colorado, Montana, and Nebraska, and possibly Texas and Florida.

Order Artiodactyla

Members of the Order Artiodactyla, the even-toed or paraxonic ungulates in which the axis of the hand and foot passes between the third and fourth digits, comprise the most common hooved mammals in terms of taxonomic diversity and number of individuals in the modern fauna. A primary characteristic of the order is that the plane of symmetry of the front and hind foot passes between digits three and four, which are the largest digits (Fig. 5.31). Although being even toed is considered a primary characteristic of the order, some primitive forms may have a first digit, but it is usually vestigial, so it is the axis of symmetry that is the primary character. Digits 2 and 5 may be present and functional in some groups, such as peccaries, pigs, and hippopotamus, but in most groups, they become reduced in size, are vestigial, and are often lost early in the evolutionary history of these groups. In many forms, the third and fourth metapodials fuse in adults to become a single bone, often referred to as the cannon bone. Another primary skeletal character shared by all members of the order is the structure of the astragalus, which is a double pulley with both the proximal and distal ends having a similar shape. The calcaneum has a prominent convex articular surface for the distal fibula. Artiodactyls are primarily herbivores, although some, such as pigs, may be omnivorous, including scavenging; some extinct groups, such as entelodonts, are thought to have been scavengers as well. The order includes both browsers and grazers, and these differences in diet are often reflected in the dentition, with the former having brachyodont (low crowned) teeth while the latter are hypsodont (high crowned). In some advanced forms, the roots of the cheek teeth may be lost and the teeth are ever-growing, or hypselenodont. There are two primary types of dentition: a bunodont form with distinct cusps, such as is seen in peccaries and entelodonts (and humans), and a form referred to as selenodont in which the cusps form a semicircular or crescentic shape, and which is present in some White River taxa like the oreodonts and camelids, as well as modern deer, antelopes, and bison. Premolars are generally smaller than the molars and only rarely become molariform except occasionally the fourth premolar. The ulna and radius remain separate in some forms but may fuse in others, with the radius being the primary support of the body weight and the ulna becoming reduced to the proximal and distal ends. The femur never has a third trochanter. In primitive forms the fibula is complete, but in many groups it becomes reduced, losing the shaft so only the proximal and distal ends remain. The distal fibula articulates with the calcaneum and fits into a distinct facet in the distal tibia to serve as a lock to assist holding the astragalus in place. Although in most forms the ungual phalanx is modified into a hoof, there are some exceptions, as in the agriochoerids, in which the ungual is a claw.

Suborder Suina
Superfamily Entelodontoidea
Family Entelodontidae

Entelodonts are an extinct group of artiodactyls phylogenetically related to the living pigs and peccaries, as well as the hippopotamus. The family first appears in the late Eocene (Duchesnean) and became extinct in the early Miocene (late Hemingfordian). A possible earlier form, *Brachyhyops*, is known, but its phylogenetic relationships are unclear. The family was never diverse in North America, with only four known genera. One of these, *Archaeotherium*, is known from the White River Group and a second, *Daeodon*, is known from the Arikareean.

Entelodonts are distinguished by the proportionately large head for their body size. In *Archaeotherium* it is about 27 percent of the head–body length (Joeckel, 1990). The rostrum of the skull is greatly elongated and exceeds the postorbital length of the skull, and the braincase is small relative to the size of the animal. The zygomatic arches have expanded flanges that extend laterally and ventrally. The expansion of the squamosals gives the occiput a broad appearance. The brain cavity is small in comparison to the overall size of the skull, but the olfactory lobes were large, suggesting that smell was important to the animal. The frontal and parietal sinuses are enlarged and extend over the brain cavity. On the mandible there are two mandibular tubercles on the ventrolateral edge of the horizontal ramus (Fig. 5.32B) (Joeckel, 1990).

All three upper and lower incisors are pointed and round to subtriangular in cross section and increase in size laterally. The canines are large with serrations on the posterolateral and posteromedial margins in unworn teeth, but these are commonly worn off in older individuals. The cheek teeth are bunodont; all premolars and molars are present; and the premolars are large. The premolars are separated by diastema reflecting the elongation of the rostrum. The upper molars are brachyodont with six cusps, except for the third molar, which has a reduced posterior margin. The lower molars have four cusps, are square in outline, and increase in size posteriorly (Fig. 5.32B) (Effinger, 1998).

The neck is short and massive, with well-developed processes for the attachment of the neck musculature and ligaments, as would be expected given the disproportionately large head of the animal. The limb bones are relatively slender for their body size. The humerus is long and massive.

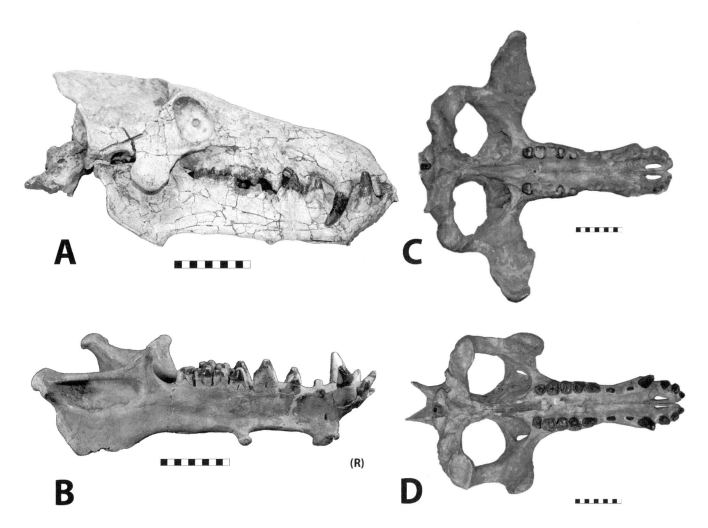

5.32. *Archaeotherium* sp. (A) Skull and jaws, right lateral view, UCM 19161. (B) Jaws, right lateral view, DMNS EPV 400. (C) Skull (male), occlusal view, DMNS 12720. (D) Skull (female), occlusal view, DMNS 1607. Scales in centimeters. Photo A by Katie McComas; photos (B–D) by the authors of specimens from DMNS and UCM. UCM 19161, University of Colorado Museum of Natural History, Boulder, Colorado, U.S.A. DMNS EPV 400, DMNS 12720, and DMNS 1607, Denver Museum of Nature and Science, Denver, Colorado, U.S.A. All rights reserved.

The ulna and radius are co-ossified, and the ulna has a prominent olecranon process. The fibula is separate from the tibia in *Archaeotherium*, but in later forms the two bones become fused. The tibia is shorter than the femur but more massive. Like pigs and peccaries, the entelodonts are functionally didactyl, with large but short third and fourth metapodials and vestigial second and fifth metapodials (Effinger, 1998).

Archaeotherium Leidy, 1850

Systematics and Evolution The type species of *Archaeotherium* is *A. mortoni* from the Oligocene of the Badlands of South Dakota. Currently six species are considered valid. *Megachoerus* Troxell, 1920, was originally considered a distinct genus but is now considered a subgenus of *Archaeotherium* (Foss, 2007).

Distinctive Characters The body mass of *Archaeotherium* is estimated at about 150 kg (Joeckel, 1990). The skull morphology of *Archaeotherium* is extremely variable, reflecting pronounced sexual dimorphism and differences indicating changes with ontogenetic growth, making selection of characters to provide a consistent taxonomy difficult. The jugal flange in *Archaeotherium* is relatively straight and increases in thickness distally so that the lateral margin is swollen. The orbit is completely enclosed with a complete postorbital bar. The coronoid process of the mandible is small, suggesting they had a wide gape when opening the mouth. The mandibular symphysis is elongated, and the two halves are solidly fused together even in young individuals. The angular process of the mandible flares laterally. The upper first and second molars are subquadrate in outline, and the first has a distinct notch on the labial side that separates the tooth into anterior and posterior halves. The carpus retains the trapezium, which is lost in advanced entelodonts. The metatarsals are unfused but tightly articulated. Both the radius and ulna and the tibia and fibula are fused. The limb segment ratio suggests they had a cursorial terrestrial mode of locomotion (Figs. 5.31, 5.32, 5.33).

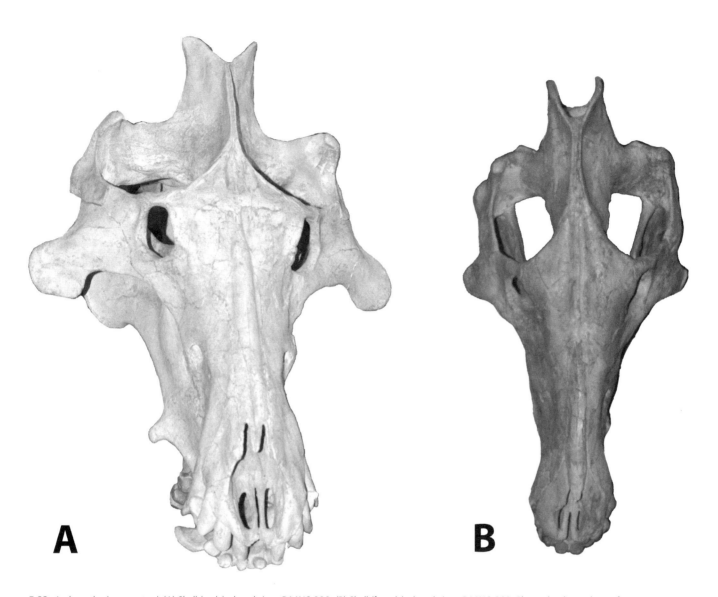

5.33. *Archaeotherium mortoni.* (A) Skull (male), dorsal view, DMNS 900. (B) Skull (female), dorsal view, DMNS 902. Photos by the authors of specimens from DMNS. DMNS 900, and DMNS 902, Denver Museum of Nature and Science, Denver, Colorado, U.S.A.

Stratigraphic and Geographic Distribution This genus has the longest time range of all the North American entelodonts, first appearing in the early Chadronian and becoming extinct in the mid-Arikareean at the Oligocene–Miocene boundary.

Archaeotherium mortoni is known from the Ahearn Member of the Chadron Formation, the Crazy Johnson Member of the Chadron Formation, and the Scenic Member of the Brule Formation in the Big Badlands. *Archaeotherium coarctatum* is known only from the Ahearn Member of the Chadron Formation. *Archaeotherium wanlessi* is reported from the Scenic and Poleslide members of the Brule Formation, while *A. lemleyi* is known only from the Poleslide Member. The genus is also known from faunas in Texas, Wyoming, Nebraska, Montana, Colorado, and Saskatchewan.

Natural History and Paleoecology The presence of grooves worn along the alveolar margins of the lower canines

and upper incisors has been suggested as indicative of rooting by entelodonts by Scott and Jepsen (1940), although Peterson (1909) had previously interpreted them as being scavenging omnivores, and perhaps in some cases active hunters. Evidence of scavenging by entelodonts is a skull of *Merycoidodon* with a fragment of a deciduous lower incisor of *Archaeotherium* embedded in the temporal bone of the skull (Mead, 1998). An analysis of the jaw mechanics and tooth wear by Joeckel (1990) indicates they were capable of ingesting large food items that included both mechanically resistant food items like roots and tubers and relatively hard materials like hard fruits and nuts, and possibly bones. Many specimens have broken teeth with subsequent polish on the remaining surface, possibly caused by biting on bone. It is possible that some coprolites found in the Big Badlands attributed to carnivores may have been produced by *Archaeotherium*. It appears that the preferred habitat of entelodonts was savanna–woodland or grassland.

Our knowledge of the paleobiology of *Archaeotherium* has recently been greatly enhanced by the discovery of a bone bed in Badlands National Park dubbed the Big Pig Dig, named for the numerous *Archaeotherium* fossils found at the site. A total of 19 genera were found at the site, with the minimum number of individuals for the four most common taxa as follows: eight *Subhyracodon*, 29 *Archaeotherium*, 11 *Leptomeryx*, and eight *Mesohippus* (Shelton et al., 2009). In all, a total of 19,290 specimens were collected from the site, and 187 m² were excavated. Both *A. mortoni* and *A. wanlessi* were identified at the Big Pig Dig site, which is in the lower portion of the Scenic Member of the Brule Formation, about 5.5 m above the contact between the Chadron and Brule formations (Miller, 2010). Many of the bones of *Subhyracodon* preserve evidence of having been scavenged by *Archaeotherium*.

Hunt (1990) determined that the younger Agate bone bed represents the short-term, attritional accumulation of animals around a watering hole during drought conditions. The Agate bone bed also preserves evidence of bone processing and trampling (Hunt, 1990). The descendent of *Archaeotherium*, *Daeodon*, is well represented in the Agate deposits, and just as at the Big Pig Dig, many of the gouges and puncture marks found on the bones of the rhinoceros *Menoceras* closely match the teeth of *Daeodon* in size and shape, supporting the idea that entelodonts had a long history as scavengers.

Entelodonts appear to have displayed aggressive behavior. Many skulls display marks and pathologies that appear to have been the result of battles between males either fighting for territory or establishing dominance for mating. It is also possible some of the marks were made during mating (Sinclair, 1921a). Sexual dimorphism is inferred but not fully established. However, differences in the degree of development of the jugal flange, size of the canines, and size of the mandibular tubercles are highly suggestive that sexual dimorphism was present (Effinger, 1998).

Daeodon Cope, 1879 (=*Dinohyus*) Peterson, 1909

Systematics and Evolution The type species of *Daeodon* is *D. shoshonensis* from the John Day Formation of Oregon (Lucas, Emry, and Foss, 1998). This animal is better known in the literature as *Dinohyus* based on *D. hollandi* recovered from bone beds in the Anderson Ranch Formation (late Arikareean) at Agate Fossil Beds National Monument in Nebraska. *Daeodon* is the terminal member of the family. Although five species have been described for this genus, currently only a single species is recognized (Lucas,

Emry, and Foss, 1998). While *Daeodon* is presumed to have evolved from *Archaeotherium*, Lucas, Emry, and Foss (1998) suggested that the genus emigrated from Asia via Beringia in the early Arikareean. The two genera overlap during the early Arikareean.

Distinctive Characters With an estimated body mass of 750 kg, *Daeodon* is the largest member of the family, greatly exceeding in size all other genera (Joeckel, 1990). Despite its large size and a skull that was 25 percent to 35 percent of the head plus body length, the jugal flange is proportionately small for the size of the skull compared to other entelodonts. The mandibular tubercles are small, and the anterior pair is smaller than the posterior. The alveolar border of the premaxillae is relatively short, resulting in a crowding of the incisors. The tibia and fibula are co-ossified. The manus lacks the trapezium, and the fifth metatarsal may be absent in some individuals (Effinger, 1998).

Stratigraphic and Geographic Distribution This monotypic genus first appears in the early early Arikareean and became extinct at the end of the early Hemingfordian. It is only known from the Arikareean Cedar Pass fauna in the Sharps Formation of the Big Badlands (Parris and Green, 1969). By the late Arikareean the genus was widely distributed across North America and has been found in California, Oregon, Texas, South Dakota, Wyoming, Nebraska, New Jersey, Mississippi, Alabama, South Carolina, and Florida.

Natural History and Paleoecology Like *Archaeotherium*, *Daeodon* was predominately a scavenger, and many of the *Menoceras* bones in the Agate bone beds have gouges that closely match the premolars and canines of *Daeodon* in shape and size. These teeth on the skulls and jaws also often have extreme wear, suggesting they were used for crushing or biting bone. The preferred habitat appears to have been savanna–woodland or grasslands.

Family Leptochoeridae

The relationship of this family to other artiodactyls has not been resolved. Traditionally it has been included as a subfamily within the Old World family Dichobunidae, but recent work indicates this creates an artificial paraphyletic family. It is a primitive group with bunodont (cusped) teeth in which the cusps tend to be inflated. The upper molars are triangular and tritubercular, while the lowers are quadritubercular. All of the cheek teeth from the third premolars to the third molars are robust. The first molar is the largest molar, and the molars decrease in size posteriorly. These are small taxa ranging in size from rabbit to about a modern peccary. Two genera are known from the White River Badlands, *Leptochoerus* and *Stibarus*.

The primitive tooth morphology suggests that the leptochoerids were omnivores and probably fed on a wide variety of foods, including fruits, seeds, and leafy vegetation. It is inferred, given their size, that they lived in dense undergrowth of woodlands. Their eventual extinction may have been the result of the loss of this habitat with the decline of forest and the spread of more open habitat in the Oligocene.

Leptochoerus (?=*Nanochoerus*) Leidy, 1856

Systematics and Evolution The type species of *Leptochoerus* is *L. spectabilis* from the Brule Formation of the Badlands of South Dakota. Three other species are considered valid: *L. elegans*, *L. supremus*, and *L. emilyae*.

Distinctive Characters The skull has a short rostrum. The lower molars are more massive than in its contemporary, *Stibarus*. The upper and lower second premolar and lower third premolar tend to be tall, single-cusped, elongate teeth. The ulna and radius are separate, and the ulna is well developed. The manus had four functional digits. The fibula is reduced, and the distal end is fused with the tibia. The pes has four functional digits, and the cuboid and navicular are co-ossified, a feature characteristic of later cervids and bovids.

Stratigraphic and Geographic Distribution The genus first appears in the Chadronian and becomes extinct at the end of the late early Arikareean, approximately 23 Ma. *Leptochoerus spectabilis* is known from both the Ahearn and Crazy Johnson members of the Chadron Formation. Of the other three species described, *L. elegans* is described from the Scenic Member and *L. supremus* from the Poleslide Member of the Brule Formation. The genus is also known from Wyoming, Colorado, Nebraska, Montana, and North Dakota.

Stibarus (=*Menotherium*, Cope, 1873; *Nanochoerus* Macdonald, 1955) Cope, 1873

Systematics and Evolution The type species of *Stibarus* is *S. obtusilobus*, and two other species are known, *S. quadricuspis* and *S. yoderensis*.

Distinctive Characters The skull differs from *Leptochoerus* in having a pinched rostrum. There are two infraorbital foramina in *Stibarus*. As the bunodont lower molars become worn, the cusps develop a subcrescentic shape. The upper and lower second premolars and lower third premolar are low and have three cusps, with the medial cusp being the largest. The upper molars are triangular with a sharp lingual apex. Nothing is known of the postcranial skeleton of this animal.

Stratigraphic and Geographic Distribution *Stibarus* first appears in the early Chadronian and becomes extinct at the end of the Whitneyan. Only *Stibarus obtusilobus*, the type species, is known from the White River Group in the Scenic Member of the Brule Formation.

Superfamily Suoidea
Family Tayassuidae

Members of the Tayassuidae, popularly known as peccaries or javelinas, originated in North America and underwent their primary evolution on this continent with later dispersals into South America in the Pliocene and to Eurasia. They first appear in the late Eocene (Chadronian). Today they are represented by the collared peccary in the southwestern United States, and their range extends southward into South America, which has three living species, the collared, white-lipped, and Chacoan. Modern peccaries are social animals and live in mixed-sex herds; this is believed to have been the case in the fossil forms as well.

Peccaries have prominent straight upper and lower canines with the uppers buttressed by massive lateral processes. These may be sexually dimorphic in some forms. Later Tertiary taxa may have large, laterally flaring zygomatic processes reminiscent of the entelodonts. The frontal is marked by deeply excavated supraorbital canals. The occiput is broad, and the external auditory meatus is enclosed in an elongate tympanic process of the squamosal. The cheek teeth are bunodont to zygodont, with some forms having teeth that approach being bilophodont. The ulna and radius may fuse in some forms. The third and fourth metapodials are large, and the second and fifth are reduced in size and may be lost. The metacarpals may remain separate, although the metatarsals fuse in some forms (Wright, 1998).

Perchoerus Leidy, 1869

Systematics and Evolution The type and only species of *Perchoerus* is *P. probus*, collected by F. V. Hayden in the Badlands of South Dakota.

Distinctive Characters As a primitive member of the family, it is mostly distinguished by lacking characters seen in later forms. Features of the braincase that distinguish the genus include a dorsally directed external auditory meatus; the articular surface of the glenoid fossa extends slightly ventral to the basioccipital but is still above the occlusal plane; and there is a robust occipital crest. Like modern peccaries, the dentition is brachyodont, with quadrate molars having four cusps that are connected by small lophs. In the upper molars, the paracone and protocone are separated by a paraconule (Fig. 5.34) (Wright, 1998).

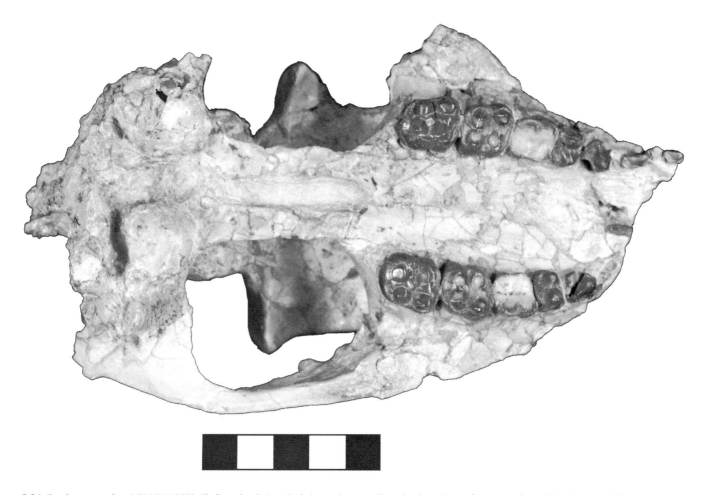

5.34. *Perchoerus probus.* YPM PU 12579. Skull, occlusal view. Scale in centimeters. Photo by the authors of specimen from YPM. Courtesy of the Division of Vertebrate Paleontology, YPM PU 12579, Peabody Museum of Natural History, Yale University, New Haven, Connecticut, U.S.A.

Stratigraphic and Geographic Distribution *Perchoerus probus* is known only from the Poleslide Member of the Brule Formation (Whitneyan).

Natural History and Paleoecology Fossils referable to *Perchoerus* are rare in the White River beds; only a few skulls and no skeletons have been found. Because of its rarity, little is known about the paleoecology of this genus. Presumably, like modern peccaries, *Perchoerus* was probably primarily herbivorous, but it may have occasionally consumed animals. Although modern peccaries are social animals and live in mixed-sex herds, it is difficult to determine if this was the case for *Perchoerus*, given its rarity and given that most of the known specimens consist of scattered remains.

Thinohyus Marsh, 1875

Systematics and Evolution A second genus of peccary, *Thinohyus*, has been reported from the Big Badlands of South Dakota but has not been identified to species (Wright, 1998). The type and only described species is *T. lentus* is from the Arikareean sediments of John Day, Oregon.

Distinctive Characters *Thinohyus* is a rare taxon, and its diagnostic features are based essentially on the type specimen (Wright, 1998) and are consequently confined to features of the skull. It lacks features that distinguish the *Perchoerus–Tayassu* clade so can be considered the sister group to all other peccaries (Wright, 1998). In its dentition, the protocone and paracone are separated by a paraconule. Features of the skull include the postglenoid canal passing between the posttympanic process of the squamosal and the postglenoid process of the jugal; the tympanic process of the squamosal is dorsoventrally shallow; and the articular surface of the glenoid fossa is approximately in the plane of the basioccipital (Wright, 1998).

Stratigraphic and Geographic Distribution *Thinohyus* in the Big Badlands is known only from the Scenic Member of the Brule Formation. It is also known from the Wounded Knee fauna in South Dakota; Nebraska; and Oregon.

Natural History and Paleoecology Given its rarity, nothing has been written about the natural history of this taxon.

Superfamily Hippopotamoidea
Family Anthracotheriidae

Members of the Anthracotheriidae are a group of primitive artiodactyls that seem to be somewhat convergent with modern hippos in their habits and appear to have been semiaquatic, although not all researchers agree on this interpretation (Kron and Manning, 1998). Also known from Europe and Africa, their diversity in North America is small, with six known genera, four of which are found in the White River Group of South Dakota. The earliest record is from the Duchesnean, and they became extinct at the end of the early Hemingfordian. The greatest diversity of anthracotheres occurs in the White River, with three genera, *Heptacodon*, *Aepinacodon*, and *Bothriodon*, present in the Chadronian. *Heptacodon* persisted into the Orellan and Whitneyan, where it overlapped with *Elomeryx*, which survived into the earliest Arikareean. The history of the anthracotheres in North America seems to be one of ecological replacement resulting from intercontinental dispersal rather than evolution in place, at least at the generic level (Kron and Manning, 1998). In the White River Group, *Aepinacodon* was replaced by *Bothriodon*, which was then replaced by *Elomeryx*.

The skull of anthracotheres tends to be long and slender, including an elongate rostrum and a narrow braincase with a well-developed sagittal crest. The internal nares are positioned posteriorly behind the upper third molar elongating the palate, supporting the idea that they may have been semiaquatic. As a general rule, they retain the primitive eutherian dental formula with the occasional loss of the first premolars. The cheek teeth are brachyodont. As the skull becomes more elongated in later forms, diastema develop between the premolars. The upper molars are large and square, with five cusps and a W-shaped ectoloph. The primary resemblances to hippos in the postcranial skeleton are the short, stout limbs. The body was not as massive as in hippos, and the skeleton lacks any graviportal modifications. The distal limb bones indicate they did not have a fully unguligrade stance. The manus retains five digits, but the pes has only four. Both the metacarpals and metatarsals are unfused (Kron and Manning, 1998). Sexual dimorphism has been documented in anthracotheres on the basis of the canines, with males having larger canines than the females, although the degree of difference is not as pronounced as in hippos or pigs. Differences in the anterior of the rostrum in *Elomeryx armatus* have also been interpreted as indicative of sexual dimorphism, with males having a premaxilla that is indented anterior to the canine, whereas in females there is no indentation (Macdonald, 1956).

Aepinacodon Troxell, 1921

Systematics and Evolution The type of *Aepinacodon* is *A. deflectus* from the Big Badlands of South Dakota, and a second species, *A. americanus*, is recognized. The name *Aepinacodon* was proposed by Troxell (1921:334) for the North American Oligocene anthracotheres not referable to *Elomeryx*, *Octacodon*, or *Heptacodon*. Among these he included *Hyopotamus americanus* Leidy, *H. deflectus* Marsh, and *Ancodon rostratus* Scott. Scott rejected this name in his White River monograph and referred the species *A. americanus* and *A. rostratus* to the European genus *Bothriodon* and ignored Marsh's species *H. deflectus*. Macdonald (1956) revived Troxell's genus in his review of the North American anthracotheres.

Distinctive Characters The genus is distinguished from *Bothriodon* by its shorter facial region of the skull, resulting from the shorter diastema between the canine and first upper premolar. The internal nares open just behind or at the level of the third molars. The canine is incisiform. The upper third premolar has a more prominent protocone.

Stratigraphic and Geographic Distribution *Aepinacodon deflectus* is known only from the Peanut Peak Member of the Chadron Formation, and *A. americanus* is known only from the Crazy Johnson Member of the Chadron Formation. Besides South Dakota, the genus is known from Nebraska and northern Colorado.

Bothriodon Aymard, 1846

Systematics and Evolution The type of *Bothriodon* is *B. velaunus* from the early Oligocene of Western Europe. There are two species recognized in North America, *B. rostratus* and *B. advena*.

Distinctive Characters The diastema between the canine and first upper premolar is more pronounced than in *Aepinacodon*, so the rostrum of the skull is more elongate. The protocone of the upper third molar is reduced compared to *Aepinacodon* (Kron and Manning, 1998).

Stratigraphic and Geographic Distribution *Bothriodon rostratus* is found in the Scenic Member of the Brule Formation and the Orella Member of the Brule Formation in Nebraska, and *B. advena* is known from the Cypress Hills of Saskatchewan. The genus has also been reported from Wyoming (Kron and Manning, 1998).

Elomeryx Marsh, 1894

Systematics and Evolution The type species of *Elomeryx* is *E. armatus* from the Badlands of South Dakota, and a second species, *E. garbanii*, has also been described.

5.35. *Elomeryx armatus.* YPM 10177. Skull, occlusal view. Scale in centimeters. Photos by the authors of specimens from YPM. Courtesy of the Division of Vertebrate Paleontology, YPM PU 10177, Peabody Museum of Natural History, Yale University, New Haven, Connecticut, U.S.A.

Distinctive Characters *Elomeryx* was about the size of a wild boar. The rostrum of the skull in *Elomeryx* is shorter than other anthracotheres such as *Aepinacodon*, and the diastema between teeth are less pronounced. The orbits are set high on the skull, another feature that has suggested the possibility of hippolike habits. The palate has a tubular extension that opens behind the last molar in *Elomeryx*, which also supports the idea of amphibious habits. The canine is large and extends ventrally in contrast to *Bothriodon*, which has small canines. The manus has five digits while the pes has four (Fig. 5.35) (Kron and Manning, 1998).

Stratigraphic and Geographic Distribution *Elomeryx armatus* is known from the Poleslide Member of the Brule Formation. The Orellan "*Metamynodon* channels" and the Whitneyan "*Protoceras* channels" of the Big Badlands contain the larger concentrations of its remains. The genus is also known from California, Nebraska, and Saskatchewan.

Natural History and Paleoecology The earliest records of *Elomeryx* are from the middle Eocene of Asia. It appears in Europe in the latest Eocene and is presumed to have dispersed into North America in the Oligocene. Although remains of *Elomeryx* are generally rare, the concentrations of its remains in the Orellan "*Metamynodon* channels" and Whitneyan "*Protoceras* channels" of the Big Badlands have been cited as evidence that anthracotheres were hippolike in their habits. On the basis of their association with nonaquatic taxa, Wilson (1975) proposed that the *Protoceras* Channel

fauna actually represented a stream border habitat and that many of the taxa were in fact forest dwellers that just happened to be preserved in the channel deposits.

Heptacodon Marsh, 1894

Systematics and Evolution The type of *Heptacodon* is *H. curtus* from the Whitneyan *Protoceras* channels of the Badlands of South Dakota, and four other species are recognized: *H. gibbiceps,* *H. pellionis,* *H. occidentale,* and *H. quadratus.*

Distinctive Characters The genus *Heptacodon* is quite distinctive. Its rostrum is short compared to other anthracotheres. The orbit is positioned more anteriorly on the skull, with its anterior margin almost to the first molar. The zygoma is deep. The mandible has an upturned symphysis. *Heptacodon* differs from the other contemporary anthracotheriids in having a fused mandibular symphysis without trace of the suture, while in most anthracotheriids the two halves of the mandible remain unfused. The canines are shifted laterally, and the upper and lower anterior premolars are not separated by diastema. The upper molars are heavily styled, and the anterior premolars are simple and generally rounded with nonangular cusps on both the upper and lower teeth (Macdonald, 1956). It differs from the other North American genera *Bothriodon, Aepinacodon,* and *Arretotherium* and Euro-American *Elomeryx* in the presence

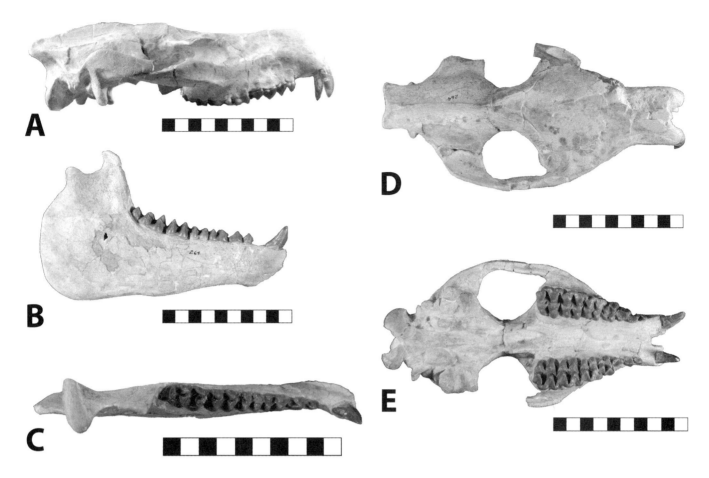

5.36. *Agriochoerus antiquus,* SDSM 264. (A) Skull, right lateral view. (B) Right mandible, right lateral view. (C) Right mandible, occlusal view. (D) Skull, dorsal view. (E) Skull, occlusal view. Scales in centimeters. Photos by the authors of specimens from the Museum of Geology, South Dakota School of Mines and Technology, Rapid City, South Dakota, U.S.A.

of a strong postprotocristid on its teeth, and the tooth rows lack a significant diastema between the canine and first lower premolar or between the first and second lower premolars (Kron and Manning, 1998).

Stratigraphic and Geographic Distribution *Heptacodon* first appears in the Duchesnean and became extinct at the end of the Whitneyan. *Heptacodon curtus* is known from the Poleslide Member of the Brule Formation in the Big Badlands and from the Orella and Whitney members of the Brule Formation in Nebraska. *Heptacodon occidentale* and *H. quadratus* are known from the Scenic Member of the Brule Formation and *H. gibbiceps* from the Poleslide Member of the Brule Formation. Stratigraphic distribution is the Duchesnean of Saskatchewan, Oregon, Texas and Utah; the Chadronian of Wyoming, South Dakota, and Colorado; and the Orellan and Whitneyan of South Dakota (Holroyd, 2002).

Natural History and Paleoecology According to Macdonald (1956), there is sexual dimorphism in the size of the canines, as in *Elomeryx*.

Suborder Tylopoda
Superfamily Merycoidodontoidea

The families Agriochoeridae and Merycoidodontidae are closely related forms often referred to by the general term oreodonts. The Agriochoeridae is considered the more primitive of the two families.

Family Agriochoeridae

The Agriochoeridae appear in the middle Eocene (late Bridgerian), and most genera became extinct in the late Eocene (Chadronian) with one genus, *Agriochoerus*, surviving until the end of the Oligocene. Eight genera are recognized, although most have not been formally named (Lander, 1998), leaving *Agriochoerus* as the only genus in the White River that has been formally described.

Members of the family retain many primitive features in the skull and the postcranial skeleton that have been lost in the Merycoidodontidae. Features that distinguish the agriochoerids from the merycoidodonts include the reduction and loss of the upper incisors; presence of a diastema between the

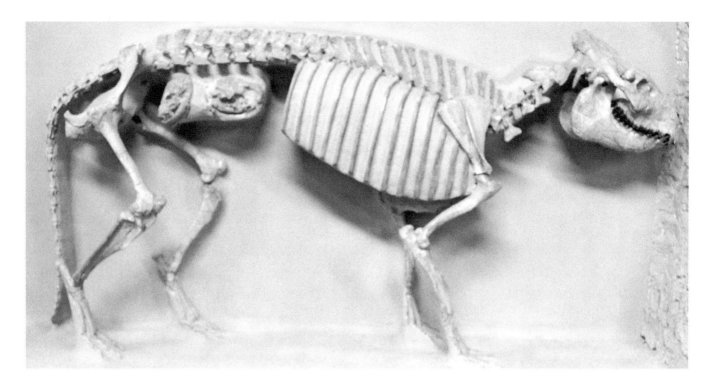

5.37. *Merycoidodon culbertsoni,* skeleton, female found with two associated juveniles, right lateral view. SDSM 28129, photo by the authors of specimen from the Museum of Geology, South Dakota School of Mines and Technology, Rapid City, South Dakota, U.S.A.

canines and premolars; bothriodonlike molars; brachyodont molars and molariform third and fourth premolars; absence of a lachrymal pit; an incompletely closed orbit; a much heavier and more carnivorelike skeleton; and a far longer and thicker tail. The manus has five digits retaining the pollex, which is lost in the merycoidodonts, but the pes consists of four digits and unguals that are claws in some of the larger taxa.

Agriochoerus Leidy, 1850

Systematics and Evolution The type species of *Agriochoerus* is *A. antiquus* from the Brule Formation of the White River Badlands, South Dakota. Numerous species and subspecies have been described. Lander (1998) in his review of the genus reduced this to four subspecies of *A. antiquus*, and a second species, *A. matthewi*, with two subspecies.

Distinctive Characters *Agriochoerus* is larger than the other contemporary members of the family. The rostrum is constricted below the canine and second premolar. The upper incisors and first premolar are reduced or lost. The diastema between the upper first and second premolar is reduced or lost, while the one between the lower first and second premolars is expanded (Fig. 5.36). The ungual phalanges are clawed. Generally the skeleton of *Agriochoerus* resembles that of the carnivorous creodonts or fissipeds.

The cervical vertebrae, especially the axis, with its great hatchetlike neural spine, and the posterior thoracic and lumbar vertebrae are strongly built; the tail is extremely long and heavy; and the limb bones and clawed feet are all suggestive of a predator (Lander, 1998).

Stratigraphic and Geographic Distribution This genus has the longest biostratigraphic range of all members of the family, first appearing in the early Uintan, becoming extinct at the Oligocene–Miocene boundary. Given its long period of existence, numerous species and even subspecies have been described, many of which have been subsequently synonymized. The genotypic species, *A. antiquus*, is the only species recognized from the White River. It is known from the Peanut Peak Member of the Chadron Formation and the Scenic Member of the Brule Formation.

Natural History and Paleoecology There is no consensus on the habits of *Agriochoerus*. W. D. Matthew interpreted it as a tree climber, and the skeleton in the American Museum of Natural History is mounted so as to suggest this habit. Alternatively, it has been suggested that they were fossorial, although this many have been limited to the smaller forms. It has also been suggested they may have dug up roots and tubers as a principal part of their food.

Family Merycoidodontidae

The second oreodont family is the Merycoidodontidae, which first appears in the Duchesnean and became extinct at the end of the Miocene.

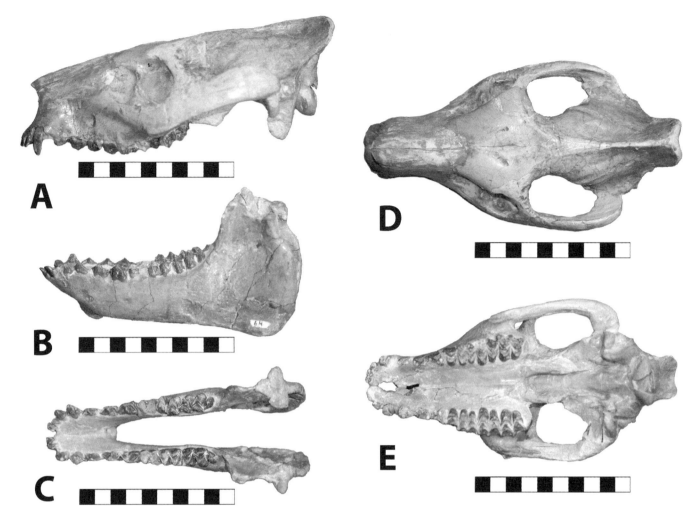

5.38. *Merycoidodon culbertsoni,* DMNS EPV 64 (A) Skull, left lateral view. (B) Mandibles, left lateral view. (C) Mandibles, occlusal view. (D) Skull, dorsal view. (E) Skull, occlusal view. Scales in centimeters. Photos by the authors of specimens from DMNS. DMNS EPV 64, Denver Museum of Nature and Science, Denver, Colorado, U.S.A. All rights reserved.

Members of the Merycoidodontidae are smaller than most of their ungulate contemporaries. This suggests a shorter life span and generation time, resulting in a high reproductive turnover. Oreodonts are the most common mammalian taxon found in the White River Badlands, suggesting they must have been a common animal on the landscape. On exhibit at the South Dakota School of Mines and Technology Museum of Geology is a skeleton of a female of *Merycoidodon culbertsoni* with two juveniles that were found in association with her, suggesting that twins may have been the normal litter size for the species (Fig. 5.37).

Merycoidodon (=*Oreodon* Leidy, 1852) Leidy, 1848

Systematics and Evolution The type species of *Merycoidodon* is *M. culbertsonii* from the Brule Formation of the Big Badlands of South Dakota, described by Leidy in 1848. It is one of the earliest fossil taxa described from the White River. The taxon is based on two cotypes in the Academy of Natural

Sciences in Philadelphia, a partial jaw with three molars and a maxilla with the second and third molars. Leidy (1852) later replaced his own genus, *Merycoidodon*, with a new genus *Oreodon* for *culbertsoni*, which has contributed to taxonomic confusion regarding this animal. The validity of this taxon is not accepted by some recent workers such as Lander (1998), who preferred the later name *Prodesmatochoeus periculorum*, as he considered Leidy's name a nomen vanum. As Leidy's first generic name is well established in the literature, and as some subsequent workers (Stevens and Stevens, 1996) have chosen to retain it, we have done likewise, although we recognize that there are many nomenclatural issues with its retention. While Leidy's second proposed genus, *Oreodon*, has no taxonomic validity, the term oreodont is still used as the popular or common name for this group of artiodactyls. In addition to *M. culbertsoni*, Stevens and Stevens (1996) recognized *M. presidioensis*, *M. bullatus*, and *M. major*.

Distinctive Characters *Merycoidodon* was about the size of a sheep. The skull has a complete postorbital bar. The

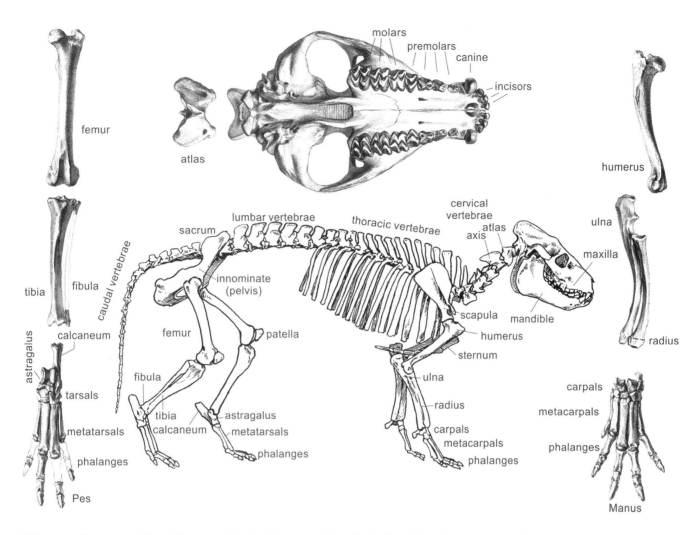

5.39. Bones of the oreodont *Merycoidodon* depicting individual elements including teeth and bone showing various features. Figures compiled from Scott and Jepsen (1940:fig 133, plate 69, fig. 2; plate 71, figs. 2, 9a, 10, 11, 12, 14, 15).

dentition forms a continuous series without any diastemas. The first lower premolar is caniniform. The other premolars are simple and trenchant, but the upper fourth premolar is more molariform in its shape. The molars are selenodont, formed by four crescents in two transverse pairs but are low crowned. The facial portion of the skull is short, and the orbit is enclosed by a complete postorbital bar. There is a well-developed fossa in front of the orbit. The auditory bulla is small (Fig. 5.38). The neck and limbs are short in proportion to the body, and it retains a long tail, which in proportion to its body size is longer than in any other member of the family. Although an artiodactyl, one of its primitive features is that the manus has five digits, and while the pollex is reduced in size, it is complete with all phalanges (Figs. 5.38, 5.39) (Lander, 1998).

Stratigraphic and Geographic Distribution *Merycoidodon culbertsoni* is found in the Peanut Peak Member of the Chadron Formation and the Scenic Member of the Brule Formation. *Merycoidodon major* is found in the "lower *Leptauchenia* beds," now known as the Poleslide Member of

the Brule Formation in the Big Badlands and the Whitney Member of the Brule Formation in Nebraska. *Merycoidodon bullatus* is known from the Scenic and Poleslide members of the Brule Formation of the Big Badlands and from the Whitney Member of the Brule Formation of Nebraska and approximately equivalent beds in Montana, North Dakota, and Colorado. The genus is also known from faunas in Wyoming and Texas.

Natural History and Paleoecology *Merycoidodon* is the most common oreodont found in the White River, and its abundance far exceeds that of any other known taxa from the region or as Scott (1962:368) phrased it, "In the middle and lower White River (lower Brule and Chadron substages) the chief oreodont genus is *Merycoidodon*, the bones of which so far outnumber those of all other mammals, that the collector comes to regard them as a nuisance." Given the abundance of this taxon, it must have been a herd animal.

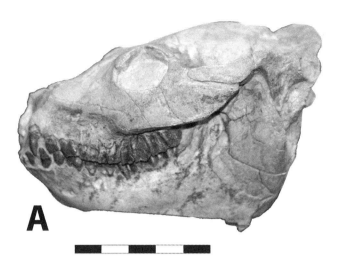

5.40. *Leptauchenia* sp. UCM 17448 (A) Skull and jaws, left lateral view. (B) Skull, dorsal view. Scales in centimeters. Photos by the authors of specimen from UCM. UCM 17448, University of Colorado Museum of Natural History, Boulder, Colorado, U.S.A.

Leptauchenia Leidy, 1856

Systematics and Evolution The type of *Leptauchenia* is *L. decora* from the Brule Formation, White River Valley, South Dakota; two additional species, *L. major* and *L. lullianus*, are also recognized. *Leptauchenia major* is found in the White River Badlands, and *L. lullianus* is found in Montana (Prothero and Sanchez, 2008).

Distinctive Characters Compared to other contemporary oreodonts like *Merycoidodon*, *Leptauchenia* has a smaller body size. The skull has a reduced rostrum. The skull is dorsoventrally flattened. The orbits are small and positioned high up on the skull with their tops projecting above the level of the forehead. There are large antorbital vacuities that extend from above the orbits almost to the premaxillae and a small, shallow lacrimal fossa. The size of the antorbital vacuities has resulted in a reduction of the nasals to mere splints (Fig. 5.40). Prominences at the anterior margin of the nasal opening in some specimens along the nasal–maxillary suture, which may have formed attachment points for the soft tissue of the nose, although Janis (1982) has proposed they may have supported nasal horns. That these structures are not present in all individuals suggests they may have been sexually dimorphic. The tympanic bulla is highly inflated, and the external auditory meatus is an elongated tube located high and posteriorly on the skull. The occiput is expanded laterally. The first upper incisor is reduced or absent. The cheek teeth are hypsodont but with thin enamel. The fore- and hind limbs are of similar length. The ulna and radius are separate, and there is a complete fibula. The cuboid and navicular are separate bones (Lander, 1998).

Stratigraphic and Geographic Distribution *Leptauchenia* first appears in the late Orellan and becomes extinct in the late Arikareean. *Leptauchenia decora* appears in the early Orellan and continues to the early Arikareean. This species is found in the Poleslide Member of the Brule Formation. Besides South Dakota, the species also occurs in faunas in Nebraska, Wyoming, Colorado, and Montana. *Leptauchenia major* appears in the late Whitneyan and continues to the late Arikareean. The earlier *L. decora* is smaller than the later *L. major* (Prothero and Sanchez, 2008). It is also found in the High Plains and western Montana. *Leptauchenia lullianus*, which ranges within the early to middle Arikareean, is larger than *L. major* (Prothero and Sanchez, 2008). Prothero and Sanchez (2008) consider all three species to have been contemporaries in the early Arikareean. *Leptauchenia lullianus* is also found in the High Plains and eastern Montana.

Natural History and Paleoecology Because of its high-set eyes and ears, and large nasal-facial vacuities, *Leptauchenia* was considered to have been semiaquatic or amphibious (Cope, 1884; Matthew, 1899; Scott, 1937; Scott and Jepsen, 1940). Schultz and Falkenbach (1956) argued that Leptaucheniines were specialized for a desert mode of life. They based their interpretations on the eolian type sediments in which many specimens were found. They were compared to desert bovids, which have similar large bullae, large facial vacuities, and specialized nasal areas. In contrast, they are almost absent in the *Protoceras* channel deposits of the Poleslide Member of the Brule Formation. Leptaucheniines have hypsodont teeth with highly worn crowns, deep-set jaws, and a thick, heavy set zygomatic region – all features well adapted for an abrasive, grittier diet. Recent studies of the postcranial skeleton have suggested locomotor similarities to the living hyraxes (Wilhelm, 1993).

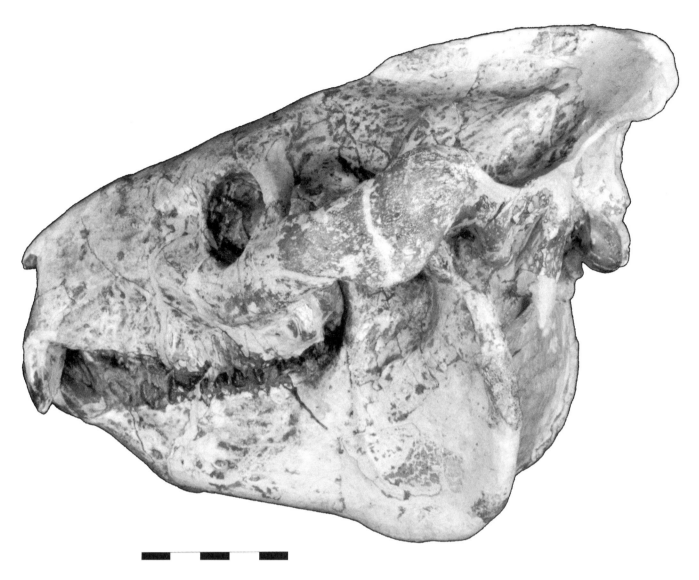

5.41. *Eporeodon major.* YPM PU 13594. Skull and jaws, left lateral view. Scale in centimeters. Photos by the authors of specimen from YPM. Courtesy of the Division of Vertebrate Paleontology, YPM PU 13594, Peabody Museum of Natural History, Yale University, New Haven, Connecticut, U.S.A.

Eporeodon Marsh, 1875

Systematics and Evolution The type of *Eporeodon* is *E. occidentalis* from the Bridge Creek Locality of the John Day Valley, Oregon. Thorpe (1937) recognized 15 species within the genus, most of which have now been synonymized with *E. occidentalis.* Currently only a second species, *E. major*, known only from Arikareean of South Dakota, is also considered valid.

Distinctive Characters *Eporeodon* is a small genus of oreodont. The skull has a short facial region and a shallow zygomatic arch, and the sagittal crest is not prominent. The dentition is brachyodont with rectangular premolars. *Eporeodon major* is larger than *E. occidentalis* (Fig. 5.41) (Lander, 1998).

Stratigraphic and Geographic Distribution *Eporeodon* appears in the Whitneyan and becomes extinct at the end of the Arikareean; most records of the genus are from the Arikareean. *Eporeodon occidentalis* is present in the Poleslide Member (Whitneyan) of the Brule Formation in the Big Badlands. The genus is known from multiple localities in South Dakota, Wyoming, Nebraska, Oregon (John Day Formation), Florida, California, Montana, Washington, and Idaho.

Miniochoerus Schultz and Falkenbach, 1956

Systematics and Evolution The type species of *Miniochoerus* is *M. gracilis* from Shannon County, South Dakota. Other species considered valid include *M. forsythae, M. chadronensis, M. affinis,* and *M. starkensis.*

Distinctive Characters *Miniochoerus* were relatively small and cranially stereotyped oreodonts and retained a minute auditory bulla. Their teeth became distinctive, with thin-enameled, quickly abraded fossettes and fossetids and foreshortened premolars (Stevens and Stevens, 2007).

Stratigraphic and Geographic Distribution The genus first appears in the early Chadronian and becomes extinct at the end of the Whitneyan. *Miniochoerus forsythae* is found in Wyoming and South Dakota from the middle Chadron Formation. *Miniochoerus chadronensis* is from the upper Chadron Formation in Wyoming and the lower Brule Formation in Nebraska. *Miniochoerus affinis* is from the Scenic Member of the Brule Formation in South Dakota and the lower Orella Member of the Brule Formation in Nebraska. *Miniochoerus gracilis* is from the Orella Member of the Brule Formation in Niobrara County, Wyoming, and the Scenic Member of the Brule Formation in South Dakota. *Miniochoerus gracilis* is also found in Nebraska and Colorado. *Miniochoerus starkensis* is found in the Brule Formation, Fitterer Badlands, North Dakota.

Mesoreodon Scott, 1893

Systematics and Evolution The type species of *Mesoreodon* is *M. chelonyx* from the Fort Logan Formation of Montana. Besides the type species, a second species, *M. minor*, is tentatively recognized by Stevens and Stevens (1996, 2007). Lander (1998) considered *Mesoreodon* a junior synonym of *Eporeodon*. The taxonomy of the genus has been complex, and the name *Mesoreodon* has been restricted to forms with wide and low skulls. Some of the confusion appears to reflect differences in the postmortem deformation of the skull. The genus appears to be the ancestor to *Hypsiops* and consequently to the *Desmatochoerus–Promerycochoerus* lineage of oreodonts.

Distinctive Characters The general form of the cranium is similar to that of *Merycoidodon* but with inclined premaxillae. The antorbital fossa is shallow and pitlike. The infraorbital foramen is positioned above the third and fourth premolars. The premolars are only slightly reduced, and all teeth are slightly narrower and higher crowned than other contemporary oreodont genera. According to Stevens and Stevens (1996, 2007), *Mesoreodon* is distinguished from *Eporeodon* by its larger size, the pitlike antorbital fossa, a higher rostrum, the slightly reduced premolars, and anteroposteriorly shorter molars that are higher crowned.

Stratigraphic and Geographic Distribution As defined by Stevens and Stevens (1996, 2007), the genus appears in the latest Whitneyan and became extinct in the early Arikareean. In the White River Badlands, it has only been found at Quiver Hill, south of Kadoka, South Dakota (Arikareean). The genus is also known from Oregon, Montana, Wyoming, and Nebraska.

Natural History and Paleoecology Nothing specific has been proposed regarding the natural history or paleoecology of this genus.

Superfamily Cameloidea
Family Camelidae

The family Camelidae is today represented by the Old World dromedary and Bactrian camels and the South America llamas, vicuñas, and guanacos. The family originated in North America, with the earliest records in the Uintan, approximately 45.9 Ma. The primary evolutionary history of the family is in North America, and they did not disperse into Eurasia until the late Miocene and into South American until the early Pleistocene.

Camelids lack cranial structures like horns, antlers, and ossicones. They retain a canine, and the third incisors and anterior premolars may become caniniform in their morphology. The skull has a sagittal crest. In front of the orbit there is a lacrimal vacuity and a maxillary fossa that is well developed in early camelids but may be reduced in size in later forms. The first and second upper incisors are reduced, but the lower incisors are large and spatulate. The true lower canine becomes incisiform, and some premolars are caniniform. The molars are four cusped with straight ectolophs and with fossettes that are closed both anteriorly and posteriorly after moderate wear. Camelids have long necks compared to most artiodactyls and are only exceeded by the giraffids in length. A distinctive character of the cervical vertebrae is that the vertebral artery passes inside the pedicle of the neural arch and its posterior portion is confluent with the neural canal. The ulna and radius are co-ossified. The fibula is reduced and lacks the midshaft so there is only a small proximal end, and the distal malleolus locks into the distal tibia. The cuboid and navicular remain separate even in advanced forms, and this is reflected in the distal end of the astragalus, which has a keel that fits between them. Only the second and third metacarpals and metatarsals are present; they have a distinctive distal flare, with the keel on the distal articular surfaces confined to the posterior margin. The metapodials may be fused or unfused depending on the taxon. Primitive camelids have an unguligrade stance, but many later genera are digitigrade (Honey et al., 1998).

Poebrotherium Leidy, 1847

Systematics and Evolution The type species of *Poebrotherium* is *P. wilsoni* from the Badlands of South Dakota; the genus includes two other species, *P. chadronense* and *P. eximium*. *Poebrotherium wilsoni* was the first Tertiary camelid described from North America.

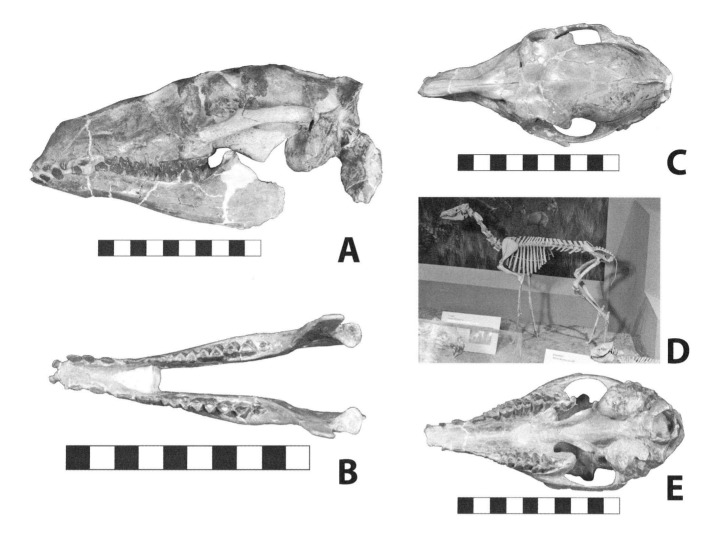

5.42. *Poebrotherium wilsoni* (A) YPM PU 12722, skull and jaws, left lateral view. (B) SDSM 2913, mandible, occlusal view. (C) SDSM 2913, skull, dorsal view. (D) USNM 15917, *Poebrotherium* sp., skeleton, left lateral view. (E) SDSM 2913, skull, occlusal view. Scales in centimeters. Photos by the authors of specimens from SDSM, USNM, and YPM. Courtesy of the Division of Vertebrate Paleontology, YPM PU 12722, Peabody Museum of Natural History, Yale University, New Haven, Connecticut, U.S.A. SDSM 2913 is from the Museum of Geology, South Dakota School of Mines and Technology, Rapid City, South Dakota, U.S.A. USNM 15917, courtesy of Smithsonian Institution, Washington, D.C., U.S.A.

Distinctive Characters Compared to later camelids, *Poebrotherium* is small. The largest species is about the size of a sheep but with a more slender build and a long neck and body (Fig. 5.42D). It maintains a complete dental formula and lacks a diastema between the anterior teeth. The teeth are less hypsodont than other camelids except *Poebrodon*. The skull lacks a postorbital bar, a feature seen in later camelids (Fig. 5.42A–C). A primitive feature seen in some *Poebrotherium* is a bifurcation of the protocone of the upper molars. The humerus has a single bicipital groove. The ulna and radius are co-ossified. The fibula has already lost the shaft and is reduced to a proximal splint that is fused to the tibia, and the distal malleolus is set in the distal tibia to lock the astragalus in place. Digit 1, is lost and digits 2 and 5 are reduced to vestigial nubbins in both the front and hind feet. The third and fourth metapodials remain unfused (Honey et al., 1998).

Stratigraphic and Geographic Distribution This is the most common Oligocene camelid and is the most primitive except for *Poebrodon*, a smaller genus from the Uintan. *Poebrotherium* first appears in the Chadronian and became extinct at the end of the early Arikareean. *Poebrotherium eximium* is found in the Peanut Peak Member of the Chadron Formation and the Scenic Member of the Brule Formation, and *P. wilsoni* is known from the Scenic and Poleslide members of the Brule Formation. The genus is also known from Wyoming, Colorado, Nebraska, North Dakota, and Montana.

Natural History and Paleoecology In the Chadronian and Orellan, camelids are common in the White River deposits of Colorado and southeastern Wyoming but rare in South Dakota and farther north, suggesting differences in habitat in these regions that limited their presence at higher latitudes. Clark, Beerbower, and Kietzke (1967) observed that in the Big Badlands, *Poebrotherium* is typically found in sediments far from channels, indicating it may have preferred drier habitat.

Paralabis Lull, 1921

Systematics and Evolution The type of *Paralabis* is *P. cedrensis* and is the only species currently recognized. The type was collected from the *Titanotherium* beds (late Chadronian) of northeastern Colorado (Lull, 1921).

This taxon has had a confusing taxonomic history. An unfortunate lapse in Matthew (1901) illustrated a lower jaw, AMNH 8969, as the type of *Protomeryx cedrensis*, but the species name *campester* was used in the text, which cited the same catalog number for the type. The assignment of the species to Leidy's genus *Protomeryx*, which had been used as a catchall name for many early camels—and McKenna (1966) considered the name a nomen dubium—was changed when Lull (1921) transferred the species to his new subgenus, *Paralabis*, within *Pseudolabis*. Both of these are now considered separate genera (Honey et al., 1998). Stock (1935) clarified the taxonomy with the recognition of *cedrensis* as the valid species name.

Distinctive Characters *Paralabis* apparently represents an advanced descendant of *Poebrotherium* in the Whitneyan. The muzzle has become elongated. The lower second to fourth premolars are shortened, and the diastema is better developed, especially between the lower first and second premolars where the depth of the jaw is at a minimum (Lull, 1921).

Stratigraphic and Geographic Distribution *Paralabis cedrensis* is known from the Poleslide Member of the Brule Formation (Whitneyan) in the Big Badlands but has also been reported from faunas in Colorado and Nebraska.

Pseudolabis Matthew, 1904

Systematics and Evolution The type and only species of *Pseudolabis* is *P. dakotensis* from the Whitneyan *Protoceras* beds of South Dakota. It was the first camel reported from the *Protoceras* beds or channel deposits in the Poleslide Member of the Brule Formation. It is the earliest representative of the Stenomyline or gazellelike camelids.

Distinctive Characters *Pseudolabis* is larger than *Poebrotherium*. The skull has a deeply depressed maxillary fossa, and the premaxilla extends posteriorly to the first premolar. It is the first camelid to have a complete postorbital bar that encloses the orbit. The teeth are relatively hypsodont, as in later members of the subfamily, and compared to *Poebrotherium*, the teeth are more elongate and transversely narrower. As the earliest stenomyline camelid, it is primitive in having a relatively short rostrum and unreduced premolars. The upper first premolar is caniniform in its morphology. The upper fourth premolar has a double internal crescent. The molars lack mesostyles (Matthew, 1904; Honey et al., 1998).

Stratigraphic and Geographic Distribution First appearing in the Whitneyan, the stratigraphic range of the genus extends into the early late Arikareean. In the Big Badlands, *Pseudolabis dakotensis* is present only in the Poleslide Member of the Brule Formation (Whitneyan). It is also found in faunas in Wyoming and Nebraska.

Family Protoceratidae

The family Protoceratidae first appears in the middle Eocene (early Uintan) and became extinct in the early Pliocene (late Hemphillian) (Patton and Taylor, 1973). The family takes its name from *Protoceras*, the first genus of the family described. The family currently includes 13 genera, and three genera, *Protoceras*, *Leptotragulus*, and *Heteromeryx*, are known from the White River Badlands.

The family is perhaps best known for the later genus, *Synthetoceras*, with its elongated slingshotlike horn on the tip of its nasals, and while many genera of the family do have well-developed horns, they are often absent in the earlier more primitive genera. There is a small facial vacuity anterior to the orbit, which is completely enclosed. Although upper incisors are present in primitive genera, they are lost in later genera—a trend that parallels the situation seen in many living artiodactyls such as camelids, cervids, and bovids. The jaw flexes ventrally at the mandibular symphysis, and the coronoid process of the jaw is one of the shortest among all ungulates. The dentition is of the primitive bunoselenodont type, with the molars widened laterally but shortened anteroposteriorly. Although the teeth are selenodont, there is no bifurcation of the lobes. There are prominent parastyles and mesostyles on the cheek teeth and a well-developed lingual cingulum (Prothero, 1998b).

Protoceratids share many features of the postcranial skeleton with camels. Among these is the vertebral canal of the cervicals passing through the neural arch and not the transverse process. One evolutionary trend in the family is a modification of the limbs for increased cursoriality, such as the fusion of the ulna and radius in later genera, but they never achieve the same level of modification seen in camels or the pecorans. The lateral metapodials are reduced in later genera, but they are not totally lost, so even the later genera have four toes and the metapodials never fuse. They also resemble camelids in the keel of the distal end of the metapodial being confined to the plantar surface of the articulation for the first phalanx. As in the camels, the distal astragalus has a distinct distal keel, reflecting the nonfusion of the cuboid and navicular (Prothero, 1998b).

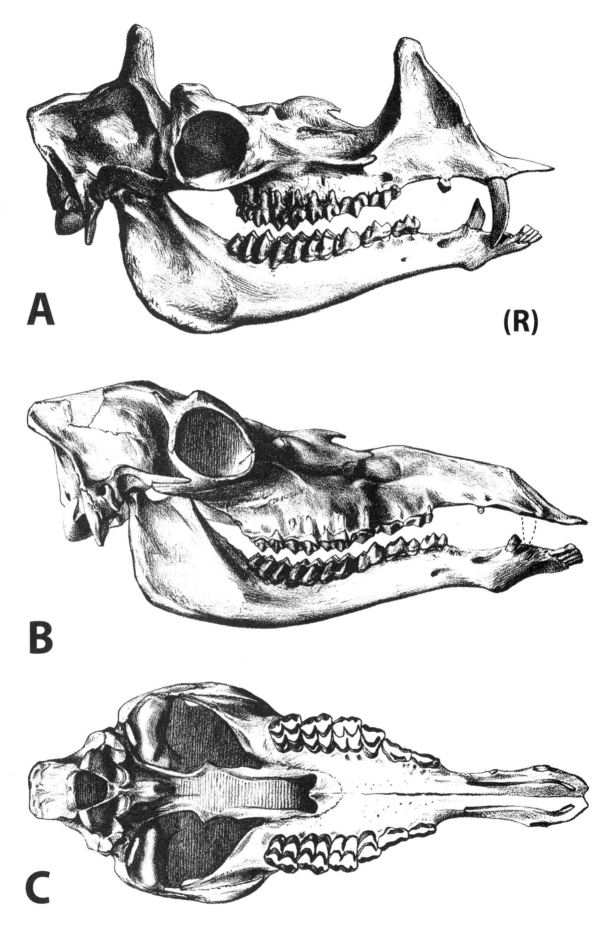

5.43. *Protoceras celer,* adapted from Scott and Jepsen (1940: plate 60, fig. 1, and plate 61, fig. 1, 1b). (A) Skull (male) reversed, left lateral view, SDSM 2814. (B) Skull (female), right lateral view, YPM PU 10655. (C) Skull (female), occlusal view, YPU PU 10655.

A

(R)

B

C

Protoceras Marsh, 1891

Systematics and Evolution The type species of *Protoceras* is *P. celer* from the upper part of the Poleslide Member of the Brule Formation of South Dakota, and two other species are recognized.

Distinctive Characters The skull has a posteriorly placed orbit, and there are strong sagittal and parietal crests. The rostrum of the skull is elongated, with shortened or retracted nasals, possibly suggesting the presence of a flexible proboscis or muzzle like a tapir or saiga antelope. The upper incisors have been lost, but in males, the upper canine is a short, strong, recurved tusk. In females, the canine is reduced and vestigial. The lower canine is incisiform, while the first premolar in the male is caniniform. The other premolars are elongated anteroposteriorly and narrow with sharp cusps, while the molars are low crowned (Fig. 5.43). The ulna and radius are starting to co-ossify but only at their distal ends; each is still a fully functional bone. The fibula is totally lost. The manus has four functional digits, while the pes is functionally two toed, but the metatarsals remain separate from each other. As in camelids, the cuboid and navicular remain separate but differ in having a fused ecto- and mesocuneiform, which occurs in later genera. The second and fifth metatarsals are long and slender splints and lack phalanges (Prothero, 1998b).

Stratigraphic and Geographic Distribution *Protoceras* first appears in the Whitneyan and became extinct in the early late Arikareean. Only one species, *P. celer,* is present in the Big Badlands from the Poleslide Member of the Brule Formation (Whitneyan). The genus is also known from faunas in Nebraska.

Natural History and Paleoecology The type specimen of *Protoceras* is based on a female skull, but with the discovery of additional specimens, the pronounced sexual dimorphism in *Protoceras* was quickly recognized (Osborn and Wortman, 1892). The male has three pairs of bony hornlike structures. There is a pair that arises from the parietals just behind the orbits, the dorsal margin of the orbit is expanded into a triangular structure, and there is a dorsal expansion of the maxilla above the narial opening. The structure and size of these seems to vary between species, but a detailed study of this variation and how much it is determined by the stage of ontogenetic development has not yet been done. In *P. celer,* the anterior protuberances on the maxillae are simple, broad recurved plates that are thickened and rugose on the margins, and the posterior horns are short and club shaped. In females, the anterior protuberances on the maxillae are missing, while the posterior pair is small and hornlike (Prothero, 1998b).

Remains of *Protoceras* are so common in channel deposits in the late Whitneyan of the Big Badlands that the deposits are known as the *Protoceras* channels. The structure of their teeth indicates folivorous (leaf-eating) habits, and with the retracted nasals, a flexible nose that is either tapirlike or mooselike suggests that they may have fed on semiaquatic plants (Janis, 1982). Their short limbs and unfused metapodials suggest that they were not fast runners and probably inhabited bushy terrain, such as riparian vegetation along watercourses rather than open country. The pronounced differences seen in the development of the hornlike structures of the skulls of males and females suggests they were important for visual display during mating and perhaps in males in establishing dominance. The horns of *Protoceras* are best seen from the side, while those in other members of the family are more readily seen from the front. Battles during rut between males may have consisted of neck wrestling and pushing and pulling, with horns of males interlocking and preventing serious damage to the contestants (Prothero, 1998b).

Leptotragulus Scott and Osborn, 1887

Systematics and Evolution The type of *Leptotragulus* is *L. proavus*, and two, or possibly three, other species are considered valid. The type is from the Uintan of Uintah County, Utah.

Distinctive Characters This is a small, primitive protoceratid. The metaconid on the lower fourth premolar is not well developed, but the parastylid is prominent on the sharply flexed anterior crest (Prothero, 1998b).

Stratigraphic and Geographic Distribution The stratigraphic range of the genus is from early Uintan to middle Chadronian. This genus is not well known from the White River Group of South Dakota. "*Leptotragulus*" *profectus* from the Peanut Peak Member of the Chadron Formation may actually belong to another genus, *Trigenicus* (Prothero, 1998b). The genus is also known from faunas in Utah, Wyoming, Texas, Nebraska, Montana, and Saskatchewan.

Heteromeryx Matthew, 1905

Systematics and Evolution The type of *Heteromeryx* is *H. dispar* from the middle Chadron Formation on Indian Creek, Cheyenne River, South Dakota. The type specimen is represented by a skull with a considerable part of the skeleton, including most of the forefeet and parts of the hind feet. It is the only species in the genus (Prothero, 1998b).

Matthew (1905) originally considered *Heteromeryx* to be in the family Hypertragulidae on the basis of the presence of a functionally tetradactyl manus with a didactyl pes. He considered the genus to be intermediate between *Leptomeryx* and *Protoceras* and a possible ancestor to *Protoceras*. The genus is now placed in the Protoceratidae (Prothero, 1998b).

Distinctive Characters *Heteromeryx* is smaller than *Protoceras*. The skull of *Heteromeryx* is rather short, with an elongate muzzle, and the orbits are enclosed by a complete postorbital bar and are located above the posterior molars. The type skull lacks horns; it is not known whether this is characteristic of the genus, or alternatively whether the skull is from a female. The nasals are reduced and separated from the maxilla by a narial notch that extends to the third premolar. It is considered to be a primitive member of the family because the skull lacks horns or protuberances. The teeth are short crowned. There are four premolars, the first small, simple, and single rooted, with long diastema before and behind it, and the others much as in *Leptomeryx*. The molars have heavy internal cingula and a rudimentary mesostyle. The ulna has a well-developed shaft and is co-ossified with a radius along its entire length. The manus is functionally tetradactyl, with four separate digits, as in *Leptomeryx*. The magnum and cuneiform are fused. The distal end of the fibula is separate from the tibia. The pes is didactyl, with separate metatarsals. The cuboid and navicular are distinct, but the ecto- and mesocuneiform are fused. The ungual phalanges are short and compressed (Prothero, 1998b).

Stratigraphic and Geographic Distribution The genus first appears in the late Uintan and became extinct at the end of the Chadronian. In the Big Badlands, it is only known from the Ahearn Member of the Chadron Formation (early Chadronian). It is also known from similar age deposits in Nebraska and Uintan to middle Chadronian faunas in Texas.

Suborder Ruminantia
Infraorder Traguloidea
Family Hypertragulidae

Hypertragulids have a skull with a narrow rostrum, orbits centrally positioned on the skull, a reduced postorbital bar so the orbit is confluent with the temporalis fossa, and an auditory bulla with a long bony external auditory meatus. There are no antlers in males, unlike their modern relatives, the cervids. They do, however, retain a large upper and lower canine, and they resemble the modern musk deer of Southeast Asia. The upper molars have strong ribs on their labial (cheek) margin of the crown. The forelimbs are shorter than the hind limbs, giving them a posture reminiscent of the musk deer. The ulna and radius are co-ossified. The manus is primitive, with five digits (pentadactyl); the metacarpals are unfused, and the lateral digits are complete with phalanges but are shortened relative to the central digits and probably functioned like dewclaws. The fibula is complete. The pes has four digits with the metatarsals unfused, and

the second and fifth metatarsals are reduced in size but still retain phalanges. A distinctive feature of the family is the "bent" astragalus, with the proximal and distal portions at an oblique angle to each other.

Hypertragulus Cope, 1873

Systematics and Evolution The type species of *Hypertragulus* is *H. calcaratus* from the Brule Formation of Colorado, and three other species are recognized.

Distinctive Characters While the overall body size of *Hypertragulus* is small, it is the largest member of the family except for *Nanotragulus*. The teeth are not as hypsodont as other members of the family. Both the upper and lower molars have prominent cingula on their anterior margins, and accessory conules are also present. Both the upper and lower second premolars are caniniform and separated from the other cheek teeth by diastema (Fig. 5.44) (Webb, 1998).

Stratigraphic and Geographic Distribution *Hypertragulus calcaratus* is the only species known from the Big Badlands, where it is found in the Scenic and Poleslide members of the Brule Formation. It is also found in California, Colorado, Nebraska, Montana, Chihuahua (Mexico), Oregon, and Wyoming.

Natural History and Paleoecology Sites containing multiple *Hypertragulus* skeletons have been found (Fig. 5.44), suggesting that this was a social animal that probably lived in herds.

Hypisodus Cope, 1873

Systematics and Evolution The type species of *Hypisodus* is *H. minimus* from the White River of Colorado, and only a single species is known.

Distinctive Characters *Hypisodus* is smaller than *Hypertragulus* and is about the size of a rabbit. It is distinguished by a narrow and shortened rostrum and an anteroposteriorly compressed braincase. The orbit is large in comparison to the size of the skull, and the postorbital bar is incomplete. The auditory bullae are greatly enlarged so that they meet at the midline of the skull (Matthew, 1902). All of the upper incisors—the canine and the first and second premolar—have been lost. The molars are hypsodont and increase in size posteriorly. The upper molars have flat labial walls. The coronoid process of the mandible is shorter and more slender than in *Leptomeryx* (Fig. 5.45). The third and fourth metatarsals are nearly fused, and the second and fifth metatarsals are reduced.

Stratigraphic and Geographic Distribution The genus appears in the Duchesnean and became extinct in the early Arikareean. *Hypisodus minimus* is only found in the Scenic

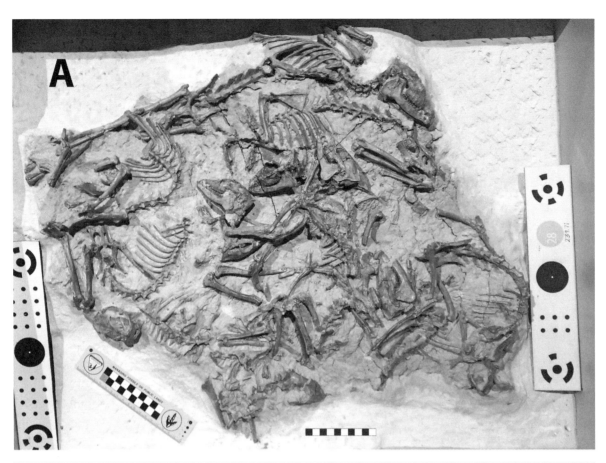

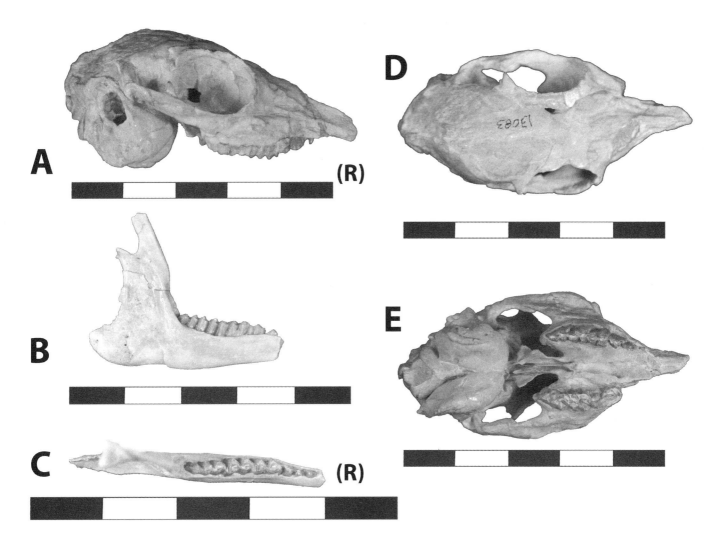

5.45. *Hypisodus* sp., DMNS 13083. (A) Skull, left lateral view, reversed. (B) Right mandible, right lateral view. (C) Right mandible, occlusal view, reversed. (D) Skull, dorsal view. (E) Skull, occlusal view. Scales in centimeters. Photos by the authors of specimens from DMNS. DMNS 13083, Denver Museum of Nature and Science, Denver, Colorado, U.S.A. All rights reserved.

Member of the Brule Formation, but the genus is also known in the Poleslide Member in the Big Badlands. The genus is widespread and is found in a number of localities in Colorado, Nebraska, and Wyoming.

Natural History and Paleoecology On the basis of the large orbit and the enlarged tympanic bulla, *Hypisodus* was probably nocturnal in its habits.

Family Leptomerycidae

Leptomerycids are another group of North American hornless ruminants. Features of the skeleton that characterize the ruminants that can be seen in fossils include the loss of the upper incisors, the presence of procumbent lower incisors,

and the fusion of the cuboid and navicular into a composite bone, the cubonavicular (Webb, 1998).

The family was originally considered to be a subfamily within the Hypertragulidae until a study by Gazin (1955) demonstrated a suite of distinctive features that distinguished the two main North American groups. Members of the Leptomerycidae can be distinguished from the Hypertragulidae by their broader rostrum and by the presence of a complete postorbital bar. As a primitive ruminant, both Leptomerycids and Hypertragulids have orbits located at the midpoint of the skull, reflecting their shorter rostrum than in later forms.

One of the most distinctive features of members of the family is an enlarged procumbent lower first incisor. Unlike the Hypertragulids, the upper canine is reduced in size, and the first upper premolar is lost. The upper first and second

5.44. *Hypertragulus calcaratus.* (A) Multiple skeletons on block, UCM 19659. (B) Skeleton, right lateral view, USNM V 16744. Scale in centimeters. Photos by the authors of specimens from UCM and USNM. UCM 19659, University of Colorado Museum of Natural History, Boulder, Colorado, U.S.A. USNM V 16744, courtesy of Smithsonian Institution, Washington, D.C., U.S.A.

premolars have protocones, and the upper fourth premolar has a strong lingual crescent. The first lower premolar is caniniform or reduced in size, and the other premolars form a continuous series of three-cusped teeth.

Postcranially, Leptomerycids can be distinguished from Hypertragulids by a number of generally more derived features. The odontoid process of the axis is spoutlike. The magnum and trapezoid in the carpus are fused, and the manus is tetradactyl. The astragalus has parallel sides and is not bent, and metatarsals 3 and 4 are fused. The midshaft of the fibula is lost, with only a vestigial proximal end, and the distal malleolus serves as a lock to aid holding the astragalus in place (Webb, 1998).

With the inclusion of *Archaeomeryx* and other Old World genera, the family Leptomeryicidae at one time geographically extended into Asia, but they have been recently placed in their own family, so the Leptomerycids are not considered to be strictly North American but rather derived from an Asian ancestor. The family first appears in the Duchesnean and became extinct in the Arikareean (Heaton and Emry, 1996).

Leptomeryx Leidy, 1853

Systematics and Evolution The type species of *Leptomeryx* is *L. evansi* from the White River Badlands of South Dakota. In addition to the type species, five other species are recognized as valid. Because *Leptomeryx* is hornless, dentition, mostly the lower (because jaws are commonly preserved), has been used to define the different species. This has resulted in the description of multiple species and has produced a confusing taxonomy as a result of the large amount of variation in the morphology of the lower premolars, especially the third. Consequently, individual species often cannot be identified on the basis of a single specimen, and a statistical sample is often needed (Heaton and Emry, 1996). The evolutionary history of *Leptomeryx* includes several important morphological transformations, including the evolution of *Leptomeryx* from *Hendryomeryx* in the latest Duchesnean or earliest Chadronian; the splitting of *L. yoderi* into *L. speciosus* and *L. mammifer* in the early to middle Chadronian; and the evolution of *L. speciosus* into *L. evansi* in the late Chadronian and early Orellan (Heaton and Emry, 1996).

Distinctive Characters This is a small artiodactyl about the size of a house cat. The general body form is similar to that of the living mouse-deer. The skull is elongate with a slender muzzle. The orbits are positioned at the midpoint of the skull. All of the upper incisors have been lost, and both the upper and lower canines are reduced in size, but the lower canine functions as an incisor and the first premolar

is caniniform in shape. The other premolars are narrow and sharp pointed, and the molars are low crowned. The hind legs and feet are longer than the front. The ulna is complete and separate from the radius, but the fibula is reduced. The manus has four digits, all separate, but in the pes there are only two digits, and the metatarsals are fused into a cannon bone (Fig. 5.46).

Stratigraphic and Geographic Distribution The genus first appears in the late Uintan and became extinct at the end of the early Hemingfordian. The type species is the only described species from the White River Group and is known from the Orellan to Whitneyan Scenic and Poleslide members of the Brule Formation of South Dakota. The genus is also found in Colorado, Nebraska, Montana, Saskatchewan, California, Wyoming, and Texas.

Natural History and Paleoecology This is one of the more common artiodactyls present in the Chadronian and Orellan faunas of the Great Plains ranging from Saskatchewan to Texas. Fossil deposits with this genus are sporadic in their distribution, making environmental and evolutionary reconstruction difficult. Important evolutionary transitions are not always recorded (Heaton and Emry, 1996).

Because *Leptomeryx* is so common, sufficiently large samples have been recovered, including accumulations of multiple skeletons that permit population studies. According to Clark and Guensburg (1970), *Leptomeryx* occurs in collections in greater numbers than any other genus except *Palaeolagus*, although this statement is difficult to believe, given the abundance of the oreodont *Merycoidodon*. Many specimens have been recovered as complete articulated skeletons, and multiple individuals have been recovered in a large block, suggesting the presence of a single herd. Scott and Jepsen (1940) cited one assemblage containing 20 individuals. These death assemblages were found near water sources associated with a swamp on a grassy or savanna plain. When wear stages of the teeth were examined, six age groups were identified in the samples. The presence of distinct age classes in the populations suggests seasonal breeding and birthing. Maximum life span was inferred to have been about 8 years (Clark and Guensburg, 1970).

Order Perissodactyla

The order Perissodactyla is represented today by horses, tapirs, and rhinoceros and by two extinct groups, brontotheres and chalicotheres. The order first appears in the late Paleocene in North America, and fossils of the group are found on all continents except Australia and Antarctica. Members of the order are characterized by having the axis of symmetry of the feet pass through the middle of the third digit

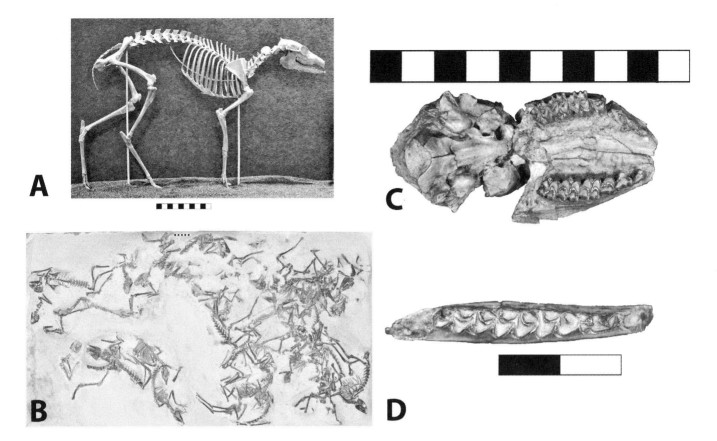

5.46. *Leptomeryx* sp. (A) Skeleton, right lateral view, PM 61157. (B) Block containing several skeletons, P 12320. (C) *Leptomeryx evansi* partial skull, occlusal view, USNM 157 holotype.

(D) *Leptomeryx evansi* partial right mandible, occlusal view, USNM 157 holotype. Scales in centimeters. Photos A and B by William Simpson, Field Museum of Natural History; photos

C and D by the authors of specimens from USNM. USNM 157, courtesy of Smithsonian Institution, Washington, D.C., U.S.A.

(mesaxonic), and while they are popularly called the odd-toed ungulates, some forms may actually have four digits, usually on the manus. Another characteristic of the order is the presence of a prominent and well-developed third trochanter on the femur. The astragalus is distinctive, with a single pulley surface for articulation with the tibia rotated so it is at an angle to the axis of the bone and a flattened and compressed navicular process. The oblique axis of the trochlea of the astragalus is also reflected in the distal end of the tibia in which the articular surfaces are oriented obliquely to the long axis of the shaft.

Family Tapiridae

The family Tapiridae originated in North America, where it underwent most of its evolutionary history. Today there is a single genus, with three species in South America and a single species in Malaysia. The family shares a common ancestry with rhinos, but the two lineages diverged in the early Eocene, and the earliest tapir is from the early Uintan.

Although tapirs have a typical perissodactyl manus and pes in which the axis passes through the third digit and the third metapodial is the largest in the series, the manus has four digits. The first has been lost and the fifth is reduced in size and nonfunctional. The pes has three digits, having lost the first and fifth, but digits 2, 3, and 4 are all functional. First appearing in the middle Eocene (Bridgerian) in the genus *Helaletes* (found in Colorado, Montana, Nevada, Utah, and Wyoming), a characteristic of tapirs is the shortening of the nasals and the development of a short, fleshy, prehensile proboscis, a feature present in living tapirs. They have a complete placental dental formula of C3/3-I1/1-P4/4-M3/3 totaling 44 teeth. The incisors are chisel shaped, and the third upper incisor is the largest in the series and is conical. It is separated by a short diastema from the smaller canine. There is a much longer diastema between the canines and premolars, the first of which may be absent. While the dentition is brachyodont, the cusps have coalesced to form lophs. In the upper dentition, the lophs form a characteristic pattern that resembles the Greek letter pi (π), with a straight ectoloph on the labial side of the tooth. The lower teeth have two unconnected transverse lophs. As in all members of the order, the femur has a well-developed third trochanter. The fibula is complete (Colbert and Schoch, 1998).

Protapirus Filhol, 1877

Systematics and Evolution The type species of *Protapirus* is *P. priscus* from the phosphorite deposits of Quercy, France (late Eocene). Five species are recognized in North America.

Distinctive Characters *Protapirus* is in the medium size range for members of the family. The third and fourth premolars are nonmolariform to submolariform. Differences in tooth morphology used to describe the different species are based on differences in the premolars, and it is possible that these differences simply reflect individual variation within a population. The genus is in need of revision. The shortening of the nasals in *Protapirus* is not as pronounced as in modern tapirs, and they extend to above the canine, so the flexible proboscis may not have been as well developed. They are separated from the maxillae by an enlarged narial incision. The manus has four digits; the first digit is lost, and the fifth digit is reduced to a similar degree as in the Malayan tapir and thus may not have borne weight (Colbert and Schoch, 1998).

Stratigraphic and Geographic Distribution The genus has the longest stratigraphic range of all the North American tapirs, first appearing in the late Uintan and becoming extinct at the end of the early Hemingfordian. *Protapirus simplex* is reported from the Scenic Member of the Brule Formation, and *P. obliquidens* and *P. validus* are reported from the Poleslide Member of the Brule Formation of South Dakota (Wortman and Earle, 1893). The genus is also known from Oregon.

Natural History and Paleoecology Modern tapirs are forest animals and closely associated with water. It appears that they filled this adaptive niche of living in humid mesothermal regions with a diversity of plants that provided large quantities of forage early in their evolutionary history. As herbivores, they feed on small branches and leaves as well as fresh sprouts along with fallen fruit and water plants. They use their flexible proboscis to feed, and it allows them to selectively pull leaves, shoots, and fruits into their mouths. The evolutionary history and major changes in the skeleton are related to the refinement of the proboscis, molarization of the premolars, and a general increase in size, but the postcranial skeleton tends to remain conservative and similar in all taxa.

Colodon Marsh, 1890

Systematics and Evolution The type species of *Colodon* is *C. occidentalis* from the Nebraska Territory. It was originally described by Leidy as *Lophiodon*, but the species was transferred to a new genus, *Colodon*, by Marsh. Currently the genus includes four other recognized species.

Distinctive Characters *Colodon* is considered a tapiroid (but not a true tapir) in the family Tapiridae by Colbert and Schloch (1998). The canines are small or absent, and the cheek teeth are short and wide. The upper first premolar has a lingual loph or cusp, but the other premolars are molariform with separated protocone and hypocone, and with the metaloph as prominent as the protoloph and connecting to the hypocone. The lower first premolar is absent, and the other premolars have wide talonids and a relatively large entoconid. The lower premolars have distinct posterior cusps that have not joined to form a loph. The third lower molar has a third lobe. The skull has a greatly enlarged narial incision and reduced nasals. Colbert (2005) noted that although *Colodon* is only about half the size of modern tapirs, the skull shared more characters of the skull with the modern genus *Tapirus* than the contemporary genus, *Protapirus*. These included several skeletal indicators that *Colodon*, like the living genus, probably had a prehensile proboscis: a retracted nasoincisive incisure similar to *Tapirus*, the occurrence of apomorphic fossae for cartilaginous meatal diverticula on the dorsal frontals and nasal, and ascending maxillae. Other similarities between *Colodon* and *Tapirus* include the telescoping of the skull, resulting from the anteroposterior shortening of the frontals, and the development of frontal sinuses; the conformation of the rostrolateral processes of the frontals; the descending processes of the nasals and the ascending process of the maxillae; and the apparent contact with the cartilaginous nasal septum by the premaxillae, as indicated by a CT scan of two species of *Colodon* from the Big Badlands (Colbert, 2005). As in *Protapirus*, the first digit is absent in the manus, but the bones of the fifth digit are more reduced in size.

Stratigraphic and Geographic Distribution The genus first appears in the early Uintan and the last definite appearance is in the Whitneyan, although possible records have been reported from the early Arikareean. In South Dakota, *C. occidentalis* is known from the Ahearn Member of the Chadron Formation, with questionable records from the Crazy Johnson and Peanut Peak members, and it is known from the Scenic and Poleslide members of the Brule Formation. *Colodon occidentalis* has also been reported from faunas in Nebraska and Saskatchewan. The genus is also known from Montana, Wyoming, California, and Oregon.

Family Equidae

The family Equidae today is represented by a single genus, *Equus*, and includes horses, zebras, and donkeys and a diverse number of fossil genera. The primary characteristics of the family are the lack of separation of the foramen ovale and

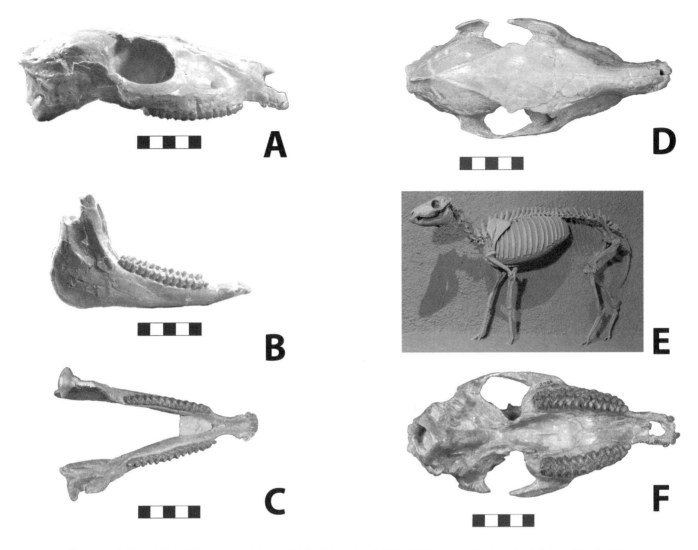

5.47. *Mesohippus bairdi,* SDSM 2920. (A) Skull, right lateral view. (B) Mandible, right lateral view. (C) Mandible, occlusal view. (D) Skull, dorsal view. (E) *Mesohippus* sp. USNM V15898, skeleton. (F) *Mesohippus bairdi,* SDSM 2920, skull, occlusal view. Scales in centimeters. Photos by the authors of specimens from SDSM and USNM. SDSM 2920 is from the Museum of Geology, South Dakota School of Mines and Technology, Rapid City, South Dakota, U.S.A. USNM V15898, courtesy of Smithsonian Institution, Washington, D.C., U.S.A.

the optic foramen from the ethmoid and anterior lacerate foramen in the base of the skull. Members of the family have a diverse range of dentition, from brachyodont and bunodont to lophodont and hypsodont, reflecting their evolutionary transition from browsers to grazers. Despite the variety of tooth morphology, all members of the family have a structure called the postprotocrista on the upper third premolar (Hooker, 1989). In early horses, the morphology of the premolars is different from the molars, but with the appearance of *Mesohippus* in the White River Group, it and all subsequent horse genera have molarized premolars. The facial region tends to lengthen to accommodate the increasing height of the cheek teeth, resulting in a shift of the position of the orbit more posteriorly on the skull. Early equids have prominent facial fossae on the maxillae, but this becomes reduced and eventually lost in later genera. Although there is a general increase in body size through time (MacFadden, 1986), this is not true for all horse lineages, and some Pliocene (Blancan) genera are small and similar in size to horses from the White River Group.

Like all perissodactyls, horses have a well-developed third trochanter on the femur. Early horses have a separate ulna and radius, but in advanced forms, the shaft of the ulna is reduced, leaving only the proximal and distal ends, which fuse to the enlarged radius. The fibula becomes reduced over time and is eventually lost in the modern genus, *Equus.* Another trend in the family is the continuous loss of the medial and lateral digits and increasing emphasis of the third metapodial to support the animal. Horses from the White River Group tend to be more primitive in their anatomy than later horses and have played a major role in understanding the evolution of horses as representing transitional forms between early and later members of the family.

Mesohippus Marsh, 1875

Systematics and Evolution The type species of *Mesohippus* is *M. bairdi*, probably collected from Bear Creek, South Dakota. Nine species are currently recognized. There have been preliminary attempts to refine the species-level taxonomy of the genus, but additional work is required.

Distinctive Characters *Mesohippus* was a small animal with an estimated body mass of 42.2 kg for *M. bairdi* and 47.7 kg for *M. barbouri* (MacFadden, 1986). There is an increase in the size of the genus from the Chadron through the Brule. The geologically older and smaller individuals have hind legs longer then the front so the rump is higher than the shoulders, while in later forms the front and hind limbs are similar in length. The orbits are farther forward on the skull than in later horses and positioned above the first molar. There is no postorbital bar, and the orbit is confluent with the temporal fossa. There is a maxillary fossa in front of the lachrymal. *Mesohippus* retains the low-crowned (brachyodont) cheek teeth present in earlier horses, but is the first horse in which the premolars have become molariform (Fig. 5.47). The humerus has an undivided bicipital groove, a feature that evolves in later horses to facilitate sleeping upright (Hermanson and MacFadden, 1992), and the greater tuberosity extends medially over the bicipital groove. The ulna, although reduced in size, remains distinct from the radius. The fibula is complete but the distal end and often portions of the shaft fuse to the tibia (Scott, 1891). Some species of *Mesohippus* have four metacarpals but have reduced the number of metatarsals to three, so it is considered to be the first fully tridactyl horse (MacFadden, 1998).

Stratigraphic and Geographic Distribution The genus first appears in the Duchesnean and became extinct at the end of the early early Arikareean, although there are some questionable records for its presence in the late early Arikareean. Most specimens of *Mesohippus* from the White River Group are referred to *M. bairdi*, which is found in all members of the Chadron Formation. *Mesohippus* from the Scenic Member of the Brule are referred to a second species, *M. westoni*. Another species, *M. exoletus*, is reported from the Peanut Peak Member of the Chadron and the Scenic Member of the Brule, and *M. barbouri* is reported from the Scenic Member of the Brule Formation (MacFadden, 1998). Outside of the Big Badlands of South Dakota, the genus is widespread and is found throughout the western United States, including Wyoming, Nebraska, Colorado, Montana, North Dakota, Oregon, and Texas, then north into Canada and Saskatchewan. It is the first equid known in the southeastern United States and Florida (Patton, 1969).

Natural History and Paleoecology The diet of *Mesohippus* has been considered to be that of a browser, and Clark, Beerbower, and Kietzke (1967) considered it to be a member of the aquatic wet forest habitat of the White River Group. They considered *Mesohippus*, along with *Merycoidodon*, to be the dominant taxa of the gallery forests along river courses during the Orellan. Isotopic analysis of tooth enamel of *Mesohippus* from the Chadron and Brule formations in South Dakota produced carbon-13 values between -10 and $-14.1^0/oo$, indicating a diet of C_3 vegetation. C_3 grasses today grow at either higher latitudes or elevations that provide cooler environments but also grow in the understory of forests (Wang, Cerling, and MacFadden, 1994). The pure C_3 signal in their tooth enamel supports the idea they were predominately browsers living in an environment dominated by C_3 vegetation, such as trees, shrubs, and herbaceous dicots.

Miohippus Marsh, 1874

Systematics and Evolution The type species of *Miohippus* is *M. annectens* from the late Oligocene/early Miocene of John Day, Oregon. Numerous species have been described for this genus. Eighteen were listed by Stirton (1940), but currently six species are recognized.

Distinctive Characters *Miohippus* and *Mesohippus* are both present throughout the White River Group. The distinctions between the two genera are subtle. *Miohippus* is about the size of a sheep and is larger than most species of *Mesohippus*, with an estimated body mass for *M. quartus* of 53.8 kg (MacFadden, 1986). The primary difference used to distinguish the two genera is *Miohippus* has an articulation between the cuboid and third metatarsal, larger hypostyles on the teeth, a more elongate face, and a deeper facial fossa. The articulation between the third metatarsal and cuboid reflects a widening of the proximal end of the former bone and reflects its increase in size to support the animal with the reduction of the other digits (Prothero and Shubin, 1989).

Stratigraphic and Geographic Distribution *Miohippus* first appears in the early Chadronian and became extinct at the late Arikareean. In South Dakota, the type species and *M. assiniboiensis* have been reported from the Poleslide Member of the Brule Formation; *M. obliquidens* is present in the Peanut Peak Member of the Chadron Formation and all members of the Brule Formation; and *M. gidleyi* is reported from the Poleslide Member of the Brule Formation. The genus is also found in California, Oregon, Wyoming, Colorado, Nebraska, North Dakota, and Saskatchewan.

Natural History and Paleoecology Given the close morphological similarities between *Mesohippus* and *Miohippus*, as well as their overlapping geography and chronology, it is difficult to determine how they were ecologically different or how they may have filled different niches. Both would have

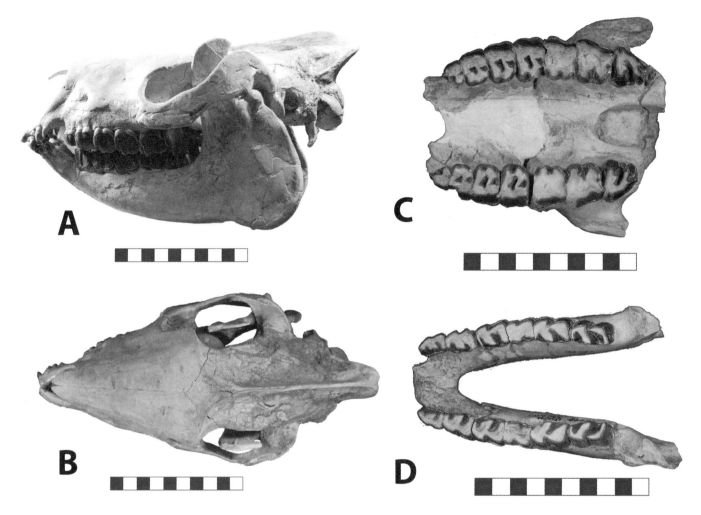

5.48. *Hyracodon nebraskensis.* (A) SDSM 28100, skull and jaws, left lateral view. (B) SDSM 28100, Skull, dorsal view. (C) USNM 336207, holotype, partial skull, occlusal view. (D) USNM 336207, holotype, left and right mandibles occlusal view. Scales in centimeters. Photos by the authors of specimens from SDSM and USNM. SDSM 28100 is from the Museum of Geology, South Dakota School of Mines and Technology, Rapid City, South Dakota, U.S.A. USNM 336207, courtesy of Smithsonian Institution, Washington, D.C., U.S.A.

been browsers but must have utilized different habitats on the landscape.

Family Hyracodontidae

This family of rhinoceros is known from Eurasia and North America from the middle Eocene to Oligocene, with five genera in North America. They are distinguished from other rhinos by their elongated limbs, suggesting a higher degree of cursoriality than their contemporaries. The skull tends to be primitive with a full dentition including all the incisors and canines, although there is incisor loss in some forms. The sagittal crest is well developed. The nasals are elongate, but there are no bosses indicative of the presence of horns. The rostrum and anterior part of the skull are laterally constricted. The dentition was more hypsodont than many of their contemporary mammalian taxa such as the horses and oreodonts. The upper third molar is reduced in size, unlike other rhinos. The pes is tridactyl, and the metapodials are elongated as part of the cursorial adaptation.

Hyracodon Leidy, 1856

Systematics and Evolution The type species of *Hyracodon* is *H. nebraskensis.* Five species are currently recognized.

Distinctive Characters *Hyracodon* is the largest member of the family. In the Whitneyan, there were two contemporary species, *H. nebraskensis* and the larger *H. leidyanus.* Although the genus is among one of the longest lived, from approximately 10 Ma, there seems to have been little evolutionary change during this time. There is no indication that it had horns and no evidence that it was sexually dimorphic. The incisors are pointed and the canines are small. The first lower premolar is lost. The paracones on the premolars and molars are not separated posteriorly from the ectoloph (Fig. 5.48) (Prothero, 1998a).

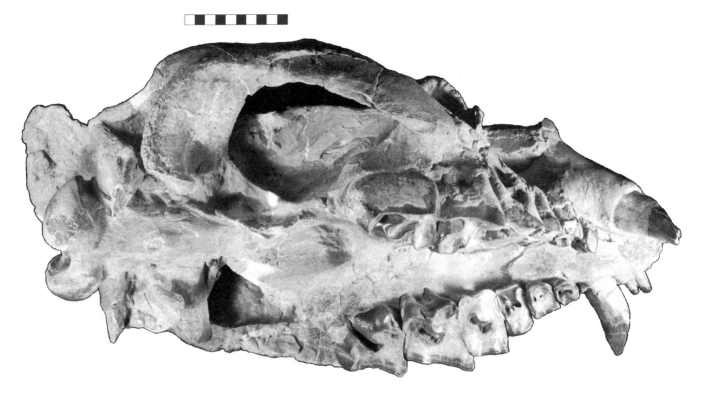

5.49. *Metamynodon planifrons,* SDSM 3634. Skull, occlusal view. Scale in centimeters. Photo by the authors. SDSM 3634 is from the Museum of Geology, South Dakota School of Mines and Technology, Rapid City, South Dakota, U.S.A.

Stratigraphic and Geographic Distribution The genus appears in the Duchesnean and became extinct in the early early Arikareean, making the last member of the family in North America. In South Dakota, *Hyracodon nebraskensis* is present in all members of the Brule Formation, and *H. leidyanus* is present in the Poleslide Member of the Brule Formation. *Hyracodon priscidens* is present in the Peanut Peak Member of the Chadron Formation. The genus is also found in Utah, Wyoming, Colorado, Nebraska, Montana, North Dakota, Saskatchewan, and Texas.

Natural History and Paleoecology Remains of *Hyracodon* are fairly common in the White River Group, making this genus the best known of the family. No large quarry samples similar to other White River Group taxa have been found, so they may not have formed herds and like modern rhinos been primarily solitary in their habits. *Hyracodon* is ubiquitous in its distribution and has been found in deposits indicative of near-stream, swampy plains and open plains habitats.

It has been generally considered to be one of the most cursorial rhinos in terms of its skeletal anatomy. This interpretation has been challenged by Wall and Hickerson (1995) because the limb indices of *Hyracodon* do not compare to modern cursorial animals and are more similar to the hippopotamus and pigs. Clark, Beerbower, and Kietzke (1967) linked the decline of *Hyracodon* in the White River Group

to the decrease of wet forest habitat and the increase in savanna and prairie open habitat. One would expect that if *Hyracodon* was cursorial in its adaptations, it would have increased in numbers with the expansion of open habitat.

Family Amynodontidae

The family Amynodontidae originated in Asia and dispersed into North America. This family of rhinocerotoids first appears in North America in the middle Eocene (late Bridgerian) and survived into the early Oligocene (Whitneyan). The family underwent a minor radiation in North America during the middle Eocene that seemed to coincide with the presence of lush tropical forest that existed in response to a homogenous equitable humid climate. They were medium to large in size and have been interpreted as semiaquatic in their habits. River channel deposits in the Orellan rocks of the White River Group have been named the *Metamynodon* channel sandstones. Their decline is attributed to the increasing aridity from the Eocene to Oligocene.

All members of the family have a large facial fossa on the skull. The size and location on the skull varies with different genera. There is a prominent sagittal crest, a primitive feature to all rhinos. All members of the family have greatly enlarged canines, and the size of the canines appears to be sexually dimorphic. There is a trend to reduce the number of premolars, and the upper and lower first premolar is absent in all forms. For most described amynodonts, the postcranial skeleton is unknown. The primitive build of the body seems to be subcursorial to mediportal, but metamynodontines are heavy limbed and barrel chested, and thus closely parallel features seen in the skeletons of hippos. Whether they had hippolike feeding habits, leaving the water to graze on land or feed on aquatic plants in the river, is unknown.

Metamynodon Scott and Osborn, 1887

Systematics and Evolution The type species of *Metamynodon* is *M. planifrons.* In addition to the type species, two other species are considered valid.

Distinctive Characters The antorbital or facial fossa characteristic of members of the family is greatly reduced in *Metamynodon.* The orbits are positioned high on the skull, similar to hippos, supporting the notion of its semiaquatic habits. The second lower premolar is lost, and the third lower premolar is small or may be absent (Fig. 5.49).

Stratigraphic and Geographic Distribution The genus appears in the late Uintan and became extinct at the end of the Whitneyan. It has the longest stratigraphic range for the family and was the last genus of the family to become extinct. *Metamynodon planifrons* is found in all members of the Brule Formation. A second species, *M. chadronensis,* is known from the Peanut Peak Member of the Chadron Formation. The genus is also known from Mississippi, North Dakota, and Texas.

Family Rhinocerotidae

The name *rhinoceros* translates directly from Greek, "nose horn." This feature is present in all five extant species of rhinos, although many extinct members of the family did not have a horn. Modern rhinos may have one horn or two aligned anteroposteriorly, growing from the top side of the nose. Unlike the horns of bovid artiodactyls, the horns of the rhinoceroses lack a bony core and are composed of compressed hair or keratin. They are attached to the skull via a bony boss, and the presence of this ossification on the nasals can be used to infer the presence of a horn in fossil forms. The family first appears in North America as immigrants from Asia, and it subsequently underwent a radiation, with 13 genera currently recognized. The family became extinct in North America at the end of the late Miocene (Hemphillian).

The skull of members of the family tends to be low and saddle shaped. The parietals support broad parasagittal crests. The nasals are long and slender and separated from the maxillae by a deep narial notch. While modern members

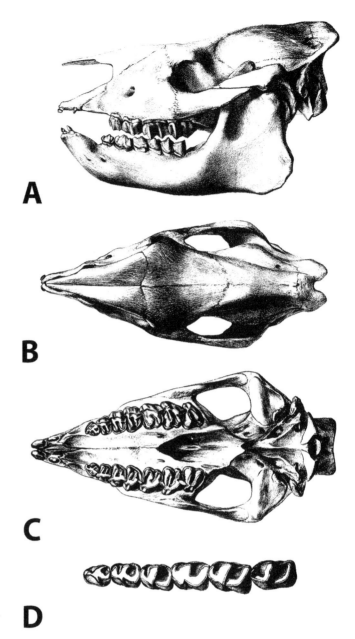

5.50. *Subhyracodon occidentalis,* adapted from Scott and Jepsen (1941:plate 85, figs. 1, 2, 2a, 3). (A) Skull and jaws, right lateral view, AMNH 534, juvenile. (B) Skull, top view, AMNH 534, juvenile. (C) Skull, occlusal view, AMNH 534, juvenile. (D) Lower cheek teeth, p2–m3, YPM PU 13138.

of the family are characterized by the presence of a prominent horn or horns on the skull, many fossil members of the family lacked horns. A distinctive dental feature of the family is a chisel-shaped upper first incisor, which occludes with a tusklike lower second incisor. Except in the most primitive members of the family, all the other incisors, along with the canines, are lost. The upper molars have the typical pi pattern of lophs with a straight ectoloph. In primitive members of the family, the premolars are not fully molarized but become more so in advanced forms. In advanced forms, the teeth are more hypsodont, and they may develop additional

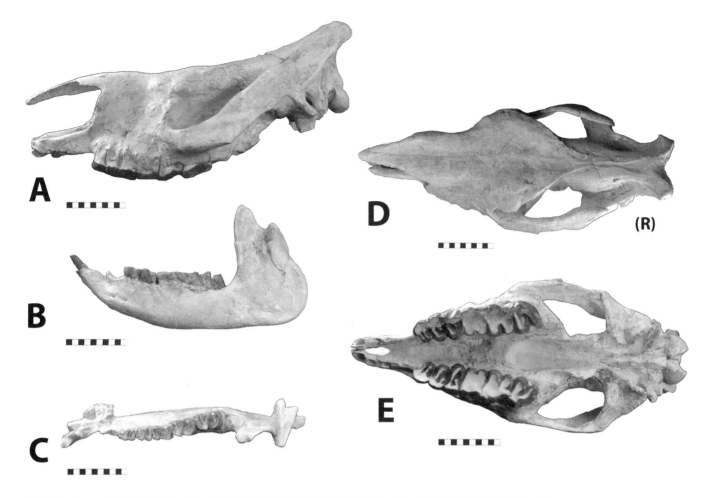

5.51. *Trigonias* sp. (A) Skull, left lateral view, DMNS EPV.953. (B) Mandible (L), left lateral view, DMNS EPV.2674. (C) Mandible (L), occlusal view, DMNS EPV.2674. (D) Skull, dorsal view, DMNS EPV.953. (E) Skull, occlusal view, DMNS EPV.953. Scales in centimeters. Photos by the authors. Denver Museum of Nature and Science, Denver, Colorado, U.S.A. All rights reserved.

internal ridges, often referred to as crochets, antecrochets, and cristae, that increase the surface area of the teeth and are an adaptation to ingesting more abrasive food.

Bones of the postcranial skeleton tend to be massive and readily preserved as fossils. The ulna and radius remain separate, and a complete and separate fibula is retained even in advanced forms. Most rhinos are tridactyl, with digit 3 being the largest, although a functional fifth metacarpal is present in some genera.

Subhyracodon Brandt, 1878

Systematics and Evolution The type species of *Subhyracodon* is *S. occidentalis* was reported as coming from the "Missouri Territory," now thought to be the Badlands of South Dakota. Two other species, *S. mitis* and *S. kewi*, are considered valid.

Distinctive Characters *Subhyracodon* is the largest of the true rhinos found in the Big Badlands and is the most common. Like many other fossil rhinos, it was hornless. It has lost the upper third incisor and canine, and the premaxillae are narrower and more delicate than in other contemporary rhinos. The premolar row is relatively shorter, and the upper second premolar has become molarized in comparison to earlier rhinos (Fig. 5.50). The limb bones are relatively heavy and more massive. The feet are tridactyl. One species of *Subhyracodon* evolved into *Diceratherium* in the Whitneyan, but the two genera coexist from the Whitneyan to early Arikareean, at which time *Subhyracodon* became extinct. One characteristic of the skull that distinguishes *Diceratherium* from *Subhyracodon* is the development of paired nasal ridges that suggest the presence of keratinous horns (Prothero, 1998c).

Stratigraphic and Geographic Distribution The genus first appears in the early Chadronian and survived until the early Arikareean. *Subhyracodon occidentalis* is present in the Scenic Member of the Brule Formation. *Subhyracodon mitis* is present in all members of the Chadron Formation. The two species are primarily distinguished by a difference in size, with *S. mitis* being the smaller of the two species. The

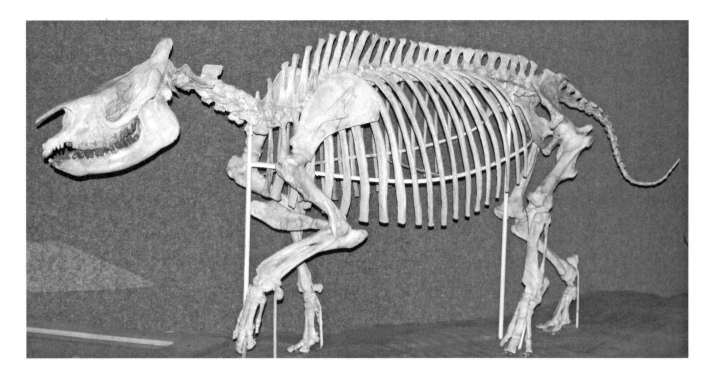

5.52. *Trigonias* sp., DMNS 872. Skeleton, left lateral view. Composite specimen. Photo by the authors. Denver Museum of Nature and Science, Denver, Colorado, U.S.A. All rights reserved.

genus is not only known from the Big Badlands of South Dakota but is also present in faunas in Wyoming, Colorado, Montana, North Dakota, Saskatchewan, and California, and it has also been found in Mississippi (Prothero, 2005).

Natural History and Paleoecology *Subhyracodon* is particularly common in the Lower Nodular zone of the Scenic Member of the Brule Formation, which is Orellan in age. Numerous individuals of *Subhyracodon* have been recovered from an ancient water hole in Badlands National Park, called the Big Pig Dig. The large number of individuals probably accumulated over a short period of time, possibly during a drought. Many of the bones of *Subhyracodon* recovered from the Big Pig Dig show evidence of bite marks that suggest scavenging by the entelodont *Archaeotherium*. Besides the Big Pig Dig, a large number of individuals were recovered from the Nuttal Rhino Quarry and the Harvard Fossil Reserve in Wyoming and the Rockerville Site in South Dakota, as well as other localities in the Black Hills, including Wind Cave National Park. Rhino tracks referred to *Subhyracodon* preserved at Toadstool Geologic Park in Nebraska include numerous parallel trackways, suggesting that the genus may have formed small herds, unlike modern rhinos, which tend to be solitary.

As a larger-bodied rhino with more massive and stocky limb bones, *Subhyracodon* probably was not as cursorial as other forms such as the smaller *Hyracodon*. Its preferred habitat seems to have been riparian habitat along rivers, where it appears to have been a mixed feeder.

Diceratherium Marsh, 1875

Systematics and Evolution The type species of *Diceratherium* is *D. armatum* from the John Day Formation of Oregon. Other recognized species of the genus include *D. tridactylum, D. annectens, D. matutinum,* and *D. niobrarense.* The genus evolved from *Subhyracodon.*

Distinctive Characters Unlike its ancestor, *Subhyracodon,* as the name suggests, *Diceratherium* had a pair of nose horns positioned side by side just before the tip of the nasals in males, as indicated by the rugosity of the bone. These are absent in the females. The skull is elongate, with a flat dorsal profile. It is also larger than *Subhyracodon.* The second to fourth premolars are more molariform than in *Subhyracodon.* The molars have strong cingula with simple crown patterns that lack either crochets or cristae. The limbs are long but the metapodials are short (Prothero, 2005).

Stratigraphic and Geographic Distribution *Diceratherium* is one of the longer-lived North American rhinos appearing in the Whitneyan; it did not become extinct until the Hemingfordian. *Diceratherium armatum* and *D. annectens* have been reported from the Peanut Peak Member of the Chadron Formation (Prothero, 1998c). The holotype of *D. tridactylum* is a complete skeleton of a female collected from the *Protoceras* beds, Poleslide Member of the Brule Formation (late Whitneyan) of South Dakota, and the species is known only from this unit in the Big Badlands. It is also known from California, Wyoming, Montana, Nebraska,

Saskatchewan, and Oregon, as well as south to Arizona and Panama.

Trigonias Lucas, 1900

Systematics and Evolution The type of *Trigonias* is *T. osborni* from the Chadron Formation of the Big Badlands. A second species, *T. wellsi*, is also known from the Big Badlands. The primary difference between the two species of *Trigonias* is size, with *T. osborni* being the smaller of the two species (Prothero, 2005).

Distinctive Characters *Trigonias* is a medium-size rhino. It retains all of its anterior teeth except the lower third incisor and canine. The upper first incisor and lower second incisor show the typical rhino chisel/tusk modifications. There is no evidence for the presence of horns. The skull is elongate, with a low, saddle-shaped profile. The sagittal and occipital crests are moderately broad (Figs. 5.51, 5.52). The manus has a functional fifth metacarpal that is absent in most other genera (Prothero, 1998c).

Stratigraphic and Geographic Distribution *Trigonias osborni* is known from the Ahearn and Crazy Johnson members of the Chadron Formation. A second species, *T. wellsi*, is from the Ahearn Member of the Chadron Formation of the Badlands of South Dakota. The genus is known also from Nebraska, Colorado, Montana, Wyoming, California, and Saskatchewan.

Natural History and Paleoecology That the two species of *Trigonias* are known only from the Chadronian and have overlapping distributions is suggestive that they may represent males and females of the same species, although *T. wellsi* is known from a fewer number of localities. This may have been a more social species than other rhinos. A quarry site in Weld County, Colorado, produced multiple individuals.

Penetrigonias Tanner and Martin, 1976

Systematics and Evolution The type species of *Penetrigonias* is *P. hudsoni* from the upper part of the Chadron Formation at Chadron Flats, north of Toadstool Geologic Park, Sioux County, Nebraska. It is also known from Wyoming and the Ahearn Member of the Chadron Formation in the Big Badlands. *Penetrigonias dakotensis* is found in Wyoming, Colorado, and the Poleslide Member of the Brule Formation in the Big Badlands, while the third species, *P. sagittatus*, is known only from the Chadronian of the Cypress Hills of Saskatchewan.

Distinctive Characters This rhino is smaller than its contemporaries, *Subhyracodon*, *Trigonias*, or *Amphicaenopus*. The premaxilla is short but still retains all of its incisors,

and it has the typical rhinocerotid incisor complex. There is some molarization of the premolars, but the degree to which it occurs is variable. The range of morphological variation present in the premolars resulted in the description of multiple species, taxonomically oversplitting the genus (Prothero, 1998c).

Stratigraphic and Geographic Distribution The genus has a long biostratigraphic range, first appearing in the Duchesnean and becoming extinct in the Whitneyan. Despite its long stratigraphic range, it has not yet been found in any Orellan faunas (Prothero, 2005). *Penetrigonias dakotensis* is known from the Poleslide Member of the Brule Formation in the Big Badlands of South Dakota and has been found in Wyoming and Colorado. *Penetrigonias sagittatus* is known only from the Chadronian of the Cypress Hills of Saskatchewan. The Duchesnean records are from Texas and Utah. Besides South Dakota, the Chadronian records are from Nebraska, Wyoming, Montana, and Colorado, and the single record from the Whitneyan is from the *Protoceras* beds of South Dakota.

Amphicaenopus Wood, 1927

Systematics and Evolution *Amphicaenopus* is only known from the type species, *A. platycephalus*, from the Whitneyan *Protoceras* channels of South Dakota.

Distinctive Characters This is a large rhino, larger than *Subhyracodon* and *Diceratherium*. The skull is broad and short with greatly expanded occipital crests. The lower second incisor is procumbent. The upper second and third premolars are not molariform, but the upper fourth premolar is nearly molariform. The nasals are short, and the notch between them and the maxilla that is commonly present in other rhinos is absent. The orbit is positioned farther forward on the skull, above the second molar, than in other rhinos (Prothero, 1998c).

Stratigraphic and Geographic Distribution The genus first appears in the early Chadronian and became extinct at the end of the Whitneyan. Like *Penetrigonias*, it has not been found in any Orellan faunas. In the Big Badlands of South Dakota, it is known from the Chadron Formation and the upper Poleslide Member of the Brule Formation (late Whitneyan).

Natural History and Paleoecology The body proportions of *Amphicaenopus* are hippolike and reminiscent of *Metamynodon* (Prothero, 2005). The lower incisors show striations due to wear, and Bjork (1978) suggested the presence of a flexible upper lip used to pull vegetation across the tusk.

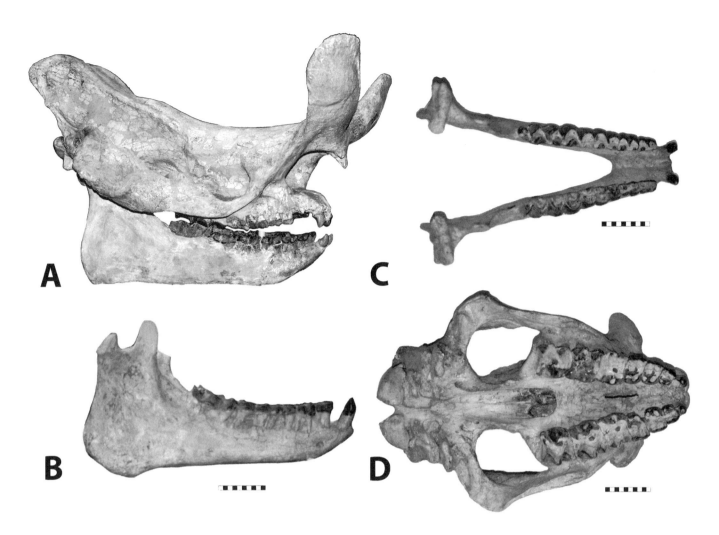

5.53. *Megacerops* sp. (A) P12161, skull and jaws, right lateral view. (B) DMNS 3748, mandible, right lateral view. (C) DMNS 3747, mandible, occlusal view. (D) DMNS 3746, skull, occlusal view. Scales in centimeters. Photo (A) by William Simpson, Field Museum of Natural History; photos (B), (C), and (D) by the authors of specimens from DMNS. DMNS 3748, 3747, and 3746, Denver Museum of Nature and Science, Denver, Colorado, U.S.A. All rights reserved.

Family Brontotheriidae

This extinct family, also sometimes referred to as titanotheres, first appears in the late Wasatchian and became extinct at the end of the Chadronian (the Eocene–Oligocene boundary). The phylogenetic relationship of this family to other perissodactyls has not been resolved. The family evolved in North America, but some forms dispersed into Asia, where some forms may have survived after the group became extinct in North America.

Early members of the family are small, but there is a trend of increasing size, and by the late Eocene, they are larger than modern rhinos and were the largest species in the fauna. Given the size of full-grown adults in the late Eocene, it is unlikely that any of the available predators, whether any of the five saber-toothed cats, creodonts, or even canids, preyed on these animals. Even juvenile brontotheres were disproportionately large compared to the contemporary predators and would have been difficult to kill.

While early forms had relatively long faces in proportion to overall skull size, in the later forms, the face is shortened while the skull posterior to the orbit has become greatly elongated. Later members of the family are characterized by their large body size and the development of prominent horns composed of bony extensions from the nasals and frontals. It is possible that the horns were covered with a keratinized skin, but there is no direct evidence for this. The size and robustness of the horns is sexually dimorphic, as is the body size, with males being significantly larger than females (Fig. 5.53A). Evidence that males used their horns on each other in in displays of dominance or during rut is provided by broken and healed ribs. In forms with large horns and skulls, the neck is short. Variation in the horns has been used as the basis for taxonomy in the group, resulting in a plethora of names being created since their initial discovery.

Brontotheres are interpreted as having been obligatory browsers on the basis of the morphology of their low-crowned dentition, which was used to either crush or shear vegetation.

5.54. The first fossil discovered in the White River Badlands. Described by H. A. Prout of St. Louis, 1846–1847. Partial mandible of a brontothere. USNM 21820. Scale in centimeters. Photo by the authors. USNM 21820, courtesy of Smithsonian Institution, Washington, D.C., U.S.A.

The incisors are small, are reduced in number, have a buttonlike crown, and do not show any signs of wear. While canines are present, they are short and do not project beyond the level of the cheek teeth. The reduction of the incisors and lack of wear suggests a prehensile lip, perhaps similar to that seen in living rhinos. The upper dentition has a characteristic W-shaped ectoloph separated from the two prominent lingual cusps. The lower molars have a distinctive M-shaped pattern formed by the fusion of the cusps into a single loph (Fig. 5.53C). The premolars never become molariform. They probably lived in warm temperate to subtropical environments that included forest to relative open woodland.

As expected from the large size and massive body of the late Eocene forms, the bones of the postcranial skeleton are massive and robust. Modifications of the skeleton include graviportal features such as columnar limbs and reduced area of articulation between limb bones, indicating a reduced range of flexion and extension. The shaft of the femur has become flattened, and the third trochanter is reduced in size compared to the overall size of the animal. The ulna and radius remain separate, and the ulna is heavy in its build. A separate stout fibula is present. The manus has four digits; the pes has three. The arrangement of the bones in the manus and pes suggests that the feet had a pad of elastic tissue, similar to that seen in modern elephants and rhinos. As in modern elephants and rhinos, the unguals are reduced and would not have carried the weight of the animal.

Megacerops (=*Menodus* Pomel, 1849 *Brontotherium*, Marsh, 1873; *Brontops*, Marsh, 1887; *Menops*, Marsh, 1887; *Allops* Marsh, 1887; *Titanops* Marsh, 1887; *Diploclonus* Marsh, 1890) Leidy, 1870

Systematics and Evolution Five genera of brontotheres, all of large body size, have been described from the Chadronian: *Megacerops*, *Brontops*, *Brontotherium*, and *Menodus/Menops*. Each of these genera in turn had multiple species, 47 described by 1929 according to Mihlbachler, Lucas, and Emry (2004), primarily based on differences in horn structure, the differences of which may be due to sexual dimorphism, stage of ontogenetic development, or simple regional differences. The variation that has been described is apparently continuous among the variables commonly used to diagnose the Chadronian brontothere species. While Mader (1998) recognized three of the four genera as valid, Mihlbachler, Lucas, and Emry (2004) considered all of them, including the well-known genus *Brontotherium*, which Mader (1998) considered to be a junior synonym of *Megacerops*, to be one genus and the same as *Megacerops*, the oldest available name created by Leidy in 1870, with *M. coloradensis* as the genotypic species. The lower jaw (Figs. 1.2A, 5.54) described by Prout (1847) as the European taxon *Palaeotherium* is in fact a jaw of a brontothere and subsequently became the holotype specimen of *Menodus giganteus* (Pomel, 1849), which some authors consider to be the valid name for the Chadronian brontotheres but is considered a nomen dubium by Mihlbachler, Lucas, and Emry (2004). Not only did these different described taxa live during the same narrow time period, but many of the specimens placed in these different genera were collected from the same localities, further supporting the idea that the names created are based on variations within a single biological species. Currently only the type species of *Megacerops*, *M. coloradensis*, and one other species included in this genus are considered valid (Mihlbachler, Lucas, and Emry, 2004).

Distinctive Characters This is the largest member of the family. The number of incisors, both the uppers and lowers, is reduced to four or fewer and are globular in shape. On the basis of the reduction of the incisors, it has been inferred that *Megacerops* may have had a prehensile lip. There is no diastema, and the premolars are partially molarized. The posterior part of the zygomatic arch is thickened. The horns are short, anteriorly directed, and round to elliptical in cross section. The placement of the horns relative to the rest of the skull gives the skull a saddlelike appearance. The horns lack any evidence of grooves for blood vessels, which suggests they were covered with tough hide and lacked a keratinous covering. As previously noted, there is a wide variety of in the shape of the horns. Part of the variation is undoubtedly due to sexual dimorphism, and skeletons considered to be females are smaller than the males. This suggests that the horns may have also served as intraspecific displays, and once the species-level taxonomy is resolved, perhaps they will serve as species identification as well (Fig. 5.53).

Stratigraphic and Geographic Distribution *Megacerops* is known only from the Chadronian and is found in all members of the Chadron Formation. Regionally, it is known from South Dakota, Nebraska, Colorado, Montana, Wyoming, and Saskatchewan.

Natural History and Paleoecology Restricted to the Chadronian, *Megacerops* was the last brontothere in North America. On the basis of the dentition, the preferred habitat was probably forest or open woodland, which existed at that time. Their extinction at the end of the Eocene may reflect the changing climate, which included a general cooling and development of more open woodland habitat. Their disappearance at the end of the Chadronian makes them one of the primary taxa used to define the Eocene–Oligocene boundary.

The most distinctive features of *Megacerops* as the last brontothere are its extreme development of the horns, and that it appears to be sexually dimorphic. Osborn (1929) was the first to study the allometric relationship of the evolutionary development of the horns in brontotheres; his data were examined by Hersh (1934), who concluded that the evolutionary trend toward larger horns in the brontotheres leading to those seen in *Megacerops* was simply the result of selection for larger body size. The positive phylogenetic allometry between horn length and skull length is among the strongest known of such relationships in any group of vertebrates, and the seemingly disproportionate lengthening of bony frontonasal horns compared to an overall increase in the size of the skull has often been used as an illustration of a vertebrate macroevolutionary trend (McKinney and Schoch, 1985). The limited data available from juvenile specimens are inconclusive but indicate a strong role for accelerated horn growth late in ontogeny. On the basis of data derived from scaling patterns in modern ungulates, the increase in brontothere size was probably accompanied by an increase in individual life spans and longer developmental stages (Bales, 1996). The horns were apparently used for fighting and further strengthening of the skull; this activity is suggested by the expanded zygomatic arches. The horns seem to have functioned for frontal attacks, and two battling individuals could have locked horns in a test of strength and endurance, similar to that seen in bison and elk today (Mader, 1998). It appears that the primary target for attack was the side of the body, as broken and healed ribs are often present in skeletons.

TRACE FOSSILS

In the White River Badlands, body fossils consist of mineralized bone and teeth. In contrast, trace fossils are burrows, tunnels, footprints, tooth marks, endocasts, nests, or egg chambers, which provide indirect evidence of animal activity (Fig. 5.55). Footprints can tell us how many animals were in an area, the direction they moved in, their size, and the speed they were going. Eggs and nests can provide information about ancient environments and climate (Lockley and Gillette, 1991). Trace fossils in terrestrial environments are uncommon because the agents that preserve them (lakes, ponds, streams, rivers) are not widespread.

Beetle Trace Fossils

In contrast to other types of trace fossils, evidence for fossil dung beetle activity is fairly common. Retallack (1984) described near spherical internal molds of silty micrite within

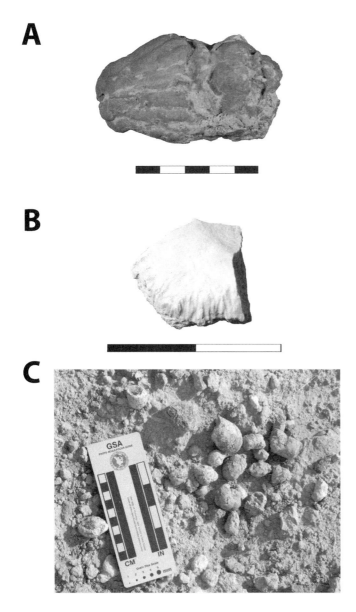

5.55. Trace Fossils from the White River Badlands. (A) Brain endocast of *Merycoidodon* sp., BADL 930. (B) Rodent gnaw marks on bone, specimen found in Badlands National Park. (C) Photos of coprolites from Badlands National Park. Scales in centimeters. Photo (A) by the authors; photos (B) and (C) by the National Park Service of specimens from Badlands National Park. All specimens are the property of the U.S. government.

is some debate regarding whether the dung beetle balls are the remains of a brood ball or are the mold of a pupal cell. Retallack (1984) was able to document deep vertical shafts 100 to 125 cm in length with short subhorizontal side passages 1 to 3 cm in length. The spherical internal molds were compared with the brood balls of modern scarabaeid beetles (Halffter and Matthews, 1966; Halffter, 1977). None of the fossils have thick, clayey shells; nor do they show any evidence of a former airspace around them. Nor is more than one near-spherical mold found at the end of each branch of the burrow (Retallack, 1984). The existence of dung beetles during the Oligocene time is indicated by the presence of the subfamily Scarabaeinae in the fossil record at other locations (Halffter, 1959; Balthasar, 1963; Matthews and Halffter, 1968; Wilson, 1977; Crowson, 1981; Grande, 1984).

Larval Cells of Bees

Although not as common as the beetle trace fossils, sweat bee cells provide important evidence of environments that were conducive to flowering plants. Retallack (1984) described tear-shaped internal molds of silty micrite from the same stratigraphic horizon as the beetle trace fossils. Upon closer examination, the tear-shaped molds had short entrance burrows of about 2 to 5 mm filled in with brown, silty claystone. The cells appear clustered around vertical shafts and were also filled with brown, silty claystone. Both the cells and the burrows had narrow crystal tubes interpreted as penetrating root traces (Retallack, 1983b). These tear-shaped molds were thought to have been created by sweat bees for the following reasons: the depth of the fossils within the paleosol, the clustering and horizontal orientation, the short entrance passages, the degree of organic coating, and the absence of spiral caps. These were all features found in other documented sweat bee nests (Apoidea, Halictidae, Halictinae) (Sakagami and Michener, 1962; Michener, 1974). Both wasps and bees were present during the Oligocene. A classic example are the bee fossils from the deposits from the Florissant Lake Beds, which are late Eocene in age.

Trackways

Fossil trackways are a rare feature in the Big Badlands of South Dakota, with only three published occurrences (Chaffee, 1943; Lemley, 1971; Bjork, 1976). Chaffee (1943) reported brontothere tracks from the Chadron Formation and camelid tracks from the upper Brule Formation. There was some debate on whether the camelid tracks belonged to *Poebrotherium* or *Pseudolabis* (Bjork, 1976). Lemley (1971) published on a trackway from the Peanut Peak Member of the Chadron Formation, consisting of two parallel trackways

the upper Scenic Member of the Brule Formation. Some of the spheres are irregular in shape and some are ellipsoidal. Many are slightly flattened. Under more powerful magnification, calcite crystal tubes were noted and were interpreted as fossil root traces. The sphere diameters ranged from 12.2 to 21.3 mm. These trace fossils are similar to the broad burrows made by scarabaeid beetles, which provision an underground nest with food for their developing larvae. The evidence for the use of dung is based on the following features: the irregular fibrous nature of the wall material, the high phosphorous content, and the abundance of large mammals that would provide the source of dung (Retallack, 1984). However, there about 17 feet in length, containing 24 tracks side by side. The prints are rounded and deep but were not assigned to a particular taxon beyond Mammalia (Lemley, 1971; Bjork, 1976).

Bjork (1976) describes a sandstone slab with six tracks representing two different mammals, proceeding in the same direction. The tracks are preserved as a negative, showing the sediment that filled the original tracks. Stratigraphically, the block was not found in place but is believed to be from the Poleslide Member of the Brule Formation, just above the *Leptauchenia* clays (Bjork, 1976). Four of the six tracks are from an artiodactyl approximately the size of a modern pronghorn. Bjork (1976) proposes that the mammal who made the tracks was a member of the Camelidae, possibly *Pseudolabis dakotensis*, *Paratylopus primaevus*, or *Paratylopus matthewi*. In addition to the camelid tracks, there are two poorly preserved carnivore tracks comprising the left front and hind foot. On the basis of the size and the extended claw marks, the track maker was believed to be *Hesperocyon*. It is not clear which animal came first. Bjork (1976) proposes the following scenario for the creation of the trackway. A shallow pool develops in the *Leptauchenia* clays, and the animals walk across the pool, leaving tracks in the mud. The tracks are buried by a sheet of sand.

In contrast to South Dakota, the Toadstool Geologic Park Trackway Site of western Nebraska is much more extensive (Nixon, 1996; LaGarry, Wells, and Terry, 1998). The trackway was discovered in the 1950s by Cecil Harvey (Harvey, 1960; Nixon, 1996; LaGarry, Wells, and Terry, 1998). The trackway consists of 75 sandstone slabs and 11 track morphologies. The site is believed to be a paleoshoreline along an ancient river system (LaGarry, Wells, and Terry, 1998). The site contains the tracks of a broad range of animals, including antelope, camels, entelodonts, birds, carnivores, rhinoceros, turtles, and invertebrates (Nixon, 1996; LaGarry, Wells, and Terry, 1998). As in South Dakota, the trackways are from the White River Group (Orella Member of the Brule Formation) and

were created by the similar process of wet clays buried by a layer of sand (LaGarry, Wells, and Terry, 1998).

Coprolites

Coprolites, or fossilized excrement (Fig. 5.55C), have been well documented in the White River Badlands since the early twentieth century (Sinclair, 1921b; Wanless, 1923; Abel, 1926, 1935; Stovall and Strain, 1936; Lemley, 1971; Edwards and Yatkola, 1974; Retallack, 1983b). Coprolites from the White River Badlands are mostly from mammals, and more exist from carnivorous mammals than from herbivores (Edwards and Yatkola, 1974), although there have been two accounts of herbivore coprolites from the Chadron Formation (Stovall and Strain, 1936; Lemley, 1971). Coprolites from carnivorous mammals are characterized by elongate, cylindrical, and usually unsegmented masses. They can be tapered on one end and rough and ragged on the other (Edwards and Yatkola, 1974). This type of coprolite often contains bone and tooth fragments from undigested prey. Herbivore coprolites can be similar to modern bovids, with a disc shape and curling on the ends (Stovall and Strain, 1936; Lemley, 1971). The abundance of coprolites is not consistent throughout the late Eocene and Oligocene. Greater concentrations can be found within the Chadron Formation and the Scenic Member of the Brule Formation. According to Edwards and Yatkola (1974), this is due to the depositional environment. It appears that coprolite preservation requires rapid burial with fine-grained sediments in a low-energy fluvial environment. The sediments within the Poleslide member of the Brule Formation and the Sharps Formation are predominately eolian. Because of a cooler and dryer climate, there were fewer low-energy streams at that time. The channel deposits that make up portions of the Sharps Formation represent a high-energy environment that would quickly destroy any type of feces.

6.1. Photographs of various taphonomic states. (A) Isolated limb bone. (B) Articulated limb bones. Pocket knife is approximately 7 cm long. (C) Tortoise (T) resting on former soil surface (S) and buried by fining upward flood deposits. Note photo scale next to shell. (D) Gnaw marks (G) on underside of a jaw bone. (E) Scratch marks (M) on a limb bone. (F) A fragment of tortoise shell that was heavily weathered before burial and fossilization. (D), (E), (F), scale in centimeters. Photos by the authors.

Death on the Landscape: Taphonomy and Paleoenvironments

6

PALEOECOLOGY IS THE STUDY OF THE INTERACTIONS, habits, and lifestyles of extinct species and the ancient communities they formed. Paleoecologists, using data from the fossil and geologic record, reconstruct these communities of plants and animals to gain a better understanding of the relationship between members of that community (Shipman, 1981) and how they responded to environmental changes. Before the paleoecology of an assemblage can be interpreted, the paleoecologist must gain a better understanding of the events that intervened between death and fossilization and what effects these events have on the retrieval of information about the past (Shipman, 1981). This process is called taphonomy. Efremov (1940) introduced the concept of taphonomy as the study of embedding or burial. It is the analysis of the transition of organics from the biosphere into the lithosphere to become part of the geologic record. The term *taphonomy* can be broken down into the Greek words *taphos* (burial) and *nomos* (laws).

Clark, Beerbower, and Kietzke (1967) made the first attempt at describing the ancient environments preserved in the White River Badlands. They proposed four different environments of deposition within the Chadron Formation. These included streams represented by channel fills of the Red River and its tributaries; limestone deposits near channels and channel margin facies; channel margin sands and silts from riverbanks; and massive to bedded clays and silts along ancient floodplains. Clark, Beerbower, and Kietzke (1967) proposed four general biotic structures, including aquatic, semiaquatic, river-border forest, and forest–savanna. They divided the paleoecology of the lower nodular zone of the Brule Formation into three different facies: near stream, open plains, and swampy plains. (See chapter 1 for a more detailed description on Clark and his colleagues' research contributions.) Refinements to geochronology (such as paleomagnetism and radioisotopic dating), the application of paleopedology to deep time records, and an expanded paleobotanical archive have since fine-tuned paleoclimatic interpretations of the Cenozoic and refined our understanding of the paleoecological changes indicated by the fossils in the White River Badlands.

FOSSIL PRESERVATION IN THE WHITE RIVER BADLANDS

During the uplift of the Black Hills, which began during the later phases of the Laramide Orogeny, Paleozoic rocks were exposed to weathering, and the resulting sediments were transported by streams and rivers flowing from west to east. These sediments are preserved within a slightly northwest–southeast-trending asymmetric basin that makes up much of what we call the White River Badlands (Fig. 4.2). The basin is bound to the north by the Sage Ridge Fault in the Badlands Wilderness Area and to the south along the Sandoz Ranch, White Clay, and Pine Ridge fault system (Clark, Beerbower, and Kietzke, 1967). A significant amount of White River Badlands sediment consists of reworked ash, which was carried via eolian and fluvial processes from volcanoes in eastern Nevada and Colorado. This combination of geologic factors promoted rapid rates of sedimentation, which aided in the fossilization process, allowing carcasses to be buried quickly after death and protecting them from destructive forces (Clark, Beerbower, and Kietzke, 1967; Shipman, 1981).

The fossilization process is a rare event and requires a series of steps to occur. The preservation potential of a fossil is the balance between the factors of preservation and destruction. Chances of preservation tend to increase if organic remains can be buried rapidly in a particular sedimentary environment (Shipman, 1981). Some of the most frequent causes of death associated with White River vertebrates include predation, flood events, starvation, and dehydration.

Regardless of the cause of death, the key ingredient in the process of fossilization is water. Once buried, and with the flesh removed, bones begin to interact with either water that is percolating downward through the soil zone (vadose water), or if the depth of burial is sufficiently deep, the bones are completely saturated and interact with groundwater (phreatic water). Most bones will experience both as they are progressively buried with additional sediment. The bone material, which is a fine-grained carbonate-substituted bioapatite (dahllite: $Ca_5(PO_4,CO_3)_3(OH)$), reacts with water, and it begins to recrystallize into a more stable form of apatite (francolite: $Ca_5(PO_4,CO_3)_3(F)$) by incorporating fluoride, uranium, and rare earth elements such as lanthanum (Metzger, Terry, and Grandstaff, 2004; Grandstaff and Terry, 2009). The proportions of these newly incorporated elements vary as a function of environmental conditions (oxidizing versus reducing and acidic versus alkaline), and can be used to aid paleoenvironmental interpretations and decipher the rate at which fossilization occurs (Grandstaff and Terry, 2009; Drewicz et al., 2011; Drewicz, 2012). In addition to the recrystallization of bone material, additional minerals, such as calcite and quartz, will precipitate into void spaces in the bone over time.

Recent National Park Service paleontological surveys have documented high concentrations of bone, often referred to as bone beds, in distinct stratigraphic layers and regions of the park. It appears that the majority of these bone beds occur in the Scenic Member of the Brule Formation. Within the Poleslide Member, bones can be found throughout the section with increased concentrations in certain layers. The bone bed accumulations of the Scenic Member represent both watering holes and fluvial channel deposits, whereas higher concentrations of bone in the Poleslide Member are associated with slower rates of eolian sedimentation and more likely are time-averaged samples that took a longer period of time to form (Benton et al., 2007, 2009). By adding information gained from paleosols associated with these bone beds, taphonomic models can be developed that provide insight into the paleoecological dynamics of the White River Badlands. Below we discuss two significant fossil sites in which taphonomic mechanisms have been studied in great detail. Both sites are located within the Scenic Member of the Brule Formation.

GENERAL TAPHONOMIC SCENARIOS IN THE BADLANDS

The characteristics of vertebrate taphonomy within the Badlands change as a function of paleoclimates throughout the Eocene and into the Oligocene, which in turn influenced the types of paleoenvironments, and thus the depositional processes that were possible. The Chadron Formation and the lower part of the Scenic Member of the Brule Formation were dominated by fluvial depositional processes. Overbank flooding and the migration of river channels was common. Episodic flooding events washed fine sediments onto broad, flat landscapes that buried the bones of animals that had been killed and scattered by predators and scavengers and occasionally trapped slower animals or those that were caught unaware.

When we examine the relationship between individual paleosol profiles and vertebrate fossils, a repetitive pattern can be detected. Fossil bones within the floodplain mudstones tend to be resting on the surface of individual paleosols (Fig. 3.5). Whether the fossil is an isolated bone or an articulated skeleton, roots are commonly seen directly beneath the fossilized remains (Figs. 3.5, 6.1). With each successive flooding event, the previous landscape would be buried, thus preserving the bones. This repetitive nature of flooding and burial explains the overall richness of vertebrate fossils within the Badlands, but it also explains why these strata are so rich in fossil tortoises: they were unable to escape the floodwaters and were easily buried by the muds and sediments that fell out of suspension onto the ancient floodplains (Fig. 6.1). On occasion, animals would also be carried by the floodwaters and end up on the banks of the rivers, similar to what is seen today on the Serengeti Plains of Africa when animals drown while attempting to cross rivers during migration. Carcasses are washed downstream and end up collecting in areas of slower water. The likelihood of the skeleton being preserved depends on the amount of time between death and burial and whether burial occurs before scavengers scatter the remains.

The type of paleosol associated with fossil bones is a clue to the relative degree of landscape stability and can be directly associated with the proximity to active river channels. Flood frequency is related to flood size. The larger the flood, the less frequent it tends to be. In modern terms we classify this relationship as 1-, 5-, 10-, 100-, and 500-year flooding events. Regardless of flood magnitude, though, areas directly adjacent to the river will always flood, and new sediments will be deposited and bury any carcasses that have accumulated on the landscape since the last flood.

In soil terms, this relationship between the magnitude of flooding and the amount of land that will be covered translates to the amount of time available for a soil to form before it is buried by new sediments. Areas closer to active rivers are characterized by soils that are weakly developed, commonly entisols, which are stacked one on top of the other as a result of frequent burial (Fig. 3.5). Farther away from the active channel, soils will not be covered as frequently by floods and new sediments, and thus they will be able to develop to a greater degree (e.g., alfisols, mollisols). Any sediment that is added tends to be in thin increments that can be easily incorporated into the active soil. Only massive flooding events will bury these soils with enough sediment to shut off any further soil development.

This relationship between proximity to an active channel and the relative degree of soil formation is also seen in the strata of the Badlands and explains why fossils tend to be concentrated within certain parts of the park. Within the lower part of the Scenic Member a definite change in the style of fluvial sedimentation and degree of paleosol development can be seen from the southwest part of the park near Scenic, South Dakota, toward the northeast into Dillon Pass. Near Scenic, the paleosols are predominantly entisols and inceptisols, and they commonly have articulated to semi-articulated skeletons resting on ancient land surfaces (cf. Fig. 3.5 and Fig. 3.9). These two features suggest a proximal position to former rivers. Moving northeast toward Dillon Pass within this same interval of strata, the paleosols become progressively better developed overall with a greater proportion of inceptisols. Fossils are still preserved at the top of individual soils, but the numbers of fossils preserved and the style of fossil preservation suggest longer periods of time between flooding events (Fig. 3.8). Bones are more commonly scattered, as if modified by scavengers, and tortoises occur in numerous orientations (including upside down and lying on their side), suggesting that they were removed from their original life position after death.

Flooding is only one part of the story. As rivers flow, they migrate across their floodplain via lateral migration: the progressive erosion of the outside bend (cut bank) in a river meander and deposition on the inside of the meander (point bar). As lateral migration proceeds, any bones lying on the active landscape or buried by recent flooding are secondarily incorporated into the channel as the riverbank is progressively eroded. Because of their size and relative density, bones tend to collect within the river channel along with the coarser sediments such as sand or gravel. Over time, large amounts of bone material can be concentrated and buried in the river channel. This process, referred to as time averaging, creates jumbled deposits of individual bones from numerous types of animals that did not necessarily live either in the same habitat or at the same time. Paleosols will not be directly associated with these ancient river channels, although they may overprint the entire channel deposit if it is abandoned by the river as it shifts the position of its channel (avulsion). Otherwise, continued lateral migration will deposit finer channel materials on top of the bones, followed by overbank muds deposited during the next flooding event. These muds can in turn be modified by ancient soil formation.

Case Study 1: The Big Pig Dig

The Big Pig Dig site was excavated for 15 field seasons and provided an incredible opportunity for park visitors to see paleontological resources and interact with scientists. The site was found by two park visitors, Steve Gasman and Jim Carney, who spotted exposed vertebrae and limb bones in a drainage ditch along Conata Road (County Road 502) southeast of the Conata Picnic Area (Plate 9; Fig. 1.1). Once it was determined that the bones were at risk to theft and erosion, the National Park Service initiated an emergency salvage operation, which turned into a long-term scientific excavation.

The excavation revealed a dense accumulation of semiarticulated to disarticulated bones with no apparent preferred orientation (Plate 10). Bone positions ranged from highly angled to horizontal. The analysis and interpretation of the lateral extent, as well as the geometry of individual sediment layers, supported the concept of a watering hole (Plates 11, 12). Sedimentological analysis suggested that the watering hole had formed from a stream channel or an oxbow lake formed by a meander loop cut off from a channel (Terry, 1996a).

The bones were intimately associated with a green mudstone layer representing anoxic/reducing conditions (Plate 10), with thicker portions of the green mudstone associated with greater concentrations of bone. The bones showed little or no weathering, although some were modified by carnivore/scavenger processing, including puncture and groove marks that were filled with surrounding sediment. Some bones were crushed along their length or on either end, possibly as a result of either burial compaction or processing. Spiral fractures, a type of crack within fresh bone that is normally associated with carnivore/scavenger processing, have also been noted on some fossils (Stevens, 1996a, 1996b).

Initially the fauna recovered from the site was not considered diverse, with a total of five taxa represented, including *Archaeotherium* (Figs. 5.32, 5.33), *Subhyracodon* (Fig. 5.50), *Mesohippus* (Fig. 5.47), *Leptomeryx* (Fig. 5.46), and cf. *Prosciurus*. However, by the close of excavations in 2008, a total of 19 genera had been discovered, with the minimum number of individuals for the four original taxa as follows: 8 *Subhyracodon*, 29 *Archaeotherium*, 11 *Leptomeryx*, and 8 *Mesohippus* (Shelton et al., 2009). A total of 19,290 specimens were eventually collected from the site, and 187 m² were excavated.

In 1995 one of the authors (D.O.T.) was invited by the National Park Service to complete a sedimentary analysis of the Big Pig Dig to gain a better understanding of the depositional environment of the site. After extensive sampling and detailed analysis, the Big Pig Dig was interpreted as the attritional accumulation of animals around a watering hole during drought conditions (Terry, 1996b). Animals may have died from thirst, with small or weakened animals becoming trapped within the soft substrate of the watering hole while attempting to reach the remaining water. This model can be broken down into five steps that explain many of the sedimentological and paleontological characteristics of the Big Pig Dig:

Step 1: Basin forms. A floodplain basin is initially filled with water. Animals are using the basin as a watering hole.

Step 2: Drying starts. With the onset of drought conditions, animals begin to concentrate around the watering hole. Animals are likely wading into the water. Some animals die, possibly floating on the surface. As animals decompose, portions of the body drop off and settle to the bottom in random orientations. Scavengers and predators are likely exploiting the abundance of dead or weakened animals.

Step 3: Drying continues. Animal mortality increases as the drought intensifies. Water level within the basin drops. Animal remains become more concentrated, possibly contaminating the water. Animals walking through the water step on remains of other animals. This trampling pushes bones down into the soft mud, orients bones at high angles, and mixes muds at the bottom of the lake. Decay of animals within the water likely induces anoxic conditions. Scavengers and predators continue to exploit dead or weakened individuals. Depending on the stability of the substrate, small or weakened animals may become trapped while heading for remaining water in the center of the basin. Some animals die while clustered together into herds (entelodonts).

Step 4: Basin dry. Continued drought eventually leads to the complete drying of the basin. Some bones are completely covered, while others protrude from the surface of the dry lake bed. Mud cracks likely develop. Muds and sediments left over are likely rich in organic matter as a result of abundant animal remains. Anoxic conditions may exist in the remaining muds and sediments.

Step 5: Basin filled and bones covered. The basin is flooded, and the bone-bearing basin is buried by an influx of new sediment.

This model explains many of the sedimentological and taphonomic characteristics of the site. The fine-grained nature of the deposits, with occasional thin stringers of silt and fine sand, suggests that fluvial action was responsible for the deposition of individual bone bed units and the eventual burial of the bones. The lenticular geometry of the bone-bearing green layer (Plate 11), the rarity of soil features within individual horizons, and the preservation of relict depositional fabric are consistent with a small pond or floodplain lake, such as an oxbow. The lower contact of the bone-bearing green layer is undulatory and has a gradual transition over 1 to 2 cm with the underlying lower red layer (Plate 11). The foot traffic of animals would have mixed the bottom of the basin, resulting in a gradational lower boundary between the bone-bearing green and lower red layers. This would also explain the occurrence of swirled zones of relict bedding. Depending on the consistency of the substrate, animals would have made hoofprints or footprints that would protrude downward into the lower red layer (Plate 11).

Trampling by animals could also explain the high-angle orientation of some bones. As animals walked across the area, some long bones or other articulated sections would have been pushed down into the soft substrate of the lake bottom. The next influx of sediment (upper red layer) would cover the bone-bearing green layer, creating a sharp contact and incorporating fragments of the bone-bearing green layer as ripped-up clasts along the contact. Depending on the flow velocity during this next influx of sediment, some bones may have been transported a short distance or partially aligned to the current. The fresh to slightly weathered state of the bones suggests that they were not exposed at the surface for a significant period of time.

The presence of bone processing/modification suggests that carcasses and/or weakened animals were accessible and that scavenging and predation was taking place. Although we cannot be absolutely sure as to what animal or animals were responsible for the bone modification, the entelodont *Archaeotherium* is a likely suspect (Figs. 5.32, 5.33). Many of the puncture marks on the bones are similar in size and shape to those that could be made by the teeth of *Archaeotherium*. Scavenging by *Archaeotherium* as well as any of the many predators that existed at the time could explain the scattered, articulated to disarticulated nature of some bones. The low diversity of the bone bed's fauna may be indicative of drought-sensitive species, though further preparation of field jackets in the laboratory may yield additional species, especially smaller ones.

Although this particular model is based on the concept of a drought-induced death assemblage, there is no direct sedimentological or stratigraphic data to support the development of drought conditions. The most likely sedimentological evidence of drought is the intricate network of cracks and veins throughout the site (Plates 10, 11). The origin of these features is problematic and should not be taken as proof of desiccation. The best evidence for drought conditions comes from comparisons of the bones within the Big Pig Dig to modern and ancient analogs. Studies of modern ecosystems in Africa suggest that animal concentrations increase as the amount of available water decreases (Western, 1975). During these dry periods, animals will cluster around remaining sources of water. Foster (1965) noted that rhinoceroses died half-submerged in rivers, possibly in the attempt to gain relief from the drought. In some instances animals remained around a drying watering hole even though flowing rivers were available several kilometers away. Foster (1965) suggested that death was due not to a lack of water but to an insufficient amount of forage available to support the animals. For example, Goddard (1970) applied the term "nutritional anemia" to explain the death of large numbers of black rhinos during a prolonged drought as a result of insufficient food sources.

Mead (1994) proposed, on the basis of an analysis of the jaw mechanics and facies associations of *Subhyracodon* versus *Hyracodon* (Figs. 5.50, 5.48), that *Subhyracodon* was a water-dependent organism that frequented riparian habitats, while *Hyracodon* was water independent and frequented open-range habitats. *Subhyracodon* is abundant in the Big Pig Dig, while *Hyracodon* has yet to be discovered. According to Mead (pers. comm.), the presence of *Mesohippus* in the Big Pig Dig also supports the concept of a water-dependent fauna (Fig. 5.47). Hunt (1990) reports the grouping of chalicotheres within the lower Miocene Agate bone bed of northwestern Nebraska. According to Hunt (1990), the Agate bone bed represents a short-term, attritional accumulation of animals around a watering hole during drought conditions. The presence of segregated animals supports this interpretation. Such segregation would not be expected to develop if the bone bed was the result of catastrophic flooding and mass death. The Agate bone bed also preserves evidence of bone processing and trampling (Hunt, 1990).

The Big Pig Dig bone bed may be one of several within this stratigraphic interval. Analysis of core samples shows that the same relict bedding structures and cracking and veining and colors similar to the bone-bearing green layer occur below the present quarry (Terry, 1996a). An oxbow lake or similar depression on a floodplain would provide a suitable setting for seasonal flooding events, eventual desiccation, and burial of animal remains. Vertebrate remains have been found nearby within greenish layers lateral to the bone-bearing green layer (Kruse, pers. comm., 1995). Several of these greenish layers are repetitive, with reddish layers similar to the upper and lower red layers.

Case Study 2: The Brian Maebius Site

The Brian Maebius site was discovered by and named after an intern employed at Badlands National Park in 1997. The site has since been studied by several institutions, culminating in four master's theses (DiBenedetto, 2000; Factor, 2002; McCoy, 2002; Metzger, 2004). The site is located in the Badlands Wilderness Area and is stratigraphically positioned in the Scenic Member of the Brule Formation just below the Hay Butte marker. The site is relevant to this discussion because of the dense accumulation of fossil bone and unique preservational environments, which range from severely deteriorated to pristine, concentrated in several stratigraphic layers that have been overprinted by pedogenesis (Factor, 2002).

At the Brian Maebius site, the base of the stratigraphic section is marked by a sandstone layer that fines upward into a mudstone and claystone (Plate 13). The site is interpreted as a former channel that was isolated by avulsion or neck cut off to form an oxbow lake (Factor, 2002). Lateral accretion surfaces suggest migration to the north and northeast with flow direction to the east. DiBenedetto (2000) describes a tree stump felled by flooding and possibly aligned with paleoflow (Plate 7a). The bones appear to have been washed in during periods of high flow, possibly during periods of reactivation of the oxbow during flooding events, in which the oxbow acted as a sediment trap.

According to DiBenedetto (2000), carnivore coprolites are common, and he proposed that the site was near a carnivore den. He based his argument on the observation that individual bones have tooth impacts and lack evidence of transport. Some of these impacts are believed to be small, needlelike tooth marks, suggesting they were made by pups or kits. They also exhibit minimal time of surface exposure. Various fossil rodent and leporid remains were collected and are believed to have been deposited as avian pellet concentrations (DiBenedetto, 2000). Unfortunately, there is no sedimentological evidence to support a carnivore den model. An alternative theory is that the site was a slack-water area during flooding events that concentrated carcasses that were later scavenged (Factor, 2002).

Over time, the river-dominated environments of the lower part of the Scenic Member of the Brule Formation gave way to progressively greater amounts of eolian influence. This switch in depositional environments influenced both the formation of paleosols and the preservation of associated fossil remains. During periods of high amounts of eolian influx, landscapes were not stable but instead were vertically aggraded as the soil incorporated this steady influx of sediment. In the rock record, this resulted in zones of massive volcanic silts that appear to be randomly sprinkled with vertebrate fossils and soil features, such as scattered roots and glaebules (Figs. 3.13, 3.14). During periods of low eolian influx, soils were able to develop to a greater degree, and fossil bones were concentrated along the ancient landscape in conjunction with denser accumulations of fossil roots and better-formed soil horizons. Once eolian sedimentation increased again, these former periods of landscape stability would be buried and preserved by the next massive zone of volcanic silts.

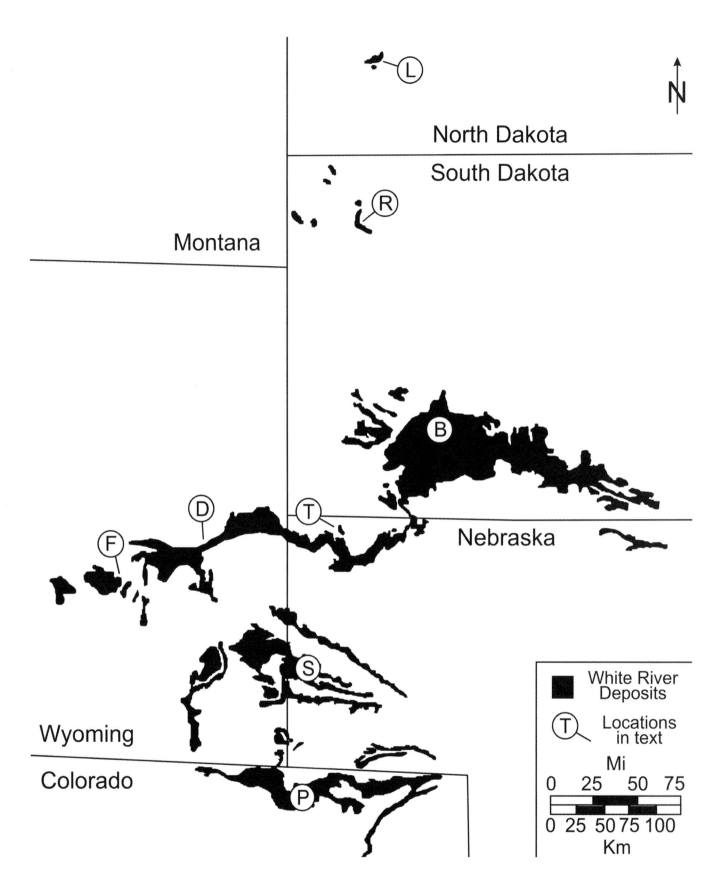

7.1. Map showing the distribution of the White River Group and correlative strata across the northern Great Plains. B = Badlands National Park, South Dakota; D = Douglas, Wyoming; F = Flagstaff Rim, southwest of Casper, Wyoming; L = Little Badlands of southwest North Dakota; P = Pawnee Buttes of northwest Colorado; R = Reva Gap, east of Buffalo, South Dakota; S = Scotts Bluff National Monument, Nebraska; T = Toadstool Geologic Park, north of Crawford, Nebraska.

THE EOCENE–OLIGOCENE BOUNDARY

The Eocene and Oligocene epochs are specific intervals of time recognized across the globe, but in order to understand their significance, it is important to understand how they are identified. The geologic history of our planet is stored within the archives of various types of rock bodies and strata, and the plant and animal remains preserved within them. Early scientists were aware that changes in plant and animal life had occurred, that the types of plants and animals preserved in a type of rock varied even if the rock type was the same, and that the existence of fossil remains in rock layers high on the tops of mountains suggested that the landscape had not always been as it appeared. Other than methods that allowed early geologists to establish the relative ages of various rock bodies, such as superposition, which states that in a sequence of horizontal strata the oldest layers of rock are on the bottom, these early scientists had no way to quantify the ages of strata and their associated fossils, when deposition started, how long it lasted, or when it stopped. Relative ages could be established across broad regions by tracing laterally continuous rock units or occurrences of particular fossils that allowed these early scientists to construct a relative geologic order of important paleontological and geological events, such as the appearance and extinction of the dinosaurs, but absolute ages for these events were unavailable. This changed with the development of radiometric dating of geologic materials and the discovery of magnetic signatures preserved in rock. The combination of radiometric dating plus paleomagnetism in sedimentary rock strata around the globe has allowed the establishment of a chronologic framework within which the patterns of evolution and geologic events portrayed in the geologic column can be placed.

The Eocene–Oligocene boundary represents the most dramatic change in global paleoclimatic conditions since the extinction of the dinosaurs. This change, also referred to as the Hothouse to Icehouse Transition (Prothero, 1994; Prothero, Ivany, and Nesbitt, 2003; Koeberl and Montanari, 2009), is associated with dramatic changes in global temperature, large drops in sea level, the establishment of permanent glacial ice on Antarctica, and extinctions in the nonmarine and marine fossil record. The causes of this dramatic climate change are still a matter of debate. Hypotheses to explain this climate change have included plate tectonic reorganization and opening of marine passageways between South America, Australia, and Antarctica; uplift of the Himalayan Mountain Range; and multiple extraterrestrial impact events during the late Eocene. Any attempt to understand the causes of this transition must take into account both terrestrial and marine records across the globe during this period of time.

Our particular interval of interest, the Eocene–Oligocene boundary, is one of several divisions of epochs within the Tertiary Period of the Cenozoic Era. For each of these divisions within the geologic column, one particular place on the globe is used as the global reference against which all other locations of the same age can be compared. These scientifically agreed-upon stratigraphic levels and locations are referred to as a Global Stratotype Section and Point (GSSP) and are formalized for many of the boundaries in the geologic column (Alvarez, Claeys, and Montanari, 2009). For the Eocene–Oligocene boundary, the global standard is a section of shallow marine limestones exposed along the Adriatic coast near Massignano, Italy, which contains several layers of impact ejecta, volcanic ashes, evidence of magnetic reversals, and abundant marine fossils (Premoli and Jenkins, 1993; Hyland et al., 2009). The Eocene–Oligocene boundary is currently dated at approximately 33.8 Ma and is associated with a particular set of normal and reversed paleomagnetic signatures referred to as 13N (normal) and 13R (reversed) polarity when compared to today (Hilgen and Kuiper, 2009). The numbers for polarity events increase with older rocks, counting backward from 1N/R for our modern record (Fig. 2.2).

Within the various epochs of the Tertiary such as the Eocene and Oligocene, smaller units of time can be identified. For the terrestrial record on a specific continent, these subdivisions are based on the mammalian fossil record. Mammals are used because they are abundantly preserved as fossils in most places, many lineages have a relatively rapid evolutionary rate, and their teeth are anatomically complex enough to allow recognition of different species. These are biostratigraphic units because they are based on living organisms and for North America are called the North American land mammal ages (NALMAS). As with any biostratigraphic unit, it is defined by a detailed knowledge of the stratigraphic ranges of the fossil taxa used to characterize the time interval. The boundaries of a NALMA may be recognized by the first appearance of a taxon in the fossil record, whether because it has dispersed into North America from another continent or evolved from an earlier ancestor in North America. It may also be defined by the absence of a genus or species that was previously present on the continent but becomes extinct. Not all taxa are equally useful as biochronologic markers. This is because some have a long geological record and so cannot be used to identify a small unit of time. Other taxa are not useful because they have a limited distribution, so the taxa cannot be identified over wide geographic regions. Taxa that have undergone relatively rapid evolutionary change, that have distinct morphologies in their dentition (the favored part of the body used by paleomammalogists) or the skeleton, and that have a widespread distribution are the most useful in terms of defining and refining land mammal ages.

Four North American land mammal ages are present in the Big Badlands: Chadronian, Orellan, Whitneyan, and the earliest Arikareean. They span the latest Eocene to early Oligocene, approximately from 37.1 to 27.7 Ma (Fig. 2.2). These land mammal ages were originally established by the Wood Committee (Wood et al., 1941), and each takes its name from the geographic locality from which a fauna with a distinctive combination of taxa was collected. The late Eocene Chadronian land mammal age is based on fauna recovered from the Chadron Formation near its type area in the vicinity of Chadron, in northwestern Nebraska and in southwestern South Dakota. These deposits are referred to as the *Titanotherium* beds in the older literature. This land mammal age was originally defined as the time when brontotheres and *Mesohippus* overlapped their geologic ranges (Wood et al., 1941). However, brontotheres and *Mesohippus* now are known to overlap in the older Duchesnean land mammal age (Mac-Fadden, 1998), and the uppermost occurrence of brontotheres may extend into the earliest Orellan land mammal age (Prothero and Emry, 1996, 2004). The taxa that mark the start of the Chadronian land mammal age include the first occurrence of the oreodonts *Bathygenys* spp. and *Limnenetes* spp., the amphicyonid *Brachyrhynchocyon dodgei*, and the enteledont *Archaeotherium coarctatum* (Prothero and Emry, 2004). The end of the Chadronian is currently marked by the transition between the Chadronian *Leptomeryx speciosus* to the Orellan *Leptomeryx evansi*. The last occurrences of several mammal species also occur at the Chadronian–Orellan boundary, including the last occurrence of the enteledont

Archaeotherium coarctatum, the horse *Mesohippus westoni*, the oreodont *Miniochoerus chadronensis*, and the rabbit *Palaeolagus temnodon* (for a complete listing, see Prothero and Emry, 2004). Brontotheres, the horse *Miohippus grandis*, and the camel *Poebrotherium eximium* have their highest occurrences just above the Chadronian–Orellan boundary (Prothero and Emry, 1996, 2004). Radiometric dates for sediments containing Chadronian faunas place its beginning at 37.1 Ma and ending at 33.4 Ma (Fig. 2.2).

The first NALMA in the Oligocene is the Orellan, originally based on the fauna from the Orella Member of the Brule Formation (Wood et al., 1941). This land mammal age takes its name from a type locality near the town of Orella in northwestern Nebraska but includes southwestern South Dakota and eastern Wyoming as well. The term replaced the older names "turtle–*Oreodon* beds" or "*Oreodon* beds" used for rocks of this age. Many of the taxa used to define the Orellan land mammal age by Wood et al. (1941) can no longer be used because of subsequent taxonomic revisions. While *Eporeodon*, *Eumys*, and *Ischyromys* were identified as first appearing in the Orellan, *Eumys* and *Palaeolagus haydeni* have been questionably identified in the upper Chadron Formation. Currently the base of the Orellan is marked by the first occurrence of *Leptomeryx evansi* (Prothero and Emry, 2004). *Hypertragulus calcaratus* first appears at the start of the Orellan within the White River Group, although outside the White River Group *Hypertragulus* has been reported from the Chadronian of Mexico (Ferrusquia-Villafranca, 1969). A large number of taxa have last occurrences at the end of the Orellan land mammal age. These include the hyaenodont *Hyaenodon crucians*, the rabbits *Palaeolagus intermedius* and all species of *Megalagus*, the enteledont *Archaeotherium mortoni*, the oreodont *Miniochoerus gracilis*, the rhinoceros *Subhyracodon occidentalis*, the small artiodactyl *Hypertragulus calcaratus*, and all the species of the horse *Mesohippus*. In addition to the large rodent *Ischyromys typus*, 12 genera of rodents may have become extinct at the end of the Orellan (for a complete listing, see Prothero and Emry, 2004). Radiometric dating places the beginning of the Orellan at 33.9 Ma and ending at 32.4 Ma (Fig. 2.2).

The Whitneyan land mammal age is based on the fauna from the Whitney Member of the Brule Formation (Wood et al., 1941). The type locality is near the town of Whitney in northwestern Nebraska, but this member extends into southwestern South Dakota and eastern Wyoming. The Whitneyan replaced the older term "*Protoceras–Leptauchenia* beds" for faunas from this time interval. While *Protoceras* is restricted to this time interval, its usefulness as a biostratigraphic indicator is limited because it is restricted to a limited depositional setting, the *Protoceras* channels.

The traditional start of the Whitneyan land mammal age is the first occurrence of abundant *Leptauchenia* oreodonts. *Leptauchenia major* does have a first occurrence at the start of the Whitneyan, but *Leptauchenia decora* that is abundant in the basal Whitneyan rocks has its first occurrence in Orellan rocks. Other first occurrences at the start of the Whitneyan land mammal age include first species of the nimravids *Nimravus* and *Eusmilus*, the rhinoceros species *Hyracodon leidyanus* and *Diceratherium tridactylum*, the oreodont *Miniochoerus starkensis*, the camels *Paratylopus primaevus* and *Paralabis cedrensis*, the tapir *Protapirus obliquidens*, and the beaver *Agnotocastor praetereadens*. The end of the Whitneyan land mammal age is marked by the last occurrences of the tapir *Protapirus obliquidens*, the rabbit *Palaeolagus burkei*, all species of the camels *Paratylopus* and *Paralabis*, and the nimravids *Dinictis* and *Hoplophoneus* (for a complete list, see Prothero and Emry, 2004). Radiometric dates places the beginning of the Whitneyan at 32.4 Ma and ending at 30.0 Ma (Fig. 2.2).

Originally the Arikareean land mammal age was based on faunas found in the Arikaree Group of western Nebraska, and the fauna preserved at Agate Fossil Beds National Monument was considered to be the most typical of the time period (Wood et al., 1941). Recent research has greatly modified the definition of the Arikareean, and it is now the longest of the NALMAS, beginning at 30.0 Ma and ending at 18.6 Ma, and transgresses the Oligocene–Miocene boundary (Fig. 2.2). This land mammal age is defined by the first appearance of a number of Eurasian immigrants including the Talpinae (talpid moles) and the rodents *Plesiosminthus*, *Leidymys*, and *Allomys*. Other first appearances in this land mammal age include the oreodonts *Desmatochoerus megalodon*, *Promerycochoerus* spp., and *Sespia nitida*, the camels *Miotylopus* and *Nanotragulus*, and the dog *Sunkahetanka geringensis*. Taxa whose last occurrence is in the early early Arikareean include the primate (or dermopteran) *Ekgmowechashala*, the rodent *Eumys*, the creodont *Hyaenodon*, the nimravid cat *Nimravus*, and the rhinoceros genera *Hyracodon* and *Subhyracodon*.

TERRESTRIAL RECORD

The White River sequence across the northern Great Plains preserves the best terrestrial record of this Eocene–Oligocene climate transition anywhere in the world. The term "sequence" is used herein, instead of "group," to recognize the different stratigraphic rankings of these rock units across the region. Understanding the spatiotemporal record of this paleoclimatic change within the terrestrial deposits of the White River sequence provides important information to establish linkages with the rate and timing of paleoclimatic changes in ancient marine environments. In turn, this can provide insights into ancient global climate dynamics, which can be used to understand our current episode of global climate change.

Although Badlands National Park is the best-known part of the White River sequence, these deposits stretch for hundreds of miles across the northern Great Plains (Fig. 7.1). From west to east and north to south, these deposits vary in their sedimentary composition and age. Some locations preserve an almost complete record of the Eocene–Oligocene transition (EOT); others are condensed or missing large spans of time. The reasons for this wide disparity in composition and preservation include proximity to sediment sources and tectonic influences on basin accommodation, or later erosion of parts of local sequences. In general, the White River sequence is more completely preserved in northwestern Nebraska and east-central Wyoming; it is less completely preserved toward the north in northwestern South Dakota and southwestern North Dakota. When put in a temporal framework, these widespread deposits provide a level of detail and insight into the geologic history of the White River sequence that would be unattainable by looking at any one location.

The Badlands of Northwest Nebraska

The best-known exposures of the White River sequence in northwest Nebraska can be found approximately 20 km north-northwest of the town of Crawford at Toadstool Geologic Park (Fig. 7.1). This exposure of badlands, while extremely small when compared to the Big Badlands of South Dakota, preserves a key interval of strata not found in Badlands National Park. In essence, Toadstool Geologic Park is a Rosetta stone for the White River sequence: it ties together the deposits of the White River sequence in southwest South Dakota to those in east Central Wyoming.

The stratigraphic nomenclature of the Nebraska Badlands has undergone significant revision over the recent past. Before the work of LaGarry (1998), Terry (1998), and Terry and LaGarry (1998), the hierarchy of Schultz and Stout (1955) was used to categorize the stratigraphy of the Toadstool Geologic Park region. This hierarchy was based on subdivisions of the White River sequence into distinct lettered intervals that corresponded to a combination of changes in rock type, fossil content, and marker beds (volcanic ashes and paleosol horizons). These lettered subdivisions were then interpreted as direct correlations to rock units in the Big Badlands of South Dakota. The stratigraphic revisions of LaGarry (1998), Terry (1998), and Terry and LaGarry (1998), which were based only on changes in lithology in the Toadstool Geologic Park region, resulted in dramatically different correlations to deposits in the Big Badlands of South Dakota. These revisions and correlations are followed herein, with references made to the hierarchy of Schultz and Stout (1955) as needed for clarification.

The base of the badlands stratigraphic sequence around Toadstool Geologic Park is similar to that already established in Badlands National Park. After the retreat of the Western Interior Seaway at the end of the Cretaceous, exposure, weathering, and ancient soil formation modified the marine Pierre Shale into bright hues of red, yellow, and purple (Plates 2, 14). This episode of weathering and soil formation persisted for approximately 30 million years to produce paleosols similar to those of the Yellow Mounds Paleosol Series in South Dakota before the first phases of fluvial deposition began. Just as in South Dakota, these first pulses of late Eocene fluvial sedimentation were also pedogenically modified into bright red paleosols similar to those described in the Chamberlain Pass Formation (Terry and Evans, 1994) that laterally correlate to bright white bodies of channel sandstone up to 7 m thick (Chadron A of Schultz and Stout, 1955). Pedogenic modification of the Chamberlain Pass Formation in northwest Nebraska was not as intense as that in South Dakota, which resulted in the preservation of rare vertebrate fossils of brontotheres and tortoises (Vondra, 1956; LaGarry, LaGarry, and Terry, 1996) in channel sandstones.

The Chamberlain Pass Formation in northwest Nebraska is overlain by the Peanut Peak Member of the Chadron Formation (Chadron B of Schultz and Stout, 1955), which is exposed as the same greenish-gray, smectite-rich, haystack-shaped outcrops seen in the Big Badlands. The Peanut Peak Member is in turn overlain by the Big Cottonwood Creek Member (Chadron C and Orella A of Schultz and Stout, 1955), a unit of the Chadron Formation not seen in Badlands National Park. The Big Cottonwood Creek Member has a greater silt content than the underlying Peanut Peak Member; it is recognized by a change in erosional relief to more cliff-forming exposures (Plate 14).

Just as in South Dakota, the Chadron Formation is overlain by the Brule Formation in northwest Nebraska, but the names of individual members differ, as do lithologies. The Orella Member of the Brule Formation (Orella B–D of Schultz and Stout, 1955) directly overlies the Chadron Formation and is recognized by a change to more silty sediments, multiple thin sheets of sandstone, and an overall browner coloration (Plate 14). The Orella Member is also noted for downcutting of ancient river channels into underlying strata by up to 10 m. This downcutting, referred to as the Toadstool Park Channel Complex (Terry and LaGarry, 1998), is easily seen at Toadstool Geologic Park as thick units of cross-bedded sandstone that incised the Chadron Formation (Plate 14) to form a prominent paleovalley. This paleovalley was subsequently backfilled with additional sediments of the Orella Member. The lower part of the Orella Member is most similar to the Scenic Member of the Brule Formation in the Big Badlands, whereas the upper part is similar to the Poleslide Member.

The Orella Member is overlain by the Whitney Member in northwest Nebraska (Whitney A–C of Schultz and Stout, 1955). It is recognized by a change from brownish-beige fluvial sheet sandstones and overbank mudstones to pinkish, volcaniclastic eolian siltstone with abundant potato-ball concretions of calcium carbonate (Plate 14). The Whitney Member has no direct lithologic counterpart in the Big Badlands of South Dakota. The Whitney Member is in turn capped by the informally designated, beige Brown Siltstone Member of the Brule Formation (LaGarry, 1998) and is recognized by a change to a greater concentration of volcanic ash. The Brown Siltstone Member, which is capped by coarse cobbles of the Arikaree Group, are considered to be equivalent to the lower Sharps Formation of southwest South Dakota (J. Swinehart, pers. comm. to E. Evanoff, 2010).

With a lithostratigraphic framework in place between Toadstool Geologic Park of northwest Nebraska and the Big Badlands of South Dakota, geologic dating of these deposits, either through correlation of paleomagnetic zonations, radioisotopic dating of discrete volcanic ash beds, and to a lesser extent biochronology, provides a spatiotemporal model of deposition for the White River sequence across this region (Fig. 7.1). On the basis of thickness alone, the White River sequence is more complete at Toadstool Geologic Park compared to the Big Badlands of South Dakota, which in turn allows for a more detailed temporal model using the exposures at Toadstool Geologic Park as a reference.

The Chamberlain Pass and Chadron formations of Nebraska were deposited during the late Eocene through the early Oligocene based on magnetostratigraphy and radioisotopic dating of several volcanic ashes within the Big Cottonwood Creek Member of the Chadron Formation (Prothero and Swisher, 1992; Prothero, 1996; Sahy et al., 2012). The time represented by the deposition of the Big Cottonwood Creek Member in Nebraska is not represented by deposition in the Big Badlands but is instead manifested as an erosional unconformity of up to 1.05 million years, which created paleotopography on top of the Peanut Peak Member of the Chadron Formation before burial by the Scenic Member of the Brule Formation (Evanoff in Benton et al., 2007).

The Eocene–Oligocene boundary occurs within the upper 10 m of the Big Cottonwood Creek Member, approximately 2 to 4 m above the Upper Purplish White Layer, a volcanic ash dated at 33.907 ± 0.032 Ma (Sahy et al., 2010). In the Big Badlands of South Dakota, the Eocene–Oligocene boundary is lost within the 1.05 my unconformity at the Chadron–Brule contact. Biochronological data suggest that the Chadron Formation is primarily Chadronian, with the exception of the upper 6 to 8 m of the Big Cottonwood Creek Member, which is Orellan.

The Brule Formation of northwest Nebraska was deposited during the early Oligocene based on magnetostratigraphy and radioisotopic dating of several volcanic ashes within the Whitney Member (Prothero and Swisher, 1992; Prothero, 1996; Sahy et al., 2012). Although the Orella Member contains two distinct ash beds (the Horus and Serendipity ash beds) within the Toadstool Park channel complex paleovalley backfill sequence (Kennedy, 2011), analysis of these ashes has yet to provide a reliable radiometric date. When compared to the Badlands of South Dakota, the timing of deposition for the Orella Member is similar to the Scenic Member of the Brule Formation, but precise tie points are elusive. Evanoff (in Benton et al., 2007) reports the presence of a monazite-bearing volcanic ash within the lower Scenic Member, but no radiometric dates have been produced. Terry (unpubl. data) has noted distinctive pockets of bentonite (altered volcanic ash) within the Scenic Member along BIA 2 just west of the White River Visitor Center, and southwest of Sheep Mountain Table within the Indian Creek drainage (Fig. P.1). Whether or not these bentonites correlate with previously recognized ash beds is unknown. Biochronological data for the Orella and Scenic members are identical (both are Orellan); they provide no additional refinement to ages.

The Whitney Member of the Brule Formation in Nebraska is temporally equivalent, on the basis of magnetostratigraphy, to the Poleslide Member in the Big Badlands of South Dakota, but not lithologically. The upper part of the Orella Member in Nebraska is a better lithologic match to the Poleslide Member of South Dakota. In Nebraska, the Orella and Whitney members are transitional across their boundary, suggesting a gradual change from one depositional regime to another. In the Big Badlands, the Scenic and Poleslide members are separated by a recently recognized thin, ashy layer, the Cactus Flat marker bed (Stinchcomb, 2007; Stinchcomb, Terry, and Mintz, 2007). Although no attempts have been made to obtain radiometric dates, its stratigraphic position and magnetostratigraphic signature suggest that it matches the Lower Whitney Ash of Nebraska at 31.85 ± 0.02 Ma. Biochronologic data for these units suggest a Whitneyan age but provide no additional refinement.

The uppermost member of the Brule Formation in Nebraska is the informally designated Brown Siltstone Member. The Brown Siltstone has up to three recognized ash beds in its lower part, the Nonpareil Ash Zone of Swinehart et al. (1985), which collectively yield a date of 30.05 ± 0.19 Ma. The most likely ash equivalent to the Nonpareil Ash Zone is the Rockyford Ash in the Big Badlands. The White River sequence in northwest Nebraska is uniformly overlain by coarse conglomerates of the Arikaree Group.

The disparity in direct lithologic correlations for some units between Toadstool Geologic Park and the Big Badlands, in addition to variations in timing of deposition across the region (such as the Peanut Peak Member), can be attributed to several processes. Simply put, deposition is not instantaneous across an entire region. Climates and environments are diachronous—that is, they migrate through space and time. For example, modern-day reductions in Arctic sea ice volume are attributed to increases in global temperature, a trend that has been active since the last major continental glaciers in North America began their retreat approximately 20,000 thousand years ago. Areas once dominated by periglacial features 20,000 years ago are now dominated by prairie and forest ecosystems through the Midwest United States. Deserts do not deposit sand across an entire region at once. Eolian environments migrate and can become repetitively stabilized and reactivated depending on changes in climate conditions and the activity of stabilizing vegetation, like the Sandhills of Nebraska. Likewise, the fluvial and eolian environments that were responsible for deposition of the Chadron and Brule formations of Nebraska and South Dakota generated sedimentary deposits that accumulated across large areas over time, but deposition across the region was diachronous.

In some instances, deposition of a particular lithologic unit did not extend into the Big Badlands. All rock units must taper and end laterally because the particular environment responsible for their deposition eventually gives way to a new environmental setting. This seems to be the case with the Big Cottonwood Creek, Whitney, and Brown Siltstone members of the Chadron and Brule formations in Nebraska. Direct lithologic equivalents of these deposits have yet to be formally recognized in the North Unit of Badlands National Park, but units similar to these are present within the South Unit of the park and can be seen on the southwest corner of Cuny Table near the intersection of BIA Highways 2 and 41. At this location, a 3 to 5 m interval of strata similar to the Big Cottonwood Creek Member rests on the Peanut Peak Member of the Chadron Formation and is capped in turn by an interval of strata tens of meters thick that is similar to the Orella Member, followed by a thicker interval of strata similar to the Whitney Member in color, erosional relief, and concentration of potato-ball concretions. This suggests that at least some of these lithologic units are wedge shaped and taper to the north and east from Nebraska into South Dakota, although detailed measurements are needed for better characterization of these units. The most critical observation from measurement of these units is that the physical rock record of the EOT from hothouse to icehouse conditions is better preserved in northwest Nebraska and east central Wyoming than in the Big Badlands of South Dakota as a result of 1.05 million years of missing rock record between the Chadron and Brule formations in Badlands National Park.

When taken as a whole, the paleoclimatic data from the White River sequence suggest an overall trend of cooling and drying. This trend was not a sharp event at the EOT proper but was the result of gradual changes that began in the late Eocene and carried over across the Eocene–Oligocene boundary to culminate in a dramatic change in paleoclimates and paleoenvironments of the early Oligocene. The clues to this overall change from warmer and wetter conditions of the late Eocene to cooler and drier conditions of the early Oligocene is based on the analysis of the types of sediment that were deposited and the ancient environments that they represent, a change in the types of fossil soils, and changes in the types of animals that survived into the Oligocene (Plates 15, 16).

Sediments in Badlands National Park change from clay-dominated, greenish-blue haystack mounds of the late Eocene Chadron Formation to the progressively more cliff-forming mudstone units of the early Oligocene Brule Formation (Plate 4). The clays of the Chadron Formation represent floodplain muds that were deposited as rivers overflowed their banks, burying former landscapes and preserving bones of animals that were either caught up in the flooding or that had accumulated on the ancient landscape as a result of predation. The rivers that deposited these muds were similar to the meandering rivers seen today in the eastern half of the United States. Rivers such as these are characterized by an abundance of water, with finer sediments transported in suspension and coarser sand and gravel transported as bed load at the bottom of the river channel. The type of clay that was deposited on the floodplains (smectite) is derived from the chemical weathering of volcanic ash, which we know existed in this region in great abundance because we see occasional layers of volcanic ash from large eruptions, as well as shards of ash in rock samples. The particular type of weathering that creates smectite from volcanic ash is termed hydrolysis, which represents the alteration of ash by water. As hydrolysis proceeds, additional atoms are stripped from the smectite to produce kaolinite clay. Kaolinite is abundant in the Chamberlain Pass Formation at the base of the Badlands (Plate 2). Hydrolysis is favored under humid climate conditions.

As we move closer in time to the Eocene–Oligocene boundary, the proportion of smectite clay to volcanic glass changes in favor of the glass shards. Although smectite and kaolinite are still present in the rocks of the Badlands, they are becoming progressively less abundant. As this change to progressively greater concentrations of silt-size ash occurs, the rivers of the Badlands change from deeper gravel- and sand-rich channels of the Chamberlain Pass and Chadron formations to channels with finer grain size, sedimentary structures, and bedding suggestive of drier, and possibly ephemeral, conditions. The amount of available water appears to be decreasing across the Eocene–Oligocene boundary, which results in a decrease in the amounts of hydrolysis, and in turn less smectite and kaolinite. By the time we reach the Poleslide Member of the Brule Formation in the Big Badlands, we see a strong influence of wind-dominated deposition. Windblown volcanic dust deposits steadily accumulate but are interrupted by periods of landscape stability represented by greater concentrations of fossil remains and well-expressed paleosols (Figs. 3.13, 3.14).

The paleosols dramatically change leading up to and across the Eocene–Oligocene boundary in Badlands National Park, and at Toadstool Geologic Park in northwest Nebraska. Paleosols of the Chamberlain Pass and Chadron formations are similar to modern soils that would be found in forests today (Plates 6, 7, 15). Fossil roots are suggestive of treelike vegetation, and occasional petrified logs and wood fragments have been found in the Chamberlain Pass Formation (Fig. 5.2). Fossil roots in paleosols of the early Oligocene Brule Formation are dramatically smaller and suggest an open rangeland with occasional trees (Plate 16). Higher up in the Brule Formation, roots are commonly no larger in diameter than a pencil, with most much smaller. Paleosols in the upper part of the Brule Formation are similar to those found on modern prairies (Retallack, 1983b; Mintz, 2007; Stinchcomb, 2007).

As for fossil vertebrates, the most dramatic change is the extinction of the large brontotheres at the end of the Eocene. Brontotheres were browsers, according to their tooth structure, and they probably inhabited a forest environment and likely could not adapt to the progressively open landscapes of the Oligocene. An extinction of many types of reptiles is also associated with the EOT and suggests a change to progressively cooler and drier conditions (Hutchison, 1992). This is also supported by a change in fossil land snails from the more humid-adapted varieties of the Eocene to forms indicative of the drier and cooler environments of the Oligocene (Evanoff, 1990b; Evanoff, Prothero, and Lander, 1992).

Although all of these indicators suggest a change toward progressively cooler and drier conditions from the Eocene to Oligocene, not all data sets agree. Isotopic data from fossil bones at Toadstool Geologic Park, Nebraska, suggest that temperatures dropped by as much as 8°C across the Eocene–Oligocene boundary, but that the amount of available moisture stayed relatively constant (Zanazzi, Kohn, and Terry, 2009). In contrast, Boardman and Secord (2013) analyzed isotopes from fossil tooth enamel that record diet and water intake and concluded that precipitation decreased slightly across the EOT in Nebraska but that temperatures remained relatively unchanged. It is also possible that the changes we see in the rock record are nothing more than an environmental response to an increase in the rate at which sediment was being introduced into the region. The Oligocene experienced sedimentation rates almost double that of the late Eocene in this region (Zanazzi et al., 2007), suggesting a greater amount of sediment input. This pattern is also seen in Badlands National Park (Terry, 2010). Increased sedimentation could produce the changes seen in the change from river- to wind-dominated environments and associated paleosols of the Brule Formation.

The Badlands of Eastern Wyoming

Further to the east in the Wyoming badlands, strata become thicker, but some parts of the rock record that were preserved in Nebraska are absent. As in Nebraska, the White River sequence rests on Cretaceous strata. The zone of intense pedogenic alteration at the base of the sequence similar to the Yellow Mounds Paleosol Series is present in the far eastern part of the state between Lusk, Wyoming, and the Nebraska–Wyoming border, but is not visible toward Douglas, Wyoming (Terry and LaGarry, 1998). Fluvial units similar to the Chamberlain Pass Formation are also present between Lusk and the state line, but also are not visible further to the west (Terry and LaGarry, 1998). Units similar to the Peanut Peak and Big Cottonwood Creek members of the Chadron Formation are present, as is the Orella Member of the Brule Formation.

One of the most striking differences between the Nebraska and Wyoming exposures is the dramatic increase in the number of distinct volcanic ash beds that are preserved in Wyoming, which is the result of the western volcanic sources responsible for the voluminous amounts of volcanic ash in the White River sequence (Larson and Evanoff, 1998). These numerous ashes have provided key radiometric dates that have enhanced our ability to correlate the deposits, magnetozones, and biochrons of the White River sequence across various exposures in Wyoming and to the east in Nebraska (Evanoff, Prothero, and Lander, 1992; Prothero and Swisher, 1992; Prothero, 1996; Larson and Evanoff, 1998).

The most detailed studies of the White River sequence in Wyoming have been carried out on deposits preserved in paleovalleys (Evanoff, 1990a), such as in and around the town of Douglas (Fig. 7.1). Similar to the deposits of northwestern Nebraska, the exposures at Douglas reflect a change upward from fluvial-dominated deposits during the Eocene within units similar to the Peanut Peak and Big Cottonwood Creek members to progressively more eolian conditions during the Oligocene within strata similar to the Orella Member of the Brule Formation. The White River sequence at Douglas is unconformably capped by conglomeratic strata similar to the Arikaree Group. Vertebrate fossil assemblages at Douglas mirror those in Nebraska, with Chadronian fossils in units similar to the Peanut Peak and Big Cottonwood Creek members of the Chadron Formation and fossils of the Orellan within strata similar to the Orella Member of the Brule Formation. The Eocene–Oligocene boundary in the Douglas area falls within the uppermost part of strata similar to the Big Cottonwood Creek Member at approximately the same level as a volcanic ash, referred to as the 5 tuff, that has been dated to 33.59 ± 0.02 Ma to 33.91 ± 0.06 Ma (Evanoff, Prothero, and Lander, 1992; Prothero and Swisher, 1992).

Further to the west and south of Douglas, the White River sequence is preserved within additional paleovalleys that were cut into underlying Mesozoic, Paleozoic, and Archean bedrock (Evanoff, 1990a). The most detailed studies have been carried out at Flagstaff Rim, approximately 30 km southwest of Casper, which has been developed for extraction of petroleum resources from strata underlying the White River sequence. The base of the White River sequence at Flagstaff Rim is marked by large boulders and cobbles that are overlain by strata similar to the Peanut Peak Member of the Chadron Formation farther to the east (Emry, 1973), but at Flagstaff Rim, this interval is intermixed with vibrant red mudstone and white channel sandstone (Plate 14). This is overlain by cliff-forming greenish-gray strata similar to the Big Cottonwood Creek Member, but enriched in volcanic ash compared to strata further to the east. Numerous volcanic ash beds are also preserved and help to date the rock strata and associated vertebrate fossils (Emry, 1973; Prothero and Swisher, 1992; Sahy et al., 2012). Although this is a thick section of strata, the amount of time preserved is limited to only the late Eocene. These strata are unconformably overlain by conglomeratic units of the Split Rock Formation of late Oligocene and possibly early Miocene age (Lillegraven, 1993).

From a sedimentological point of view, the strata at Flagstaff Rim follow the same pattern seen in Badlands National Park: clay-rich units at the base of the section become increasingly dominated by silt-size volcanic ash up section. The big difference is that this change occurs totally within the late Eocene. Just as in the Badlands of South Dakota and Nebraska, preliminary data from paleosols at Flagstaff Rim suggest that climatic conditions were becoming progressively drier and that temperature was decreasing (Griffis, 2010; Griffis and Terry, 2010). The vertebrate fossil record from Flagstaff Rim is extensive (Emry, 1973) but is of limited utility in deciphering paleoclimatic conditions.

Other Exposures of the White River Sequence

To the south of Toadstool Geologic Park, the White River sequence is exposed at Scotts Bluff National Monument in west central Nebraska (Fig. 7.1). The exposures are similar to the Brule Formation, and dated volcanic ash beds place the majority of the strata and vertebrate fossils in the Oligocene (Prothero and Swisher, 1992). The Eocene–Oligocene boundary is not exposed at Scotts Bluff National Monument. In northeast Colorado, strata of the White River sequence are preserved in isolated outcrops and buttes, the most notable of which are the Pawnee Buttes. These buttes rest on the same zone of intense weathering seen at the base of the Badlands in South Dakota, but the strata that comprise the buttes are relatively similar to those seen at Scotts Bluff National Monument.

North of Badlands National Park, the strata of the White River sequence thin and disappear just beyond the town of Wall before appearing again in isolated patches and buttes in the northwest corner of South Dakota near Buffalo (Fig. 7.1). Outcrops are thin compared to the Big Badlands, and they rest on a variety of Paleocene and Eocene strata. The base of the White River sequence is marked in places by white channel sandstones similar to the Chamberlain Pass Formation, referred to as the Chalky Buttes Member of the Chadron Formation in North Dakota (Hoganson, Murphy, and Forsman, 1998). The Chalky Buttes Member is overlain by the South Heart Member of the Chadron Formation, which is identical to the Peanut Peak Member seen in the Big Badlands (Plate 14). The Chadron Formation is in turn overlain by the undifferentiated Brule Formation. The exact position of the Eocene–Oligocene boundary is uncertain, but it likely falls near the contact between the Chadron and Brule formations, according to magnetostratigraphy and biochronology (Prothero, 1996; Hoganson, Murphy, and Forsman, 1998). A volcanic ash referred to as the Ash Creek tuff is described from the Brule Formation in North Dakota (Hoganson, Murphy, and Forsman, 1998), but it has yet to be correlated with any of the ashes found elsewhere in the White River sequence. As in the Big Badlands of South Dakota, rock units across the Eocene–Oligocene boundary in North Dakota change from claystone-dominated units in the Eocene to silt-dominated units in the Oligocene. The paleosols of this region have yet to be investigated.

THE EOT ACROSS THE GLOBE

Although the strata of the White River sequence across the northern Great Plains represent the best-preserved nonmarine archive of the EOT, other sites across the globe, both marine and nonmarine, provide important regional-scale data that can be used to derive an overall synthesis of this critical time in Earth's history. Our comparisons of ancient paleoclimatic conditions are, by default, referenced to our modern-day climate conditions across the globe. Just like today, climatic and environmental conditions during the Eocene and Oligocene varied across the globe.

In North America, the other notable records important to our understanding of the of the terrestrial EOT are preserved in isolated pockets and buttes of strata and associated fossil assemblages extending from Saskatchewan to New Mexico, and Oregon to California (Prothero, 1994; Prothero and Emry, 1996). Although none of these areas has the spatio-temporal resolution of the White River sequence, each of these sites preserves important floral and faunal evidence that documents local paleoenvironmental and paleoclimatic conditions associated with the EOT. Only a few are discussed here. Prothero and Emry (1996) provide detailed descriptions of these other areas.

The strata of John Day Fossil Beds National Monument are several hundred meters thick and are noted for their preservation of a rich vertebrate fossil record, as well as unique assemblages of fossilized plant material. Just as in Badlands National Park, the John Day beds also contain numerous paleosols, some of which are brightly colored (Retallack, Bestland, and Fremd, 2000). Numerous volcanic ashes are also preserved at John Day Fossil Beds National Monument, which allows the record of paleoclimatic change to be tightly constrained in time and compared to other parts of the world. Just as in the Big Badlands of South Dakota, the EOT in Oregon is interpreted as a change from warm and humid conditions to cooler temperate conditions, based on a combination of data from paleosols and paleobotany (Sheldon and Retallack, 2004; Retallack, 2007).

The paleobotanical record of the Pacific Northwest indicates several fluctuations in temperature and precipitation during the last 5 million years of the late Eocene, culminating with a major episode of cooling of 3 to 6°C, increased seasonality, and a decrease in precipitation (Myers, 2003). According to Wolfe (1992, 1994), a pronounced changeover in terrestrial flora occurred over approximately 1 million years just after the Eocene–Oligocene boundary. Warm, subtropical vegetation was replaced by deciduous, temperate forms, especially at high and midlatitudes.

At Florissant Fossil Beds National Monument, volcanic activity to the west during the late Eocene blocked a stream and formed lakes, similar to the catastrophic events surrounding the eruption of Mount Saint Helens in 1980. Within the lakes, numerous insects, plants, and occasional small animals were preserved in the layers of ash and sediments that washed into the lake. Surrounding forests were also buried, preserving the stumps of redwood and other trees as large petrified edifices. Although the deposits of Florissant Fossil Beds National Monument do not preserve the Eocene–Oligocene boundary proper, the fossil record of plants and insects provides evidence of forested conditions during the late Eocene (Meyer, 2003) and can be used to better understand regional variations in paleoclimatic conditions during the EOT.

In Montana, numerous small patches of terrestrial strata are preserved in lowlands between mountain blocks (Sheldon and Retallack, 2004; Retallack, 2007). As in the Big Badlands of South Dakota, these fluvial strata preserve vertebrate fossils and paleosols, the analysis of which suggests that paleoclimatic change associated with the Eocene–Oligocene boundary was minimal, if any. On the basis of chemical data from paleosols, overall conditions during the Eocene and Oligocene were wetter in Oregon and drier in Nebraska (Sheldon and Retallack, 2004).

The paleoclimatic record of the EOT in South America, which is based on stable isotopic records from fossil mammalian tooth enamel from the Patagonia region, also shows no discernible change in climate (Kohn et al., 2004; Kohn, Zanazzi, and Josef, 2010). In comparison to other regions across the globe, the typically observed change from forested to open conditions across the EOT had already occurred in Argentina by the late Eocene.

Records of the nonmarine EOT in Europe are variable in both their degree of preservation and the interpretations that have been drawn from them. Paleobotanical records from Germany suggest cooling (Roth-Nebelsick et al., 2004), whereas isotopic records from mammalian tooth enamel, mollusks, and fish fossils suggest fluctuations in climate but no clear trend in either temperature or humidity (Grimes et al., 2005). According to Hren et al. (2013), the EOT in southern England does indeed record a transition to cooler and likely drier conditions. On the Isle of Wight in southern England, paleoclimatic conditions were static, with consistent temperatures and an overall increase in precipitation, although there may have been an increase in seasonality (Sheldon et al., 2009). Records of paleoclimatic conditions preserved in paleosols from the Ebro basin of Spain also suggest that conditions were unchanged in this region across the EOT.

As in North America, the most dramatic evidence of paleoenvironmental disturbance associated with the EOT in Europe is the extinction of numerous lineages of fossil vertebrates. This event, referred to as the Grande Coupure, marks the extinction of endemic European fossil communities and their replacement with new species from Asia, and is coincident with the first major pulse of glaciation on Antarctica (Costa et al., 2011).

In Africa, the Eocene–Oligocene boundary is difficult to define because of the lack of datable material, but estimates of the position of the boundary based on mammalian fauna assemblages suggest that climate change, extinctions, and turnovers were minimal, and that they were dominated by wet, tropical conditions (Rasmussen, Bown, and Simons, 1992).

The record of the terrestrial EOT in China is sparsely preserved (Xiao et al., 2010). On the northeast corner of the Tibetan plateau, the EOT is preserved as a thick, stacked sequence of ancient lakes, with limited data from ancient soils. The lake record across the EOT in China is one of aridification associated with a reduction in temperature based on evidence provided by the sedimentological record and paleobotanical evidence (Xiao et al., 2010). In the late Eocene these lakes are dominated by couplets of mudstone and gypsum, with the gypsum representing evaporitic phases of lake hydrology. The gypsum beds disappear as deposition continues into the Oligocene, suggesting a reduction in available moisture that would have mobilized the gypsum into these former lake basins. This reduction in moisture and temperature may have been due in part to the uplift of the Himalayan Mountain system, initiated by the collision of India with Southeast Asia earlier in the Paleogene, which altered weather patterns and helped to reduce the amount of CO_2 in the atmosphere as a result of increased rates of rock weathering (Ruddiman, Kutzbach, and Prentice, 1997). On the basis of the fossil record of pollen and calculation of paleoaltimetry for this region (Dupont-Nivet, Hoorn, and Konert, 2008), the uplift of the Himalayan Range began at least 4 million years before the EOT and would have been a significant contributor to climate change. More recent research by Licht et al. (2014) suggests that monsoonal intensity during the late Eocene in southeast Asia was as strong as today, and that elevated levels of CO_2 in the late Eocene compensated for the decreased elevation of the Himalayan Range at that time. Their data also suggest a weakening of the Asian monsoon during the EOT.

MARINE RECORD OF THE EOT

The majority of paleoclimatic and paleoenvironmental data that is used to interpret the dynamics of the EOT has been recovered from marine settings, in particular from analysis of various fossil assemblages and stable isotopic data of carbon and oxygen from these assemblages and associated sediments (Pagani et al., 2005; Pearson et al., 2008). In brief, the EOT in the marine realm was a time of major reorganization. The magnitude of reorganization and response to the EOT varied across the globe as a function of latitude and water depth. Data from marine records show dramatic shifts in oxygen isotope signatures of paleotemperatures, atmospheric shifts in greenhouse gases, shifts in ocean currents, dramatic drops in sea level, and the extinction or proliferation of particular fossil lineages.

The oxygen-stable isotopic record of ocean water conditions is preserved within the calcareous shells (tests) of single-celled planktonic and benthic foraminifera. As individual foraminifera grow, they extract calcium, carbon, and oxygen from seawater to create their tests of calcium carbonate. Although foraminifera generally incorporate slightly greater amounts of O^{18} into their shell structure, cooler ocean conditions favor even greater amounts of O^{18}. With greater cooling there is a greater discrepancy of O^{16} versus O^{18} uptake by the foraminifera. Because the ratio of O^{16} to O^{18} present in water molecules is related to temperature, measurements of this ratio in fossil foraminifera over time can show relative trends. For the EOT, the oxygen isotope record shows an overall cooling trend during the late Eocene, with the first pronounced step of cooling at approximately 34.2 Ma and the second, most pronounced step within the earliest Oligocene at 33.6 Ma, which is coincident with a pronounced increase in Antarctic ice volume (Coxall et al., 2005; Miller et al., 2009; Xiao et al., 2010).

Records of carbon isotopes can be extracted from organic sediments. In particular, certain types of organic molecules preserve ratios of carbon isotopes that can be used to calculate atmospheric levels of CO_2, which is a greenhouse gas. The carbon isotopes within organic marine sediments over the last 2 million years of the late Eocene show several punctuated drops in the amount of atmospheric CO_2 within a trend of gradual declining values leading up to the Eocene–Oligocene boundary (Pagani et al., 2005). By the early Oligocene, levels of CO_2 reached modern levels. Such a decline in atmospheric CO_2 would promote global cooling.

These stepwise shifts in the isotopic record of stable isotopes in marine environments culminate with a dramatic shift at the EOT. Individual smaller shifts correspond with periods of marked cooling with subsequent periods of gradual and stable warming (Coccioni, Frontalini, and Spezzaferri, 2009). With respect to the EOT, there is a dramatic increase in the abundance of O^{18} within foraminifera tests, indicating a substantial drop in seawater temperatures of 5°C on average (Liu et al., 2009). This isotopic shift is coincident with both a dramatic drop in global sea levels and the appearance of the first permanent glaciers on Antarctica. Leading up to the EOT, the record of glaciation on Antarctica is sporadic (Miller et al., 2009).

It was also at this time that oceanic circulation patterns were significantly changed by reorganization of tectonic plates. The collision of Arabia with Eurasia, combined with a reduction in global sea level, resulted in the eventual blockage of the subtropical Eocene Neotethys current from the Indian Ocean into the Atlantic (Jovane et al., 2009). The EOT also saw the opening of Drake Passage between Antarctica and South America and between Antarctica and Tasmania, thus allowing the establishment of the Antarctic Circumpolar Current and progressive cooling of ocean waters (Diester-Haass and Zahn, 1996; Exon et al., 2002). According to Miller et al. (2005), the rate of oceanic crustal spreading slowed during the EOT, which would have resulted in a reduction in the size of midocean ridges and would have also allowed ocean levels to fall. This combination of falling sea level and reorganization of oceanic circulation patterns led to numerous chemical changes in the ocean, including greater amounts of calcium carbonate precipitation, increased amounts of nutrient-rich upwelling bottom waters, increased abundances of diatoms (photosynthetic microorganisms with siliceous tests), and a reduction in overall ocean water acidity, especially in deeper ocean settings (Miller et al., 2009).

The overall paleobiological response in the marine realm to the EOT shows a complex combination of appearances, extinctions, and changes in size and diversity of particular lineages depending on latitude and water depth. This is not surprising, given the effects of falling sea levels on habitats due to glaciation and associated drops in global temperature. Prothero, Ivany, and Nesbitt (2003) and Koeberl and Montanari (2009) provide an in-depth treatment of individual marine fossil groups and their response to the Eocene–Oligocene boundary.

WHAT TRIGGERED THE EOT

The underlying cause of the EOT is still debated. More than likely its causes are similar to the proposed scenarios of climate change and extinction at the Permian–Triassic and Cretaceous–Tertiary boundaries, which were the result of a multitude of factors acting together and influencing each other. In contrast to the Permian–Triassic and Cretaceous–Tertiary boundaries, the EOT in North America did not see a major extinction in the terrestrial realm. Certain lineages went extinct, such as the brontotheres, but most lineages survived. In the marine realm, the extinction of certain microfossil lineages was offset by the origination of others. Although the EOT itself may not have been a period of extreme paleobiological change, the preceding 2 million years witnessed at least three separate asteroid or comet impacts (Poag, Mankinen, and Norris, 2003), as well as the separation of Antarctica from South America and subsequent establishment of the Antarctic Circumpolar Current (Diester-Haass and Zahn, 1996; Exon et al., 2002), the uplift of the Himalayas (Ruddiman, Kutzbach, and Prentice, 1997), and an overall drop in the amount of CO_2 in the atmosphere (Pagani et al., 2005).

Late Eocene Impact Events

Evidence of comet and asteroid impacts are preserved throughout the geologic record. The degree to which an impact event can possibly influence environmental conditions is variable and depends on the size, composition, and speed of the impactor, as well as the chemistry of the target rock (land versus ocean). Comets are essentially dirty snowballs with variable proportions of frozen materials, such as methane and water, mixed with rock. Asteroids and meteors are more variable in composition; they include stony bodies and almost pure metal.

During an impact, the velocity of the impactor is transferred to the target rock as kinetic energy in the form of heat and pressure. This results in the vaporization of the target rock for a defined distance away from the site of impact, as well as the destruction of the impactor. Vaporization is accompanied by melting of target materials and excavation of a crater. The material created by the impact (ejecta) will vary as a function of distance away from the target site. Areas proximal to the impact site will preserve coarser material as an ejecta blanket from the excavation of the crater and the size of the ejecta decreases with distance. Materials that were vaporized or melted eventually cool and condense into glassy spherules. These layers of impact debris (impactites) can sometimes traverse the globe, depending on the magnitude of the event (e.g., the end-Cretaceous impact at Chicxulub). In addition to glassy spherules, evidence of impacts in ejecta layers includes enrichments in certain elements that are not found in great abundance in the Earth's crust (such as iridium), as well as shocked minerals (most commonly quartz) that show microcrystalline lines of deformation that result from the pressure of impact at the target site.

At least three separate impact events are reported over 2 million years of the late Eocene based on records from Italy (Poag, Mankinen, and Norris, 2003), and at least three separate impact craters have been reported. The two largest craters are Popigai in northern Siberia (100 km wide) and the Chesapeake Bay crater (85 km wide), located at the tip of the Delmarva Peninsula along the Atlantic coast of the United States (Coccioni, Frontalini, and Spezzaferri, 2009). A third, smaller crater (20 km) is found on the continental shelf east of Toms River, New Jersey. The relationship of these impacts to particular ejecta layers is unknown, although links to distinct proximal ejecta deposits have been established, such as the correlation between the North American strewn field and the Chesapeake Bay impact (Poag, Mankinen, and Norris,

2003). Some issues of correlation are related to ambiguity of geologic dating of individual ejecta layers, as well as dating between sections.

Ejecta layers have not yet been reported from terrestrial EOT deposits, even though several locations preserve strata of the correct age (such as the Flagstaff Rim section of Wyoming). Numerous EOT marine core sections with ejecta have been reported (Poag, Mankinen, and Norris, 2003), which is likely due to the enhanced preservation potential of marine settings. When placed within a temporal framework of impacts versus extinctions and paleoclimatic perturbations at the GSSP in Massagnano, Italy, impacts appear to have had no direct effects on planktonic foraminifera, although changes in the proportions of particular types of warm-water versus cool-water foraminifera are seen (Coccioni, Frontalini, and Spezzaferri, 2009). The sizes of the Popigai and Chesapeake Bay impacts suggest that some form of environmental perturbation would be expected. For comparison, fragments of comet Shoemaker-Levy 9, which impacted Jupiter in July 1994, were 1 km or less in diameter, but the atmospheric blemishes that were created were larger than the planet Earth. Although these individual late Eocene impacts did not generate punctuated extinction events, the last 2 million years of the late Eocene may have been subjected to an increased rate of overall cosmic bombardment based on elevated amounts of extraterrestrial helium found in sediments (Koeberl, 2009).

All of these events likely contributed to the transition from a hothouse to icehouse paleoclimate, but what was the tipping point? Was it a gradual trend that eventually reached a threshold? What were the drivers? In order to answer these questions, the rate and timing of paleoenvironmental and paleoclimatic changes in terrestrial and marine realms need to be correlated and compared. The paleoclimatic records of the EOT do not always agree, which is not unexpected, given the current global variation in localized climates. The key is to understand the global variation and regional trends of paleoclimatic conditions during the EOT. With a terrestrial–marine chronologic framework of paleoclimatic data, global climate models can be developed in order to decipher the EOT. Such models are important for critical periods of paleoclimatic transition because they can be used to better understand our current patterns of climate change and to predict future climate responses. Once the drivers have been identified, it may be possible to link these to the patterns of extinction of some lineages and changes in the North American fauna across the EOT.

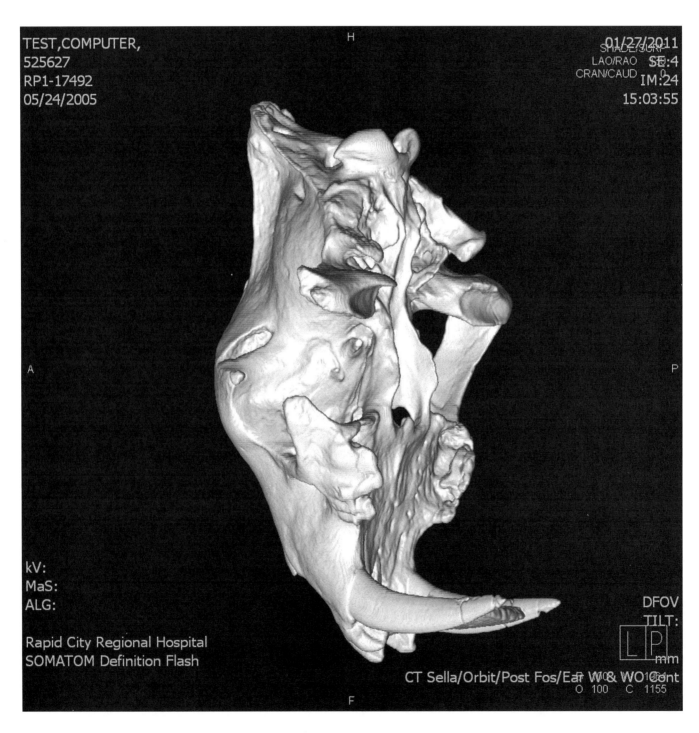

8.1. CT scan of *Hoplophoneus* sp., BADL 59490. Scan courtesy of the Rapid City Regional Hospital and the National Park Service. Specimen is the property of the U.S. government.

National Park Service Policy and the Management of Fossil Resources

THE NATIONAL PARK SERVICE (NPS) PROTECTS OUR natural and cultural heritage for future generations. At Badlands National Park, this natural heritage includes fossil vertebrates and invertebrates, fossilized plants, and trace fossils. The report accompanying the enabling legislation for Badlands National Monument in 1939, which was designated a national park in 1976, describes the future park as "a vast storehouse of the biological past." Careful protection of fossils and their context over the years has allowed researchers in many different disciplines in the earth sciences to conduct research and improved our understanding of that biological past. This in turn has produced important data about the geology and paleontology of the park, which have been published in scientific journals and which have been included in this book.

The guidelines set up by the NPS Geological Resources Program (Santucci, 2009) serve as the core of the paleontology program at Badlands National Park. The guidelines include providing inventory and monitoring, enhancing visitor understanding and enjoyment of the fossil resource, and ensuring adequate technical capacity for the management, protection, interpretation, curation, and ultimately protection of the park's paleontological resources so they are available to future generations for enjoyment and education. Since 1994, with the hiring of the first park paleontologist, Badlands National Park has developed a program that supports each of these components. Early on, it was recognized that these tasks could not be accomplished alone and that the park needed to develop partnerships with other federal agencies, tribes, museums, and universities. Over the past 20 years, in cooperation with these other institutions, the park has surveyed several hundred acres of land, collected several thousand fossils, and developed a database for over 300 fossiliferous localities. However, much work remains to be done.

On March 30, 2009, President Barack Obama signed the Paleontological Resource Preservation Act (P.L. 111–11) into law. This legislation serves as the primary authority for the management, protection, and interpretation of paleontological resources on federal land. The combination of annual visitation levels of over 1 million people and the great abundance of fossils exposed by erosion at visitor-use areas results in both positive and negative impacts to these resources. Badlands National Park staff members face many unique challenges to meet the goals of the 1916 National Park Service Organic Act, "to conserve the scenery and the natural and historic objects and the wildlife therein and to provide for the enjoyment of the same in such manner and by such means as will leave them unimpaired for the enjoyment of future generations." The protection of a portion of the White River Badlands as a national park has ensured that the scientific information provided by this world-class paleontological and geological resource can be made available to everyone in perpetuity.

PARTNERSHIPS

Badlands National Park staff work in close partnership with many outside institutions. Before the development of a paleontology program in 1994, park management relied completely on the expertise of paleontologists and geologists at museums and universities to assist with emergency salvage collections and provide input on major management decisions related to fossils in the park. The park still relies on this advice and input today, but the park is now also able to pursue more in-house projects. Museums and universities still provide important expertise and laboratory and curatorial space, and they are actively engaged in many research projects in the park. In return, Badlands National Park not only provides protection of a world-renowned resource but also ensures access to qualified researchers.

The Oglala Sioux Tribe is also a close partner with Badlands National Park. In 1968, under Public Law 90–468, Congress approved a revision of the boundaries of Badlands National Park to include a portion of the Pine Ridge Indian Reservation, creating the South Unit of Badlands National Park. While the Oglala Sioux Tribe retained ownership of the land, the management of the South Unit is by both the NPS and the tribe. One of the main reasons for establishing the South Unit was to protect "lands of outstanding scenic and scientific character." At the time of this writing, a new general management plan (GMP) has been drafted, with the goal of creating the first tribal national park in the United

States. Among the many goals of the GMP is to promote a mentoring program between NPS employees and tribe members as part of a gradual transition from primary management by NPS employees to a staff of all fully trained tribe members. There is also a strong educational emphasis, which promotes the understanding of Oglala Sioux history, culture, and land-management principles. Now that the new GMP is fully approved, the final step to make this new park a reality will require Congress to provide the required legislation.

VISITOR UNDERSTANDING AND ENJOYMENT

The park has a unique situation regarding how it manages its paleontological resources because of the combination of sheer abundance and easy access. The park provides the opportunity for even the most casual visitors to encounter fossils during their visit, either while hiking or simply making a road stop. Visitors can have the experience of being the first to witness the exposure of a fossil that has been buried for 30 million years—something that does not happen in many other national parks. On the other hand, without careful and active management, this can also pose a serious threat to the resource and can result in the loss of thousands of fossils each year that are illegally removed from the park.

To educate visitors about the significance and scientific value of the fossils themselves, as well as the equally important context in which they are found, the park has provided visitors with the opportunity to witness the discovery, excavation, and preparation of fossils in two different settings. The first was the Big Pig Dig, one of the most exciting paleontological discoveries within the White River Badlands, named for the numerous piglike entelodont *Archaeotherium* fossils found at the site (Plates 9, 10). Badlands National Park worked in partnership with the South Dakota School of Mines and Technology to excavate, prepare, and curate over 19,000 fossils from the site. The site was active for 15 field seasons and provided an incredible opportunity for park visitors to see paleontological resources and interact with scientists. On average, between 5000 and 10,000 visitors would stop by the site each summer.

In 2012 the park opened the Saber Site, located next to the Ben Reifel Visitor Center. In May 2010, a 7-year-old girl, while participating in a NPS Junior Ranger program, discovered the back of a *Hoplophoneus* skull just becoming exposed from the surrounding rock (Fig. 8.1). The girl's family reported the discovery to the ranger leading the Junior Ranger program. It was this discovery that led to the development of the Saber Site quarry and fossil preparation lab at the Ben Reifel Visitor Center. Visitors were exposed to the discovery of fossils firsthand, and they were able to witness the excavation and documentation of fossils. Inside the visitor center, visitors could see the preparation of fossils for exhibit and study, along with interesting displays and discussions of important paleontological topics.

Visitor Site Reports

It is often tempting to pick up a bone fragment as a souvenir and take it home. We all have heard the adage that if everyone collected fossils, there would be none left. This adage is actually true; paleontological surveys within the park have found that once fossil-rich outcrops near visitor use areas often contain fewer fossils than areas further away (Benton et al., 2007). Many of these areas were once important research areas but now have little scientific value. Park visitors can play a large role in making important fossil discoveries. While both the Big Pig Dig and the Saber Site were discovered by park visitors, these are just two examples. On average, the park staff receives between 100 and 150 fossil reports from park visitors each year. In order to ensure that the visitor provides accurate information, a special form has been designed to guide the visitor through the reporting process (Fig. 8.2). These forms can be obtained at each of the major visitor centers and park entrance stations. The most crucial information involves instructions on how to help the park staff relocate the fossil found by the visitor. Park visitors are encouraged to include GPS readings and copies of park maps showing the location of the fossil discovery. Photographs of the specimen and the specimen location are also helpful. An e-mail address is also valuable for follow-up discussions and further descriptions of the site. Visitors are strongly discouraged from tampering with or collecting a fossil. Such activities could result in a citation, no matter how well intentioned. When a fossil is removed from its context without proper documentation, it loses most of its scientific significance. With 244,000 acres of sharply eroded buttes, pinnacles, and spires formed by Cenozoic sediments that contain fossils from two epochs and four land mammal ages, the many eyes provided by park visitors, along with careful reporting of their discoveries, provide an important paleontological management tool.

PALEONTOLOGICAL SURVEYS AND THE PARK PALEONTOLOGICAL LOCALITY DATABASE

In 2000 and in 2003 the park received funding for 6 years of paleontological surveys within the Scenic and Poleslide members of the Brule Formation (Plate 4). The purpose of the surveys was to document fossil-rich areas within the park and to encourage further work to mitigate these areas. Other outcomes included providing important information to resource management and law enforcement divisions of the park on the location of sensitive sites for further protection,

Visitor Site Report

Paleontological Field Identification

Date _____

Visitor Center _____

Employee Name _____

Site Location

Please mark the location of the fossil on this map of the Cedar Pass Area. You can also attach a copy of another map showing the fossil location.

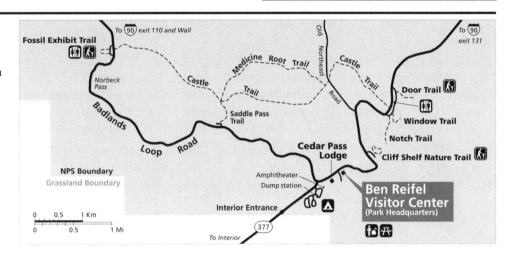

GPS Coordinates

Please note latitude and longitude or Universal Transverse Mercator coordinate (UTM).

Help us locate the fossil by answering these questions about the site.

_____   _____   _____
Latitude   Longitude   UTM

Is the site near a road or visitor use area?   YES   NO

If yes, which one? _____

How far is the site from a major landmark? _____

Include any additional information you feel would be helpful.

_____

_____

_____

_____

_____

_____

Site Description

Help us locate the fossil by answering these questions about the terrain.

In what type of terrain was the fossil found?

Butte   Wash   Sod   Table   Flat   Other _____

What is the color and texture of the surrounding rock?

Color: Banded   Gray   Green   Other_____

Texture: Clay   Popcorn/Rough and broken   Sandy   Nodule

8.2. Visitor site report form that can be used by visitors to document their fossil discovery. Form is the property of the U.S. government.

and to ensure that these areas were not negatively affected when future park development was considered. All fossil occurrences were correlated with stratigraphic position and depositional environment. During the 6-year project, the survey team documented a total of 4113 fossil occurrences. All locality information was recorded in the park Geographic Information System (GIS) database.

The development of the park paleontological locality database began in 2004. Much of the data recorded during the 6 years of paleontological surveys were integrated into individual localities. A database was created that included locality descriptions (longitude and latitude, stratigraphic position, depositional environment, and associated faunal lists), and all site locations were entered into the park GIS database. At the time of this writing, over 300 localities have been documented within the park. Most recently, the park has extended the database into the Badlands Wilderness Area (Fig. P.1). Surveys are needed for the younger Sharps Formation as well (Plate 4).

PALEONTOLOGICAL SALVAGE
AND QUARRY OPERATIONS

The majority of paleontological occurrences in the park consist of a single bone. However, during survey work, field crews sometimes discover extensive bone accumulations that require larger-scale excavation (Plate 9). The larger excavations require a meter grid system and the use of highly accurate survey tools to document the exact location and position of individual fossils within a bone bed (Plate 10).

NPS RESEARCH PERMIT AND
REPORTING SYSTEM

Although casual collecting in the park is not allowed, fossil collecting by qualified researchers as part of a scientific study is allowed with a research permit signed by the park superintendent. A research permit application requires a proposal with a valid research question.

In 2001 the NPS made a concerted effort to standardize the research permitting system. After several complaints from the research community about the inconsistency of permitting programs between individual national parks, the NPS Permit and Reporting System was consolidated into a central online database to address these issues. To apply for a permit, the researcher simply logs on to the NPS Web site (https://irma.nps.gov/RPRS/), at which point the NPS research coordinator for that particular park is notified of the permit application via automatic messaging. The park staff then can begin the permit evaluation process, and if the application is acceptable, they can issue a research permit.

The permitting process requires the drafting of a summary report of all the work done within the park, known as the Investigator's Annual Report, or IAR. All fossils collected under the research permit remain federal property and must be curated and stored following the NPS curatorial guidelines. However, because of limited curatorial space, the park does allow fossils to be curated at nonfederal repositories. The majority of fossils collected from Badlands National Park are housed at other institutions. The park does have a collections storage facility, but it is limited in size. The park has a loan agreement with institutions that have collected and curated fossils under a valid research permit. The Museum of Geology at the South Dakota School of Mines and Technology in Rapid City is just one example of the many partner repositories curating fossils from the park. Many fossils were collected from the White River Badlands before the park was established, and in order to help the park staff fully understand the paleontology of the park, efforts are being made to try and document these collections as a way of achieving this goal.

NPS CURATORIAL DATABASE

Congress mandates that all NPS collections be entered into a centralized database, and NPS management guidelines stipulate that all related information be recorded for each museum specimen. The Interior Collection Management System (ICMS) is the database for museum collections used by all agencies within the Department of the Interior, including the NPS. Because of stringent NPS curatorial guidelines, an extra responsibility is added to the collection of each fossil specimen. When each fossil is collected, relevant information such as proper locality and stratigraphic information must be recorded. This is the same type of information recorded for specimens collected by professionals in all museums. Curators in the NPS make a special effort to record the correct identification and all associated information. Because the deaccession process can be lengthy, special care must be taken when deciding to collect and curate a fossil. Fossils in partner repositories are cataloged into ICMS, and this information is maintained by the park curator. This allows the curator to keep track of the location of all specimens collected in the park and to ensure that they are being maintained in proper curatorial conditions.

NATIONAL ENVIRONMENTAL
POLICY ACT COMPLIANCE

The National Environmental Policy Act (NEPA) was passed by Congress in 1969 and took effect on January 1, 1970 (Directors Order 12, 1985). NEPA mandates that every federal agency prepare an in-depth study of the impacts of major federal actions that can have a significant effect on the environment. Each agency must propose alternatives to these actions. They also must use this information as an integral part of the decision-making process. Finally, they must involve the interested and the affected public. Part of the environment that can be affected includes paleontological resources. This includes resources in the field, museum collections, undiscovered resources, and unauthorized collections. At Badlands National Park, all proposed major actions within the park are carefully evaluated for their impact on paleontological resources.

During busy years with a strong economy, Badlands National Park can receive over a million visitors per year. A park with such a high visitation rate requires a strong infrastructure to meet visitors' needs. Construction projects that provide this infrastructure can include roads, trails, and visitor centers. Fossils may not only be encountered during the initial construction phase but also during maintenance, and even when structures are removed or replaced.

The Badlands Loop Road (Highway 240) was mandated through the 1929 legislation that established Badlands National Monument. Peter Norbeck, a United States senator from South Dakota, drafted the legislation to ensure the prompt development of the monument with full participation from the state of South Dakota (Shuler, 1989). The legislation required that in order for the monument to be established, a road must be built by the state of South Dakota to show off the grandeur of the Big Badlands. Because the road was actually built right along the Badlands wall, it was built in the worst possible place from an engineering perspective. The soft strata that form the Badlands features erode quickly and are not stable enough to support a road. The Badlands Loop Road crosses several major passes: Cedar Pass, Norbeck Pass, Dillon Pass, and the Ancient Hunters/Pinnacles area (Fig. P.1). Each of these passes contains active landslides, along with fault activity. The Badlands Loop Road, especially near the major passes, is often under construction. Because of the high density of fossils in the park, every construction project to remove landslide debris or to repair the road can potentially result in the discovery of fossils and must be closely monitored in order to be in compliance with NEPA.

Any excavation into bedrock within the White River Badlands requires careful monitoring by a paleontologist. Large excavation equipment can easily damage delicate fossils. A paleontologist monitoring a construction site must act quickly to salvage fossils that are at risk of being damaged or lost, and at the same time document their location and stratigraphic context. Ideally the paleontologist can accomplish all of these goals without delaying the overall progress of a project.

LAW ENFORCEMENT ISSUES AND THE PALEONTOLOGICAL RESOURCES PRESERVATION ACT

With easy access to fossils and the many duties of the law enforcement staff, it is difficult to keep up with the illegal poaching activity within the park and surrounding federal lands. Over the past 3 years, there have been between 1 and 3 fossil poaching cases each year in the park (Osback and Griswold, pers. comm., 2014). Fossil theft extends beyond actual law enforcement cases. The commercial trade in fossils is ever growing and international in scope. A recent estimate for a complete brontothere skull sold through Sotheby's auction house in October 2011 was between $40,000 and $50,000 (Fig. 5.53).

Fossil poaching at Badlands National Park, as on any federal land, can take on many forms. The casual park visitor may pick up a piece of fossilized bone during a hike along a park trail. This activity may seem innocuous, but multiplied by a million visitors per year, this activity can have a major impact on the resource. Many college geology field camps visit the park on their way to field-mapping projects further west. Unfortunately, some college students have returned to the park on their own, not understanding that they need a research permit to collect fossils. In 1999 four college students were caught with over 2000 fossils they had illegally collected in the span of 4 days. They all received steep fines as penalties for their actions. Professional fossil poachers also visit the park, especially the South Unit. Poachers use dirt bikes to access remote areas and collect fossils out of view of law enforcement (Fig. 8.3). In addition to the removal of the fossil, the use of the dirt bikes can result in heavy erosion of the soft strata and the destruction of geological features.

In an effort to combat fossil poaching, Badlands National Park has been working with the Department of Earth and Environmental Science at Temple University in Philadelphia, Pennsylvania, in order to develop a method to chemically "fingerprint" fossil bone material. This method will not stop the act of poaching directly, but it may serve as a means to strengthen legal cases against suspected poachers. The concept of chemically fingerprinting a fossil bone is based on the process of fossilization described in chapter 6. As bone fossilizes, it incorporates a distinct suite of elements that are characteristic to particular rock units and geographic locations within the outcrops. These chemical fingerprints are composed of rare earth elements (REEs), a group of elements that, although relatively abundant, are named for their tendency to be found in low concentrations in geological materials. REEs include the lanthanide series of the periodic table, from atomic numbers 57 to 71, as well as scandium and yttrium, with atomic numbers of 21 and

8.3. Photographs of fossils that were targeted for illegal collection, and pits where fossils were stolen. (A) Fossil tortoise wrapped with electrical tape (e). (B) Fossil skull (S) wrapped with newspaper, plastic, and tape. This specimen was recovered from suspected poachers. Photo is the property of the U.S. government. (C), (D) Poach pit with only postcranial fragments (pf) of bone remaining. Photos (A), (C), and (D) are by the authors.

39, respectively. Although REEs normally tend to occur in low concentrations, their ionic radius makes them a perfect fit for inclusion into bone that is recrystallizing during the fossilization process. The REE signature incorporated into a fossilizing bone is influenced by the type of sediment that the bone is in, the acidity and alkalinity of the environment of deposition (pH), and whether the environment is wet or dry, which influences the availability of oxygen in the system (oxidizing versus reducing). The REEs and other elements enter into the bone via groundwater that has chemically reacted with the surrounding sediments and environment. Once the REE signature is incorporated into the recrystallized bone, it is locked in and cannot be easily altered. Chemical signatures are determined by mass spectrometry and statistically analyzed to establish patterns and distinctiveness (Fig. 8.4).

At the time of this writing, 100 individual fossils from the Chadron Formation and from the Scenic and Poleslide members of the Brule Formation in Badlands National Park

have been analyzed for their REE signatures (Fig. 8.4). These fossil samples stretch over 50 km from Sheep Mountain Table southwest of Scenic, South Dakota, to the Door and Window Overlooks just northeast of Cedar Pass (Fig. P.1). Results to date suggest that fossils from different rock units have unique REE signatures. This is similar to the results of Metzger, Terry, and Grandstaff (2004), Grandstaff and Terry, 2009), Lukens et al. (2010), and Terry et al. (2010), who found that individual bone accumulations within various rock units of the Badlands could be statistically distinguished. As new fossil sites are discovered and chemically analyzed, a library of geochemical fingerprints will be created that can be used by law enforcement officers to establish the point of origin of fossils recovered from poachers.

The Paleontological Resources Preservation Act (PRPA) (P.L. 111–11) provides the NPS with seven important mandates to enhance paleontological stewardship. Some of the most significant mandates involve the clarification of criminal

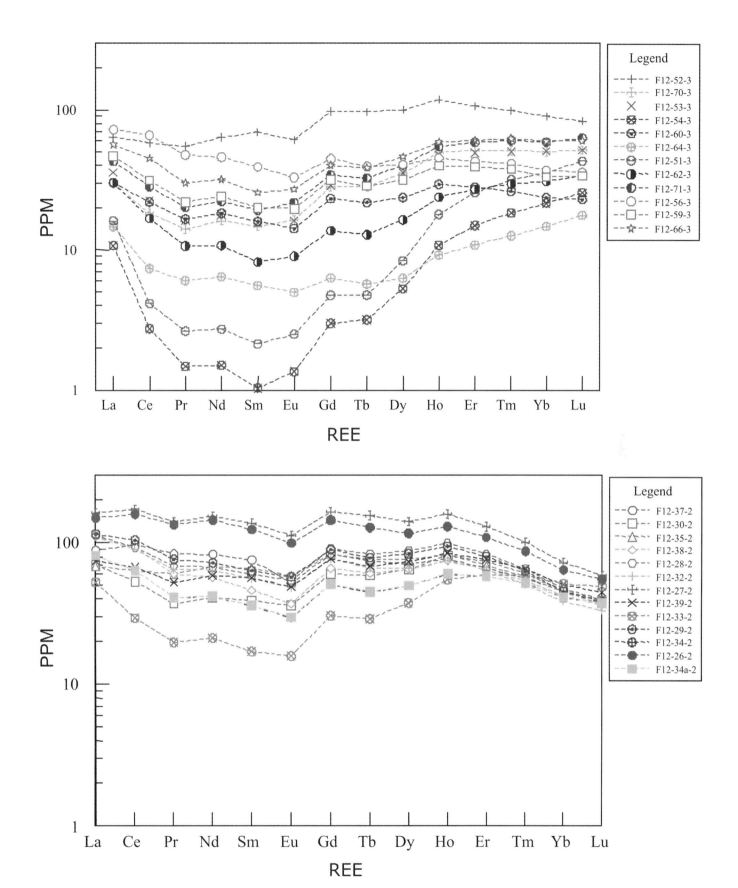

8.4. Graphs of representative rare earth element signatures from fossils in Badlands National Park. These two graphs are from different stratigraphic units in the White River Group.

penalties, which increase fines and jail time for fossil poaching offenses. A summary of these mandates, drafted by the NPS Geologic Resources Program (Santucci, 2009), is as follows:

1. PRPA Section 6302 calls for the management and protection of fossil resources using scientific principles and expertise. NPS personnel will develop inventory and monitoring programs and the scientific and educational use of fossil resources. The NPS is encouraged to partner with outside institutions and the public to achieve many of these goals.

2. PRPA Section 6303 calls for the development of educational programs to expand public awareness about the significance of paleontological resources.

3. PRPA Section 6308 provides the NPS a confidentiality provision exempting the requirement to disclose sensitive paleontological locality information.

4. PRPA Section 6305 calls for the curation of NPS paleontological resources, along with associated data or records, in approved repositories.

5. PRPA Section 6306 provides clarity regarding prohibited acts involving paleontological resources and specifies criminal penalties associated with these criminal acts.

6. PRPA Section 6307, along with other existing authorities, enables the NPS to seek civil penalties and restitution for the violation of any prohibited activities involving paleontological resources.

7. PRPA Section 6310 directs the secretaries of Interior and Agriculture to issue regulations appropriate to carry out the act.

It is with great pride that the NPS serves the people of the United States, protecting this unique, nonrenewable resource. Nearly every major natural history museum around the world houses fossils collected from the White River Badlands, demonstrating a legacy of research and education for over 150 years. Many tools are needed to protect this spectacular resource, including documentation, excavation, preparation, curation, and education. The goal is to continue this legacy for many generations to come.

Glossary

A horizon mixed zone of mineral and organic matter at the top of a soil profile. Commonly referred to as topsoil.

acidic chemical condition in soils denoted by higher concentrations of hydrogen protons than base cations, such as sodium, calcium, and potassium.

alfisol soil order in semiarid to humid areas, typically under a hardwood forest cover. It has a clay-enriched subsoil and a relatively high native fertility. "Alf" refers to aluminum (Al) and iron (Fe).

alkaline chemical condition in soils in which base cations, such as sodium, calcium, and potassium, are more abundant than hydrogen protons.

anoxic environment lacking oxygen.

anterior near or toward the front of something (such as the body)

antorbital vacuities openings in front of the orbit within a skull.

aperture opening of a snail shell.

apomorphy in cladistics, a derived character state, or one that evolved later. Any feature novel to a species and its descendants. See *synapomorphy*.

aragonite form of calcium carbonate that makes up mother-of-pearl shells.

arboreal living in or often found in trees.

argillan film of clay on the surfaces of peds that form by the translocation of clays downward in a soil profile.

argillic horizon pedogenic horizon (Bt) that is enriched in clays that have been translocated downward.

aridisol soil most commonly found in association with arid environments, such as deserts; often have accumulations of carbonate or gypsum in their profiles.

articulation degree of connectivity between various bones in a fossil deposit. Articulated bones retain their original life position on the organism, and disarticulated bones are free-floating but generally in the same area as other bones.

avulsion process by which a river changes its overall course of flow by breaching the natural levee that contains it.

B horizon primary horizon of accumulation (illuvial horizon) in a soil profile. The type of materials that concentrate in the B horizon reflect dominant climatic conditions.

bed load portion of sediment transported by a river that is in contact with the stream bed as it moves.

benthic foraminifera microscopic single-celled organisms that precipitate a calcite shell (test) and live on the seafloor.

bentonite particular type of swelling clay (smectite) named for Benton, Wyoming.

bicipital groove groove on the anterior border of the proximal humerus between the medial and lateral tuberosities.

bioapatite apatite is a group of phosphate minerals, calcium apatite is the mineral that forms bone. The individual crystals are held together by collagen, which decays upon death and burial.

biochronology correlation of rock units based solely on the presence of fossils, either individual taxa or an assemblage.

biomineralization process by which living organisms produce minerals, often to harden or stiffen existing tissues; an extremely widespread phenomenon, it occurs in all six taxonomic kingdoms, and over 60 different minerals have been identified in organisms. Examples include silicates in algae and diatoms, carbonates in invertebrates, and calcium phosphates and carbonates in vertebrates.

biostratigraphic unit rock units defined solely by their fossil content. Biostratigraphic units are not required to adhere to lithologic boundaries.

biostratigraphy branch of stratigraphy focusing on correlating and assigning relative ages of rock strata by using the fossil assemblages within the strata.

bone bed accumulation of fossil bone that has a higher-than-normal concentration within a given area relative to the volume of sediment.

brachyodont having teeth with low crowns, well-developed roots, and only narrow canals in the roots, typically found in browsing mammals.

bunodont mammal with cheek teeth with low, rounded cusps on the occlusal surface of the crown, typical of omnivores.

C horizon bottom of a soil profile; has little to no modification from its original parent material state.

C3, C4 vegetation particular photosynthetic pathways utilized by certain types of plants. C3 plants include most shrubs and trees, whereas C4 plants are dominated by grasses. These differences in photosynthetic pathways, which are a response to general climatic conditions (such as moisture availability) can be preserved in the rock record as distinct ranges of carbon isotopic values of preserved organic matter. Measurement of the amount of carbon 13 in tooth enamel can provide information about an extinct animal's diet.

caldera cauldronlike volcanic feature usually formed by the collapse of land after a huge volcanic eruption.

cambic horizon beginning stages of subsurface accumulation of translocated materials to form a B horizon. The cambic horizon (Bw) is sufficiently different from either the A or C horizon but has not reached a diagnostic stage of accumulation.

caniniform of or bearing similarity to the shape and appearance of a canine tooth.

cannon bone bone in hoofed mammals that extends from the hock to the fetlock; especially the enlarged metacarpal or metatarsal of the third digit of a horse. The term may also be applied to the fused third and fourth metatarsals in some artiodactyls.

carbonate primarily composed of the carbonate ion, CO_2; can refer both to carbonate minerals and carbonate rock (which is made of chiefly carbonate minerals)

carina angular lateral margin of a snail shell.

carnassials teeth in mammalian carnivores with sharp edges for cutting flesh, formed by the fourth upper premolar and first lower molar teeth in modern carnivores, but may be formed by other premolars and molars in primitive carnivorous mammals like the hyaenodonts.

catena laterally linked set of soils across a landscape, the result of laterally changing CLORPT conditions. See CLORPT.

caudal of the tail region.

cementation hardening and welding of clastic sediments (those formed from preexisting rock fragments) by grain compaction and the precipitation of secondary minerals in the pore spaces between individual particles.

centrocrista tooth crest that connects the paracone and metacone (more lingually).

cephalic relating to the head or the head end of the body.

cervical neck region of an animal. In a turtle shell, it refers to the portion relating to, involving, or situated near the anterior portion of the shell.

channel belt broad band of individual channel deposits formed from a stream or river that is in a relatively stable position for a long period of time, typically decades to centuries.

chemical weathering chemical breakdown of geologic materials, commonly by dissolution, oxidation, or hydrolysis.

cingulum portion of the teeth occurring on the lingual or palatal aspects that forms a convex protuberance at the cervical third of the anatomic crown. It represents the lingual or palatal developmental lobe of these teeth.

cladistics method of classification of animals and plants according to the proportion of measurable characteristics they have in common. It is assumed that the higher the number of characteristics shared by two organisms, the more recently they diverged from a common ancestor.

clay smallest particle size class (4 µm or less) commonly described by geologists; special class of minerals (phyllosilicates) defined by sheetlike arrangement of their atomic structure. Particular clay minerals in the White River Group include smectite, kaolinite, and illite.

claystone clastic sedimentary rock composed of more than two-thirds clay-size particles (grains less than $\frac{1}{256}$ mm in diameter).

claystone breccia fragments of clay eroded from their source, transported, and redeposited into a new layer. A claystone breccia is commonly created by lateral migration of river cutbanks into their floodplain.

CLORPT five factors of soil formation: climate, organisms, relief, parent material, and time. See *catena*.

concretions bodies of specific minerals in soils and rock bodies that form by precipitation from fluids. Common varieties include calcium carbonate and iron oxide.

condyle articular prominence of a bone.

continental shelf large, flat expanse of ocean floor that stretches from the beach out to the shelf-slope break.

coprolite fossilized feces. The shape of the coprolite and materials inside of it can provide clues to the animal and its diet.

costal relating to, involving, or situated near a rib.

crevasse splay thin sheet of sand deposited in a proximal floodplain position by breaching the natural levee along a river.

cumulic soil soil characterized by a gradual buildup of sediment that is incorporated into the active soil. Cumulic profiles are overthickened compared to normal soils and tend to form at the base of slopes or in places where eolian additions to the soil are common.

cursorial adapted specifically to run.

dahllite original form of bioapatite in vertebrate bones. Dahllite recrystallizes into francolite upon fossilization.

detrital zircon sediment grains of the mineral zircon ($ZrSiO_4$) that are typically derived from multiple source rocks. Zircons are resistant to weathering and contain uranium, which allows them to be radiometrically dated.

dextral coiling coiling in a snail shell in which the coil is to the right. In dextral shells, the aperture opens to the right of the spire axis when the apex of the spire is held vertically. See *sinistral coiling*.

diachronous "over time"; commonly used to refer to a rock unit that has different ages across a large area.

diapophyses part of the transverse process of a thoracic vertebra that articulates with its corresponding rib.

diastema space or gap between two teeth. Many species of mammals have a diastema as a normal feature, most commonly between the incisors and molars.

diatoms microscopic single-celled plants that precipitate opaline silica shells (tests) and live in fresh- and ocean waters.

digitigrade standing or walking on the toes (digits). See *plantigrade* and *unguligrade*.

discoidal shells coiled snail shells only slightly elongated along the axis of coiling. The resulting shell is very flat and wide, like a disk that is also called a low-spired shell.

distal location farther away from a point of reference; in vertebrates, this is usually from the midline of the body.

drab-haloed root trace greenish traces of former roots in paleosols created by the anaerobic decay of plant matter.

dung ball mass of feces rolled into a ball by a dung beetle and buried in the ground to serve as a food source for developing beetles inside the mass. Dung balls are a common trace fossil in certain layers of strata in the Badlands.

durophagous ability to break and eat the hard parts of organisms, such as bone crushing in hyenas or the ability to break the shells of snails.

E horizon soil horizon characterized by high rates of eluviation (loss) of clays and elements to lower parts of the soil profile. These horizons, also referred to as albic horizons, are commonly white in soil profiles when very well developed.

ectepicondylar process on the distal humerus (upper arm bone) or femur (upper leg bone) on the lateral side just above the articular surface. It is associated with attachment of the extensor muscles of the lower forelimb.

ectoloph one of the principal crests of a lophodont molar extending from the paracone to the metacone. See *lophodont, metaloph,* and *protoloph*.

ejecta material created by the impact of an extraterrestrial object, including glass spherules and shocked quartz.

eluviation process by which materials are removed from the upper part of a soil profile and translocated downward to accumulate in the B horizon. See *horizonation* and *illuviation*.

endothermy ability to maintain the body at a metabolically favorable temperature, largely by the use of heat set free by its internal bodily functions instead of relying almost purely on ambient heat.

entepicondyle on the distal humerus or femur, a process above the medial condyle for the attachment of muscles, that faces posteriorly (for sprawling tetrapods) or medially (for erect tetrapods).

entisol soil that does not show any profile development other than a horizon. An entisol has no diagnostic horizons and is basically unaltered from its parent material, which can be unconsolidated sediment or rock.

entocondylar medial condyle of a bone on the side next to the body.

entoconid in mammalian dentition, a major cusp on the lingual side of the talonid.

eolian pertaining to wind activity. In geology, eolian refers to features formed by the wind. Also spelled aeolian or æolian.

epiclastic sediment derived from local sources, in contrast to those derived from distant volcanoes.

epipubic bones pair of bones projecting forward from the pelvic bones of modern marsupials and of some fossil mammals.

euhedral crystals mineral with well-formed crystal faces.

exodaenodonty cusp of tooth is more pronounced than surrounding cusps.

fenestra relatively large opening through bone; literally "window."

fissiped having the toes separated from one another, as in the feet of certain carnivorous mammals.

floodplain flat area on either side of a river channel that is covered during flooding. The floodplain is built by the deposition of fine-grained sediments that settle out of the floodwaters.

fluvial processes, sedimentary deposits, and landforms associated with rivers and streams.

fontanel opening or gap in the skull.

forbes herbaceous flowering plant that is not a graminoid (grasses, sedges, and rushes); used in biology and in vegetation ecology, especially in relation to grasslands and understory.

formation widespread, lithologically distinct rock unit sufficiently thick to be easily mapped in the field. The formation is the fundamental unit of lithostratigraphy.

fossa depression in a bone, often a point of muscle attachment; fossettes and fossetids are smaller depressions or fossas.

fossilization process by which skeletal materials are preserved in the rock record. Bone fossilization includes recrystallization of bioapatite and incorporation of mineral materials.

fossorial animal adapted to digging and life underground such as a badger or mole. It is an adjective used to describe the habit of living underground even if even if the physical adaptions are minimal; most bees and many wasps are thus called fossorial, and many rodents are considered fossorial. Some organisms are fossorial to aid in temperature regulation, while others utilize the underground habitat for protection from predators or for food storage.

francolite recrystallized version of dahllite found in fossil bone. Recrystallization is accompanied by the inclusion of rare and trace elements into the bone.

geochronology science of dating and correlating distinct periods of geologic time in the rock record.

glaebule three-dimensional body of mineralized material in soils that forms by precipitation. The most common types are calcium carbonate and iron oxide.

Global Stratotype Section and Point (GSSP) precise stratigraphic point used as the global standard for the boundary between two units of geologic time.

globose conic conical snail shell with highly inflated whorls.

group stratigraphic unit composed of two or more formations.

herbivore animal that eats plants.

hermaphroditism process of self-fertilization in some types of organisms.

high-spired shells coiled shells in snail shells that are elongated along the axis of coiling. The resulting shell is very tall in comparison to its width.

horizonation process by which soils are differentiated into distinct horizons by physical, chemical, and biological processes, resulting in either the gain or loss of particular soil materials within each horizon. See *eluviation* and *illuviation*.

hydrolysis process by which water interacts with a mineral, leading to the replacement of hydrogen from the water with assorted cations on the mineral surface. Over time, this will change the original mineral structure to a clay.

hydromorphy influence of water on the development of a soil profile.

hydroxyapatite inorganic calcium phosphate mineral of bone and teeth, imparting rigidity to these structures (also hydroxylapatite).

hypercarnivore animal with a diet of more than 70 percent meat, with the balance consisting of nonanimal foods such as fungi, fruits, and plants.

hypocarnivore animal that consumes less than 30 percent meat for its diet, the majority of which consists of nonvertebrate foods that may include fungi and fruits.

hypoconulid in mammalian dentition, the buccal main cusp of the talonid on the lower molars.

hypotarsus in birds, a ridge or process located on the posterior side of the tarsometatarsus, near the proximal end.

hypselenodont high-crowned teeth with open roots and are ever-growing teeth.

hypsibrachydont teeth containing both low-crowned and high-crowned cusps.

hypsodont having teeth with high crowns. Hypsodont dentition is associated with a diet of abrasive foods.

hystricomorphous in which the deep masseter passes through the infraorbital foramen to attach to the side of the snout in front of the eye, as in porcupines. See *myomorphous, protrogomorphous,* and *sciuromorphous*.

illuviation process by which materials removed from the upper part of a soil profile (eluviation) collect in the B horizon. See *eluviation* and *horizonation*.

impactite defined body of material, such as an ejecta blanket, produced by the impact of an extraterrestrial bolide.

inceptisol soil with a defined cambic (Bw) horizon between an A and C horizon.

inflated whorls spiral whorls of a snail shell that are expanded, thus looking as if they have been inflated.

insectivore animal that feeds on insects.

intercotylar prominence ridge or process in birds located on the proximal end of the tarsometatarsus next to the hypotarsus.

iridium element in low abundance in crustal materials but relatively enriched in certain extraterrestrial objects. Iridium is commonly used as evidence of bolide impacts in the rock record.

isotopes elements with the same number of protons but different numbers of neutrons, which creates different atomic masses. For example, the ^{12}C, ^{13}C, and ^{14}C isotopes of carbon have six, seven, and eight neutrons, respectively.

kaolinite (kaolin) aluminum-rich type of clay mineral that forms under well-drained and acidic soil conditions.

lacustrine processes and features associated with lakes.

land mammal ages biochronologic units of the late Cretaceous and Cenozoic defined by mammalian faunas. See NALMA.

lanthanides group of elements on the periodic table, also referred to as rare earth elements, which range from atomic number 57 to 71 and include scandium (21) and yttrium (39).

Laramide Orogeny period of mountain building during the late Cretaceous and into the Paleogene that generated the Rocky Mountains and several isolated mountain features, such as the Black Hills of South Dakota and Wyoming.

lateral accretion process by which rivers meander via lateral construction of point bars as the meander loop excavates the opposite bank.

Laurentide ice sheet large continental glacier of North America, centered in eastern Canada, during the late Pleistocene.

levee natural buildup of sediment from floods deposited adjacent to a river channel.

lithology description of the physical characteristics of a rock unit, such as composition, color, texture, and grain size.

loess silt-size sediments formed from the deposition of wind-transported dust.

loessite siltstone formed from the lithification of eolian dust deposits (loess).

lophodont tooth form with increasing specialization for grazing, resulting in fusion of cusps into ridges (lophs). Lower molars typically have two transverse lophs, the protoloph and the metaloph. In the upper molars, these ridges are fused with a longitudinal ridge (ectoloph). See *ectoloph, metaloph,* and *protoloph*.

magnetostratigraphy science of measuring and correlating magnetic signatures in the rock record.

marker bed rock strata readily distinguishable by its physical characteristics and traceable over large distances.

master horizon primary soil horizons (e.g., A, B, C) that can be subdivided with subordinate descriptors of soil features.

meander loop bend in a river characterized by erosion on the outside of the meander (cutbank) and deposition on the inside of the loop (point bar).

member rock unit that is a subdivision of a formation.

mesaxonic having the axis of the foot passing through the third digit of the hand or foot.

metaloph crest of tooth attaching the paracone to the hypocone. See *ectoloph, lophodont,* and *protoloph*.

metapodial term used to include both the metacarpals and metatarsals, often applied when a specific identification of these bones is not possible.

micrite limestone composed of microcrystalline calcite (crystals less than 4 μm in diameter) formed by recrystallization of lime mud.

mollisol soil commonly found associated with grasslands; must have a mollic epipedon, a 20 cm thick accumulation of base-rich organic matter at the top of the soil.

monophyletic group of organisms descended from a common evolutionary ancestor or ancestral group, especially one not shared with any other group. See *paraphyletic*.

monotypic taxonomic group containing only one immediately subordinate taxon. For example, a monotypic genus has only one species.

mudstone rock that includes subequal amounts of clay and silt that is lithified.

myomorphous in which the middle masseter is attached in front of the eye and the deep masseter passes up into the orbital area and through the infraorbital foramen, as in rats and mice. See *hystricomorphous, protrogomorphous,* and *sciuromorphous*.

NALMA (North American land mammal age) intervals of time defined by distinct mammalian faunas found in North America. See *land mammal ages*.

neural portion of a turtle shell relating to the bones on the midline of the carapace formed from the vertebrae.

nodules accumulations of mineralized soil or rock, most commonly calcium carbonate or iron oxide. In petrology or mineralogy, it is a secondary structure, generally spherical or irregularly rounded in shape. Nodules are typically solid replacement bodies of chert or iron oxides formed during diagenesis of a sedimentary rock.

nomen vanum literally "vain name"; available taxonomic names consisting of unjustified but intentional emendations of previously published names. They have status in nomenclature, with their own authorship and date.

normal magnetic polarity direction of Earth's magnetic field that humans have always known, i.e., the current north–south direction shown by a compass needle.

nuchal scute located on the anterior margin of the carapace of a turtle shell above the neck.

O horizon organic horizon at the top of the soil profile, also known as leaf litter.

omnivore animal that eats both plant and animal materials.

oncolite sedimentary structure similar to stromatolites, but instead of forming columns, they form approximately spherical structures resulting from layers deposited by the growth of cyanobacteria. The oncoids form around a central nucleus, such as a shell particle, and the calcium carbonate structure is precipitated by encrusting microbes. Oncolites are indicators of warm waters in the photic zone, but they are also known in contemporary freshwater environments. These structures rarely exceed 10 cm in diameter.

ontogeny origin and development of an individual organism from embryo to adult.

orogeny forces and events resulting in large structural deformation of the Earth's lithosphere (crust and uppermost mantle) as a result of the movement of the Earth's tectonic plates.

overbank deposit alluvial geological deposit consisting of sediment that has been deposited on the floodplain of a river or stream by floodwaters that have broken through or overtopped the river's banks.

oxbow lake meander loop of a river that has been isolated by cutting through a thin neck of floodplain material.

oxidation chemical process by which an element gives up an electron, thus increasing the amount of positive charge for any given ion.

paleobotany study of ancient plants, including their taxonomy, evolution, and paleoecology.

paleocatena laterally linked set of paleosols across a paleolandscape that were the result of laterally changing paleo-CLORPT conditions. See CLORPT.

paleoclimatology study of ancient climate patterns and trends preserved in various archives of the rock and fossil record.

paleoecology study of ecological relationships and interactions of ancient plants and animals.

paleoenvironmental analysis study of ancient environments, as recorded by sediments, sedimentary structures, and fossils in the rock record.

paleomagnetism study of the geologic record of the Earth's magnetic field as preserved by magnetic minerals in rocks. These minerals record changes in the direction and intensity of the Earth's magnetic field over time.

paleopedology study of ancient soils in the rock record.

paleosol fossil soil, formed when a soil becomes preserved by burial underneath either sediments or rocks; soils that are outside the range of active pedogenesis because of burial.

paleotopography geometry of ancient land surfaces preserved in the geologic record.

paleovalley ancient valley, typically recognized by its deeply eroded bottom contact and its thick fill from river deposits.

parabolic dune eolian dune with long ridges that point upwind and connect with a tightly curved ridge on the downwind end.

paracone one of the major cusps of a tribosphenic upper molar, typically in the anterobuccal corner.

paraphyletic group of organisms descended from a common evolutionary ancestor or ancestral group, but not including all the descendant groups. See *monophyletic*.

paraxonic having the axis of the foot between the third and fourth digits.

parent material original substrate material on which a soil develops. This material is typically unconsolidated sediment.

pedogenesis sum of all physical, chemical, and biological processes that modify original materials to form a soil.

pedogenic processes that lead to the formation of soil.

peds three-dimensional units of soil material that range in size and shape from small granular shapes to large columnar and prismatic shapes. Peds form by the segregation of soil material due to rooting, shrink and swell processes, and translocation of clays onto ped surfaces.

pentadactyl having five digits on the manus/pes.

perennial streams stream or river channel that has continuous water flow throughout the year.

periglacial cold environmental conditions along the margins of glaciers.

permineralization process by which pores within a skeletal material are filled with mineral material. The most common examples are petrified (silicified) wood and fossilized bone. See *petrified wood*.

petrified wood wood that has been fossilized by the filling of pores or replacement of organic material with minerals, typically silicate minerals. See *permineralization*.

pH measurement on a scale of 0 to 14 of the acidity (low pH) or alkalinity (high pH) of a solution. A neutral pH is 7.

phreatic water groundwater below the water table in the saturated zone where all available pore space is filled with water.

physical weathering process by which geologic materials are broken down into smaller sizes and shapes.

piscivore animal that eats fish.

planktonic foraminifera floating microscopic single-celled marine organisms that precipitate a calcareous shell (test).

plantigrade standing or walking with the podials and metatarsals flat on the ground. See *digitigrade* and *unguligrade*.

pleurodont having the teeth attached by their sides to the inner side of the jaw, as in some lizards.

pneumatized presence of air spaces within bones.

postcranial bones of the skeleton that are behind the skull.

posterior further back in position; of or nearer the rear or hind end, especially of the body or a part of it.

postmortem after death.

potassium feldspar aluminosilicate mineral that contains potassium; the most common constituent of granite. K-feldspar is one of the most common minerals in continental crustal rocks.

potato-ball concretions masses of calcium carbonate that are more resistant to erosion than the surrounding rock matrix.

procoelous vertebrae vertebrae whose anterior surface is concave and posterior surface is convex.

prosobranch snail that respires using a gill.

protocone central of the three cusps of a primitive upper molar that in higher forms is the principal anterior and internal cusp.

protoloph anterior cross-crest of a tooth, as that of a tapir or rhinoceros. See *ectoloph*, *lophodont*, and *metaloph*.

protrogomorphous in which the middle and deep layers of the masseter muscle attach to the zygomatic arch, as in rodent teeth and jaws in primitive forms. See *hystricomorphous*, *myomorphous*, and *sciuromorphous*.

proximal location closer to a common point of reference; in vertebrates, this is usually from the midline of the body.

pulmonate lung-bearing organism, typically referring to types of aquatic and land snails.

radiometric dating technique used to obtain numeric dates for some types of rocks, usually on the basis of a comparison between the observed abundance of a naturally occurring radioactive isotope and its decay products, often referred to as the parent and daughter element. Also called radioisotopic dating.

rare earth element (REE) one of a set of 17 chemical elements in the periodic table, specifically the 15 lanthanides plus scandium and yttrium, taken up during diagenesis. They can be used to determine the provenance and depositional environments of vertebrate fossils. Also called rare earth metal.

reduction chemical process by which an element takes up an electron, thus reducing the amount of positive charge for any given ion.

reversed magnetic polarity direction of the Earth's magnetic field in which the current north and south magnetic poles are reversed.

rhizolith root coated by the precipitation of minerals, most commonly calcium carbonate or iron oxide, which can eventually can kill the root.

riparian area along a river that supports a distinctly different ecosystem as a result of the proximity to the river, such as a line of trees along a river that is flowing across a prairie.

saltation sediment grains, typically sand, that bounce along the ground or substrate during windstorms or during water transport.

sandstone sedimentary rock primarily composed of sand particles ranging from $\frac{1}{16}$ to 2 mm in diameter.

sandstone blankets thick accumulations of stacked sandstone sheets.

sandstone ribbons long, linear bodies of sandstone that are narrow laterally but extend for long distances along a linear trend. These ribbons have lenticular cross sections with a width-to-thickness ratio of less than 15. See *sandstone sheets*.

sandstone sheets laterally extensive but relatively thin sandstone bodies. The width-to-thickness ratio of these sheets is greater than 15. See *sandstone ribbons*.

scandium one of the rare earth elements (atomic number 21) incorporated into bone during the fossilization process. See *lanthanides*.

sciuromorphous in which the middle masseter attaches in front of the eye, as in squirrels. See *hystricomorphous*, *myomorphous*, and *protrogomorphous*.

sedimentology study of modern sediments such as sand, mud, and clay, and the processes that result in their deposition, as well as their equivalents in the rock record.

shell microsculpture very fine ornamentation that can only be seen with magnification on the surface of a snail shell. It can include small granulations or thin spiral grooves.

shocked minerals minerals in rock deformed by the pressures associated with bolide impacts. Shocked minerals display planar deformation features that represent deformation along defined atomic planes in the mineral structure.

siliciclastic clastic noncarbonate rock almost exclusively bearing silicon minerals, either as forms of quartz or other silicates.

siltstone clastic sedimentary rock composed of more than two-thirds silt-size particles, defined as grains 2 to 62 μm in size.

sinistral coiling coiling in a snail shell in which the coil is to the left. In sinistral shells, the aperture opens to the left of the spire axis when the apex of the spire is held vertically. See *dextral coiling*.

slickensides set of parallel striations on the surface of peds indicating swelling of soil materials. Also used for grooves developed between two sliding blocks of rock in a fault zone.

smectite particular type of clay mineral that expands upon addition of water. Some varieties of smectite can expand up to 50 percent of their original volume. These clays return to their original size upon drying.

soil fabric preferred alignment of soil materials created by pedogenic modification of parent materials.

soil order first level of classification in USDA soil taxonomy. Twelve soil orders are currently recognized. (See Table 3.1.)

soil profile two-dimensional exposure showing the vertical association of soil horizons.

soil structures three-dimensional bodies of soil material, such as peds and glaebules.

soil taxonomy philosophy and methodology of soil classification and identification. USDA soil taxonomy is broken down into seven levels of classification and currently recognizes over 18,000 different soils in the United States.

spherule type of impact ejecta formed by the vaporization of target rock and impactor, followed by subsequent cooling and condensation in the atmosphere to form glasslike beads that then fall back to earth.

spiral fracture particular style of break in fresh bone caused by carnivore or scavenger activity. Spiral fractures run along the length of the bone and curve around the radius of the bone shaft.

squamate scaled reptile.

steinkern internal mold of a shell, often the only indication of the animal after the shell has weathered away (German for "stone kernel").

stratigraphy study of the distribution of rocks in space and time. Stratigraphy is typically used with sedimentary rocks.

stromatolite interlayered mass of photosynthetic cyanobacteria and clay that forms through the periodic addition of a thin layer of mud and clay to the surface of the cyanobacteria mat.

subduction process that takes place at the convergent boundaries of two tectonic plates when one moves under the other and sinks into the mantle.

subfossorial adapted for digging; often seen in the hands, feet, and bone structure of some rodents, such as prairie dogs. An animal is said to be subfossorial if it shows limited adaptations to a fossorial lifestyle.

supratendinal bridge any bony bridge over a tendon, in particular the bridge over the extensor canal at the distal end of the tibiotarsus of certain birds.

suspended load sediment suspended in a water body, primarily because of the small size of the particles. See *turbidity*.

symphysis type of joint in which the apposed boney surfaces are firmly united by a plate of fibrocartilage. In vertebrates, the two main symphyses are the mandibular and pubic.

synapomorphy in cladistics, a trait shared by two or more taxa and inferred to have been present in their most recent common ancestor, whose own ancestor in turn is inferred to not possess the trait. See *apomorphy*.

talonid crushing surface of a tribosphenic molar.

taphonomy collective study of the mode of death, postmortem modification, and burial of an animal to form a fossil.

taxonomy practice and study of classification of things or concepts, including the principles that underlie such classification.

terrestrial environments of deposition on land, including lakes, rivers, dunes, and glaciers.

test microscopic shell of single-celled marine organisms, such as foraminifera.

time averaging means by which fossil bones of different stratigraphic ages can be grouped together by surficial processes, such as the incision and lateral migration of a river.

translocation process by which soil materials, either as suspended or dissolved solids, can be moved through a soil profile to accumulate at a new level.

trigonid shearing end of a molar.

tritubercular having or designating teeth with three cusps or tubercles; tricuspid.

trochlea spool-shaped area of bone providing a smooth articular area for rotation on another bone. For example, the trochlea on the distal humerus articulates with the proximal ulna.

tufa porous variety of limestone, formed by the precipitation of carbonate minerals from ambient-temperature water bodies.

tuff type of rock that is a lithified volcanic ash originally ejected from vents during a volcanic eruption.

turbidity amount of sediment suspended in a water body. See *suspended load*.

ultisol soil most commonly found in association with forests in the southeast part of the United States. Ultisols are noted for high concentrations of translocated kaolinite clay in their B horizon and a low amount of base ions.

unconformity large gap of geologic time in the rock record that forms as a result of nondeposition and erosion.

unguligrade standing or walking only on the distalmost tips of the digits, as horses and cattle. See *digitigrade* and *plantigrade*.

urostyle long unsegmented bone that represents a number of fused caudal vertebrae and forms the posterior part of the vertebral column of frogs and toads.

vadose water zone of groundwater above the water table with pore spaces filled by variable amounts of air and water; also known as unsaturated zone.

vertical accretion aggradational growth of floodplain materials by settling of sediments onto a preexisting land surface during periods of flooding.

volant capable of flight.

volcaniclasts sediment particles derived from volcanic sources.

yttrium one of the rare earth elements (atomic number 39) incorporated into bone during the fossilization process. See *lanthanides*.

zalambdodont having teeth with two ridges that meet at an angle, forming the shape of the Greek letter lambda.

References

Abel, O. 1926. Amerikafahrt: Eindrücke, Beobachtungen und Studien eines Naturforschers auf einer Reise nach Nordamerika and Westindien. G. Fischer, Jena, 462 pp.

Abel, O. 1935. Vorzeitliche Lebensspuren. G. Fischer, Jena, 644 pp.

Adkins, R. M., A. H. Walton, and R. L. Honeycutt. 2002. Higher level systematics of rodents and divergence time estimates based on two congruent nuclear genes. Molecular Phylogenetics and Evolution 26:409–420.

Adolphson, D. G. 1973. Avian fossils of South Dakota. South Dakota Bird Notes 25, 40–46.

Albright, L. B., M. O. Woodburne, T. J. Fremd, C. C. Swisher, B. J. MacFadden, and G. R. Scott. 2008. Revised chronostratigraphy and biostratigraphy of the John Day Formation (Turtle Cove and Kimberly members), Oregon, with implications for updated calibration of the Arikareean North American Land Mammal Age. Journal of Geology 116:211–237.

Aleinikoff, J. N., D. R. Muhs, E. A. Bettis III, W. C. Johnson, C. M. Fanning, and R. Benton. 2008. Isotopic evidence for the diversity of late Quaternary loess in Nebraska. Glaciogenic and nonglaciogenic sources. Geological Society of America Bulletin 120:1362–1377.

Alvarez, W., P. Claeys, and A. Montanari. 2009. Time-scale construction and periodizing in Big History: from the Eocene–Oligocene to all of the past; pp. 1–15 in C. Koeberl and A. Montanari (eds.), The Late Eocene Earth: Hothouse, Icehouse, and Impacts. Geological Society of America Special Paper 452.

Anderson, D. 2008. Ischyromyidae; pp. 311–325 in C. M. Janis, G. F. Gunnell, and M. D. Uhen (eds.), Evolution of Tertiary Mammals of North America, Volume 2: Small Mammals, Xenarthrans and Marine Mammals. Cambridge University Press, Cambridge, U.K.

Asher, R. J., M. C. McKenna, R. J. Emry, A. R. Tabrum, and D. G. Kron. 2002. Morphology and relationships of Apternodus and other extinct zalambdodont, placental mammals. Bulletin of the American Museum of Natural History 273:1–117.

Auffenberg, W. A. 1962. Testudo amphithorax Cope referred to Stylemys. American Museum Novitates 2120:1–10.

Auffenberg, W. A. 1964. A redefinition of the fossil tortoise genus Stylemys Leidy. Journal of Paleontology 38:316–324.

Auffenberg, W. A. 1974. Checklist of fossil land tortoises (Testudines). Bulletin Florida State Museum, Biological Sciences 18:121–251.

Aymard, A. 1846. Essai monographique sur un genre nouveau de mammifère fossile trouvé dans la Haute-Loire et nommé Entelodon. Annales de la société d'Agriculture, Sciences, Arts et commerce du Puy 12:227–267.

Bales, G. S. 1996. Heterochrony in brontothere horn evolution: allometric interpretations and the effect of life history scaling. Paleobiology 22:481–495.

Balthasar, V. 1963. Monographie der Scarabaeidae und Aphodiidae der Palearktischen und Orientalischen Region. Tschechoslowakische Akademie der Wissenschafte, Prague I, 391 pp.

Barnosky, A. D. 1981. A skeleton of Mesoscalops (Mammalia, Insectivora) from the Miocene Deep River Formation, Montana, and a review of the proscalopid moles: evolutionary, functional and stratigraphic relationships. Journal of Vertebrate Paleontology 1:285–339.

Barnosky, A. D. 1982. A new species of Proscalops (Mammalia, Insectivora) from the Arikareean Deep River Formation, Meagher County, Montana. Journal of Paleontology 56:103–111.

Baskin, J. A. 1998. Mustelidae; pp. 152–173 in C. M. Janis, K. M. Scott, and L. L. Jacobs (eds.), Evolution of Tertiary Mammals of North America, Volume 1: Terrestrial Carnivores, Ungulates, and Ungulatelike Mammals. Cambridge University Press, Cambridge, U.K.

Beetle, D. E. 1976. Mollusks of Badlands National Monument, SD. Bulletin of the American Malacological Union 43:49–50.

Benton, M. J. 2000. Vertebrate Paleontology. Blackwell Science Ltd., Oxford, U.K., 452 pp.

Benton, R. B., E. Evanoff, C. L. Herbel, and D. O. Terry Jr. 2007. Baseline Mapping of Fossil Bone Beds at Badlands National Park, South Dakota: Natural Resources Preservation Program Grant 2000–2002, National Park Service, 231 pp.

Benton, R. B., E. Evanoff, C. L. Herbel, and D. O. Terry Jr. 2009. Documentation of significant paleontological localities within the Poleslide Member, Brule Formation, Badlands National Park, South Dakota. National Resources Preservation Program Grant Final Report, on file at Badlands National Park, 69 pp.

Berggren, W. A., D. V. Kent, C. C. Swisher III, and M.-P. Aubry. 1995. A revised Cenozoic geochronology and chronostratigraphy; pp. 129–212 in W. A. Berggren, D. V. Kent, M.-P. Aubry, and J. Hardenbol (eds.), Geochronology, Time Scales and Global Stratigraphic Correlation. SEPM Special Publication 54.

Berry, E. W. 1926. The fossil seeds from the Titanotherium Beds of Nebraska, their identity and significance. American Museum Novitates 221:1–8.

Best, M. G., E. H. Christiansen, A. L. Dieno, C. S. Gromme, E. H. McKee, and D. C. Noble. 1989. Eocene through Miocene volcanism in the Great Basin of the western United States. New Mexico Bureau of Mines and Mineral Resources Memoir 47:91–133.

Bjork, P. R. 1976. Mammalian tracks from the Brule Formation of South Dakota. Proceedings of the South Dakota Academy of Science 55:154–158.

Bjork, P. R. 1978. The functional significance of a broken lower incisor in Amphicaenopus, an Oligocene rhinocerotid. Proceedings of the South Dakota Academy of Science 57:163–167.

Black, C. C. 1963. A review of North American Tertiary Sciuridae. Bulletin of the Museum of Comparative Zoology 130:111–248.

Black, C. C. 1965. Fossil mammals from Montana. Part 2. Rodents from the Early Oligocene Pipestone Springs Local Fauna. Annals of the Carnegie Museum 38:1–48.

Black, C. C. 1968. The Oligocene rodent Ischyromys and discussion of the family Ischyromyidae. Annals of the Carnegie Museum 39:273–305.

Black, C. C. 1971. Paleontology and geology of the Badwater Creek Area, central Wyoming. Part 7. Rodents of the family Ischyromyidae. Annals of the Carnegie Museum 43:179–217.

Black, C. C. 1972. Holarctic evolution and dispersal of squirrels (Rodentia: Sciuridae); pp. 305–322 in T. Dobzhansky, M. K. Hecht and W. C. Steere (eds.), Evolutionary Biology. Appleton-Century-Crofts, New York.

Boardman, G. S., and R. Secord. 2013. Stable isotope paleoecology of White River ungulates during the Eocene–Oligocene climate transition in northwestern Nebraska. Palaeogeography, Palaeoclimatology, Palaeoecology 375:38–49.

Boyd, C. A., M. W. Weiler, M. L. Householder, and K. K. Schumaker. 2014. The diversity of cimolestan mammals within the White River Group of South Dakota and Nebraska, Acta Palaeontologica Polonica 59(4):771–778. http://dx.doi.org/10.4202/app.2011.0045.

Brattstrom, B. H. 1961. Some new fossil tortoises from western North America with remarks on the zoogeography and paleoecology of tortoises. Journal of Paleontology 53:543–560.

Brochu, C. A. 1999. Phylogenetics, taxonomy, and historical biogeography of Alligatoroidea. Memoir (Society of Vertebrate Paleontology) 6:9–100.

Brochu, C. A. 2004. Alligatorine phylogeny and the status of Allognathosuchus Mook, 1921. Journal of Vertebrate Paleontology 24:857–873.

Brodkorb, P. 1967. Catalogue of fossil birds. Part 3 (Ralliformes, Ichthyornithiformes, Charadriiformes). Bulletin of the Florida State Museum, Biological Sciences 11:99–220.

Bryant, H. N. 1996. Nimravidae; pp. 453–475 in D. R. Prothero and R. J. Emry (eds.), The Terrestrial Eocene–Oligocene Transition in North America. Cambridge University Press, New York.

Bryant, L. J. 1984. Skeletons of the diminutive sabertooth *Eusmilus* from the Arikareean of South Dakota. Carnegie Museum of Natural History Special Publication 9:161–170.

Buchsbaum, R. 1977. Animals without Backbones. University of Chicago Press, Chicago, Illinois, 392 pp.

Bump, J. D. 1956. Geographic names for the members of the Brule Formation of the Big Badlands of South Dakota. American Journal of Science 254:429–432.

Burch, J. B. 1989. North American Freshwater Snails. Malacological Publications, Hamburg, Michigan, 365 pp.

Burkhart, P. A., J. Livingston, J. E. Rawling III, P. R. Hanson, S. Mahan, R. Benton, E. Heffron, M. Jahn, T. Anderson, and B. Page. 2008. Late Pleistocene through Holocene landscape evolution of the White River Badlands, South Dakota. Geological Society of America Field Guide 10:235–248.

Cande, S. C., and D. V. Kent. 1992. A new geomagnetic polarity time scale for the Late Cretaceous and Cenozoic. Journal of Geophysical Research 97:917–951.

Carroll, R. L. 1988. Vertebrate Paleontology and Evolution. W. H. Freeman, New York, New York, 698 pp.

Chaffee, R. G. 1943. Mammal footprints from the White River Oligocene. Notulae Nature 116:1–13.

Chandler, R. M., and W. P. Wall. 2001. The first record of bird eggs from the early Oligocene (Orellan) of North America; pp. 23–26 in V. L. Santucci and L. M. McClelland (eds.), Geologic Resources Division Technical Report NPS/NRGRD/GRDTR-01/01.

Chaney, R. W. 1925. Notes on two fossil hack berries from the Tertiary of the Western United States. Carnegie Institute of Washington Publication 349:51–56.

Cifelli, R. L. 1999. Tribosphenic mammal from the North American early Cretaceous. Nature 40:363–366.

Cifelli, R. L., and C. de Muizon. 1997. Dentition and jaw of *Kokopellia juddi,* a primitive marsupial or near marsupial from the medial Cretaceous of Utah. Journal of Mammalian Evolution 4:241–258.

Clark, J. 1937. The stratigraphy and paleontology of the Chadron Formation in the Big Badlands of South Dakota. Carnegie Museum Annals 25:261–350.

Clark, J. 1954. Geographic designation of the members of the Chadron Formation in South Dakota. Carnegie Museum Annals 33:197–198.

Clark, J., and T. E. Guensburg. 1970. Population dynamics of *Leptomeryx.* Fieldiana Geology 16:411–451.

Clark, J., J. R. Beerbower, and K. K. Kietzke. 1967. Oligocene sedimentation, stratigraphy, paleoecology and paleoclimatology in the Big Badlands of South Dakota. Fieldiana Geology Memoirs 5, 158 pp.

Clemens, W. A., Jr. 1973. Fossil mammals of the type Lance Formation, Wyoming. III. Eutheria and summary. University of California Publications in Geological Sciences 94:1–102.

Clemens, W. A., Jr. 1979. Marsupialia; pp. 192–220 in J. A. Lillegraven, Z. Kielan-Jaworowska, and W. A. Clemens (eds.), Mesozoic Mammals, the First Two-thirds of Mammalian History. University of California Press, Berkeley, California.

Coccioni, R., F. Frontalini, and S. Spezzaferri. 2009. Late Eocene impact-induced climate and hydrological changes: evidence from the Massignano global stratotype section and point (central Italy); pp. 97–118 in C. Koeberl and A. Montanari (eds.), The Late Eocene Earth: Hothouse, Icehouse, and Impacts. Geological Society of America Special Paper 452.

Colbert, M. W. 2005. The facial skeleton of the early Oligocene *Colodon* (Perissodactyla, Tapiroidea). Palaeontologia Electronica No. 8.1.12, 27 pp.

Colbert, M. W., and R. M. Schoch. 1998. Tapiroidea and other moropomorphs; pp. 569–582 in C. M. Janis, K. M. Scott, and L. L. Jacobs (eds.), Evolution of Tertiary Mammals of North America, Volume 1: Terrestrial Carnivores, Ungulates, and Ungulatelike Mammals. Cambridge University Press, Cambridge, U.K.

Cook, H. J., and W. C. Mansfield. 1933. A new mollusk from the Chadron formation (Oligocene) of Nebraska. Washington Academy of Sciences Journal 23:263.

Cope, E. D. 1873. Third notice of extinct vertebrata from the Tertiary of the Plains. Paleontological Bulletin 16:1–8.

Cope, E. D. 1874. Report on the vertebrate paleontology of Colorado. Annual Report of the Geological and Geographic Surveys of the Territories for 1873, F. V. Hayden, U.S. Geologist, Washington, D.C., 427–533.

Cope, E. D. 1884. The Vertebrata of the Tertiary Formations of the West. Report of the U.S. Geological Survey of the Territories, F. V. Hayden, Geologist in Charge. U.S. Government Printing Office, Washington, D.C., 1044 pp.

Cope, E. D. 1891. On Vertebrata from the Tertiary and Cretaceous rocks of the Northwest Territory. I. The species from the Oligocene or Lower Miocene beds of the Cypress Hills. Geological Survey of Canada, Contributions to Canadian Palaeontology 3:1–25.

Costa, E., M. Garcés, A. Sáez, L. Cabrera, and M. López-Blanco. 2011. The age of the "Grande Coupure" mammal turnover: new constraints from the Eocene–Oligocene record of the Eastern Ebro Basin (NE Spain). Palaeogeography, Palaeoclimatology, Palaeoecology 301:97–107.

Coxall, H. K., P. A. Wilson, H. Pälike, C. H. Lear, and J. Backman. 2005. Rapid stepwise onset of Antarctic glaciation and deeper calcite compensation in the Pacific Ocean. Nature 433:53–57.

Cracraft, J. 1968. A review of the Bathornithidae (Aves, Gruiformes), with remarks on the relationships of the Suborder Cariamae. American Museum Novitates 2326:1–46.

Cracraft, J. 1973. Systematics and evolution of the Gruiformes (Class Aves). 3. Phylogeny of the Suborder Grues. Bulletin of the American Museum of Natural History 151, 127 pp.

Crochet, J.-Y. 1977. Les Didelphidae (Marsupicarnivora, Marsupialia) holarctiques tertiaires. Comptes Rendus Hebdomadaires des séances de l'Academie des Sciences, Paris, series D 24:357–360.

Crochet, J.-Y. 1980. Les Marsupiaux du Tertiaire d'Europe. Editions de la Fondtion Singer-Polignac, Paris, 279 pp.

Crowson, R. A. 1981. The Biology of the Coleoptera. Academic Press, New York, New York, 802 pp.

Culbertson, T. A. 1851. Journal of an expedition to the Mauvaises Terres and the upper Missouri in 1850. Annual Report of the Smithsonian Institution 5:84–132.

Curry, W. H., III. 1971. Laramide structural history of the Powder River Basin, Wyoming; pp. 49–60 in A. R. Renfro, L. W. Madison, and G. A. Jarre (eds.), Symposium on Wyoming Tectonics and Their Economic Significance. Wyoming Geological Association Annual Field Conference Guidebook 23.

Darton, N. H. 1899. Preliminary report on the geology and water resources of Nebraska west of the one hundred and third meridian. United States Geological Survey Annual Report 19, pt. 4, 719–785.

Davies, R. J., M. Huuse, P. Hirst, J. Cartwright, and Y. Yang. 2006. Giant clastic intrusions primed by silica diagenesis. Geology 34:917–920.

Dawson, M. R. 1958. Early Tertiary Leporidae of North America. University of Kansas Paleontological Contributions, Vertebrata 6:1–75.

Dawson, M. R. 2008. Lagomorpha; pp. 293–310 in C. M. Janis, G. F. Gunnell, and M. D. Uhen (eds.), Evolution of Tertiary Mammals of North America, Volume 2: Small Mammals, Xenarthrans and Marine Mammals. Cambridge University Press, Cambridge, U.K.

DeWitt, E., J. A. Redden, D. Buscher, and A. B. Wilson. 1989. Geologic map of the Black Hills area, South Dakota and Wyoming. U.S. Geological Survey Miscellaneous Investigations Map I-1910, scale 1:250,000, 1 sheet.

DiBenedetto, J. N. 2000. Sedimentology and taphonomy of the Brian Maebius Site, Sage Creek Wilderness Area, Badlands National Park, Interior, South Dakota. M.S. thesis, South Dakota School of Mines and Technology, Rapid City, South Dakota, 136 pp.

Diester-Haass, L., and R. Zahn. 1996. The Eocene–Oligocene transition in the Southern Ocean: history of water masses, circulation, and biological productivity inferred from high resolution records of stable isotopes and benthic foraminiferal abundances (ODP Site 689). Geology 24:16–21.

Directors Order 12. 1985. Handbook Conservation Planning, Environmental Impact Analysis and Decision Making. National Park Service, Department of the Interior, Washington, D.C.

Drewicz, A. E. 2012. The effect of histology on rates of diffusion and fossilization in marine and nonmarine bones. M.S. thesis, Temple University, Philadelphia, Pennsylvania, 143 pp.

Drewicz, A. E., D. E. Grandstaff, R. Ash, and D. O. Terry Jr. 2011. Quantifying periods of fossilization in terrestrial and marine environments using rare earth elements. Geological Society of America Abstracts with Programs 43:163.

Dupont-Nivet, G., C. Hoorn, and M. Konert. 2008. Tibetan uplift prior to the Eocene–Oligocene climate transition: evidence from pollen analysis of the Xining Basin. Geology 36:987–990.

Edwards, P., and D. Yatkola. 1974. Coprolites of White River (Oligocene) carnivorous mammals: origin and paleoclimatic and sedimentologic significance. Contributions to Geology, University of Wyoming 13:67–73.

Effinger, J. A. 1998. Entelodontidae; pp. 375–398 in C. M. Janis, K. M. Scott, and L. L. Jacobs (eds.), Evolution of Tertiary Mammals of North America. Volume 1: Terrestrial Carnivores, Ungulates and Ungulatelike Mammals. Cambridge University Press, Cambridge, U.K.

Efremov, J. A. 1940. Taphonomy: new branch of paleontology. Pan-American Geologist 74:81–93.

Egi, N. 2001. Body mass estimates in extinct mammals from limb bone dimensions: the case of North American Hyaenodontids. Palaeontology 44:497–528.

Eisenberg, J. F. 1980. Biological strategies of living conservative mammals; pp.13–30 in K. Schmidt-Nielson, L. Bolis and C. R. Taylor (eds.), Comparative Physiology: Primitive Mammals. Cambridge University Press, Cambridge, U.K.

Emerson, S. B., and L. Radinsky. 1980. Functional analysis of sabertooth cranial morphology. Paleobiology 6:295–312.

Emry, R. J. 1973. Stratigraphy and preliminary biostratigraphy of the Flagstaff Rim area, Natrona County, Wyoming. Smithsonian Contributions to Paleobiology 18:1–43.

Emry, R. J., and W. W. Korth. 1996. The Chadronian squirrel "Sciurus" jeffersoni, Douglass, 1901: a new generic name, new material, and its bearing on the early evolution of Sciuridae (Rodentia). Journal of Vertebrate Paleontology 16:775–780.

Emry, R. J., and R. W. Thorington Jr. 1982. Descriptive and comparative osteology of the oldest fossils squirrel, Protosciurus (Rodentia: Sciuridae). Smithsonian Contributions to Paleontology 47:1–35.

Estes, R. 1964. Fossil vertebrates from the Cretaceous Lance Formation, eastern Wyoming. University of California Publications, Geological Sciences 49:1–180.

Estes, R. 1970a. Origin of the Recent North American lower vertebrate fauna: an inquiry into the fossil record. Forma et Functio 3:139–163.

Estes, R. 1970b. New fossil pelobatid frogs and a review of the genus Eopelobates. Bulletin of the Museum of Comparative Zoology 139:293–340.

Evanoff, E. 1983. Pleistocene nonmarine mollusks and sediment facies of the Meeker area, Rio Blanco County, Colorado. M.S. thesis, University of Colorado, Boulder, Colorado, 205 pp.

Evanoff, E. 1990a. Early Oligocene paleovalleys in southern and central Wyoming: evidence of high local relief on the late Eocene unconformity. Geology 18:443–446.

Evanoff, E. 1990b. Late Eocene and early Oligocene paleoclimates as indicated by the sedimentology and nonmarine gastropods

of the White River Formation near Douglas, Wyoming. Ph.D. dissertation, University of Colorado, Boulder, Colorado, 430 pp.

Evanoff, E., and B. Roth. 1992. The fossil land snail Helix leidyi Hall and Meek, 1855, a member of a new genus of Humboldtianidae (Gastropoda: Pulmonata). Veliger 35:122–132.

Evanoff, E., S. C. Good, and J. H. Hanley. 1998. An overview of the freshwater mollusks from the Morrison Formation (Upper Jurassic, Western Interior, U.S.A.). Modern Geology 22:423–450.

Evanoff, E., D. R. Prothero, and R. H. Lander. 1992. Eocene–Oligocene climatic change in North America: the White River Formation near Douglas, east-central Wyoming; pp. 116–130 in D. R. Prothero and W. A. Berggren (eds.), Eocene–Oligocene Climatic and Biotic Evolution. Princeton University Press, Princeton, New Jersey.

Evanoff, E., D. O. Terry Jr., R. C. Benton, and H. Minkler. 2010. Field guide to geology of the White River Group in the North Unit of Badlands National Park. South Dakota School of Mines and Technology Bulletin 21:96–127.

Evans, J. 1852. Incidental observations on the Missouri River and on the Mauvaises Terres (Badlands); pp. 196–206 in D. E. Owen (ed.), Report on the Geological Survey of Wisconsin, Iowa, and Minnesota and Incidentally of a Portion of Nebraska Territory. Lippincott, Grambo and Co., Philadelphia, Pennsylvania.

Evans, J. E. 1999. Recognition and implications of Eocene tufas and travertines in the Chadron Formation, White River Group, Badlands of South Dakota. Sedimentology 46:771–789.

Evans, J. E., and D. O. Terry. 1994. The significance of incision and fluvial sedimentation in the basal White River Group (Eocene–Oligocene), Badlands of South Dakota, U.S.A. Sedimentary Geology 90:137–152.

Evans, J. E., and L. C. Welzenbach. 1998. Episodes of carbonate deposition in a siliciclastic-dominated fluvial sequence, Eocene–Oligocene White River Group, South Dakota and Nebraska; pp. 93–116 in D. O. Terry Jr., H. E. LaGarry, and R. M. Hunt Jr. (eds.), Depositional Environments, Lithostratigraphy, and Biostratigraphy of the White River and Arikaree Groups (Late Eocene to Early Miocene, North America). Geological Society of America Special Paper 325.

Exon, N., and 20 others. 2002. Drilling reveals climatic consequences of Tasmanian Gateway opening. Eos 83:23, 253–256.

Factor, L. A. 2002. Paleopedology and taphonomy of the Brian Maebius site, Badlands National Park, South Dakota. M.S. thesis, Temple University, Philadelphia, Pennsylvania, 111 pp.

Factor, L. A., and D. O. Terry Jr. 2002. A paleosol and taphonomy comparison between the Brian Maebius site and other fossil rich localities in Badlands National Park, South Dakota. Geological Society of America Abstracts with Programs 34(1):A-71.

Falcon-Lang, H. J. 2005. Global climate analysis of growth rings in woods, and its implications for deep-time paleoclimate studies. Paleobiology 31:434–444.

Farrington, O. C. 1899. A fossil egg from South Dakota. Field Columbian Museum, Geological Series 35:193–200.

Ferrusquia-Villafranca, I. 1969. Rancho Gaitan local fauna, early Chadronian, northeastern Chihuahua. Boletin de la Sociedade Geologica Mexicana 30:99–138.

Figueirido, B., J. A. Pérez-Claros, R. M. Hunt Jr., and P. Palmqvist. 2011. Body mass estimation in amphicyonid carnivoran mammals: a multiple regression approach from the skull and skeleton. Acta Palaeontologica Polonica, 56:225–246.

Flores, R. M., and F. G. Ethridge. 1985. Evolution of intermontane fluvial systems of Tertiary Powder River Basin, Montana and Wyoming; pp. 107–126 in R. M. Flores and S. S. Kaplan (eds.), Cenozoic Paleogeography of the West-Central United States. Rocky Mountain Section of SEPM Rocky Mountain Paleogeography Symposium 3.

Flynn, J. M. and H. Galiano. 1982. Phylogeny of early Tertiary Carnivora, with a description of a new species of Protictis from the middle Eocene of Northwestern Wyoming. American Museum Novitates 22725:1–64.

Flynn, L. J. 2008a. Dipodidae; pp. 406–414 in C. M. Janis, G. F. Gunnell, and M. D. Uhen (eds.), Evolution of Tertiary Mammals of North America, Volume 2: Small Mammals, Xenarthrans and Marine Mammals. Cambridge University Press, Cambridge, U.K.

Flynn, L. J. 2008b. Eomyidae; pp. 415–427 in C. M. Janis, G. F. Gunnell, and M. D. Uhen (eds.), Evolution of Tertiary Mammals of North America, Volume 2: Small Mammals, Xenarthrans and Marine Mammals. Cambridge University Press, Cambridge, U.K.

Flynn, L. J., and L. L. Jacobs. 2008. Castoridae; pp. 391–405 in C. M. Janis, G. F. Gunnell, and M. D. Uhen (eds.), Evolution of Tertiary Mammals of North America, Volume 2: Small Mammals, Xenarthrans and Marine Mammals. Cambridge University Press, Cambridge, U.K.

Flynn, L. J., E. H. Lindsay, and R. A. Martin. 2008. Geomorpha; pp. 428–455 in C. M. Janis, C. M. Janis, G. F. Gunnell, and M. D. Uhen (eds.), Evolution of Tertiary Mammals of North America, Volume 2: Small Mammals, Xenarthrans and Marine Mammals. Cambridge University Press, Cambridge, U.K.

Foss, S. E. 2007. Family Entelodontidae; pp. 120–129 in D. R. Prothero and S. F. Foss (eds.), The Evolution of Artiodactyls. Johns Hopkins University Press, Baltimore, Maryland.

Foster, J. B. 1965. Mortality and ageing of black rhinoceros in East Tsavo Park, Kenya. East African Wildlife Journal 3:118–119.

Fox, R. C. 1983. Notes on the North American Tertiary marsupials, Herpetotherium and Peradectes. Canadian Journal of Earth Sciences 20:1565–1578.

Frailey, D. 1978. An early Miocene (Arikareean) fauna from north central Florida (the SB-1A Local Fauna). Occasional Papers of the Museum of Natural History, University of Kansas 75:1–20.

Gawne, C. E. 1968. The Genus Proterix (Insectivora, Erinaceidae) of the Upper Oligocene of North America. American Museum Novitates 2315, 26 pp.

Gazin, C. L. 1955. A review of the upper Eocene Artiodactyla of North America. Smithsonian Miscellaneous Collections 128:1–96.

Gilmore, C. W. 1928. Fossil lizards of North America. Memoirs of the National Academy of Sciences 22:1–197. Printed by the United States Government Printing Office, Washington, D.C.

Goddard, J. 1970. Age criteria and vital statistics of a black rhinoceros population. East African Wildlife Journal 8:105–121.

Goodwin, H. T. 2008. Sciuridae; pp. 355–376 in C. M. Janis, G. F. Gunnell, and M. D. Uhen (eds.), Evolution of Tertiary Mammals of North America, Volume 2: Small Mammals, Xenarthrans and Marine Mammals. Cambridge University Press, Cambridge, U.K.

Grande, L. 1984. Paleontology of the Green River Formation, with review of the fish fauna. Geological Survey of Wyoming, Bulletin 63, 333 pp.

Grandstaff, D. E., and D. O. Terry Jr. 2009. Rare earth element composition of Paleogene vertebrate fossils from Toadstool Geologic Park, Nebraska, U.S.A. Applied Geochemistry, Special Volume on the 12th International Symposium on Water–Rock Interactions 24:733–745.

Green, M., and P. R. Bjork. 1980. On the genus Dikkomys (Geomyoidea, Mammalia), Palaeovertebrata. Mémoire Jubilaire, Réné Lavocat 1:343–353.

Gries, J. P., and G. A. Bishop. 1966. Fossil invertebrates from the Big Badlands of South Dakota. Proceedings of the South Dakota Academy of Science 45:57–61.

Griffis, N. P. 2010. Late Eocene terrestrial paleoclimate record from the White River Formation at Flagstaff Rim, Wyoming, U.S.A. M.S. thesis, Temple University, Philadelphia, Pennsylvania, 56 pp.

Griffis, N. P., and D. O. Terry Jr. 2010. Vertical changes in paleosol morphology within the White River Sequence at Flagstaff Rim, Wyoming: implications for paleoclimatic change leading up to the Eocene–Oligocene Transition. Geological Society of America Abstracts with Programs 42:42.

Grimes, S. T., J. J. Hooker, M. E. Collinson, and D. P. Mattey. 2005. Summer temperatures of Late Eocene to Early Oligocene freshwaters. Geology 33:189–192.

Gunnell, G. F. 1998. Creodonta; pp. 91–105 in C. M. Janis, K. M. Scott, and L. L. Jacobs (eds.), Evolution of Tertiary Mammals of North America, Volume 1: Terrestrial Carnivores, Ungulates, and Ungulatelike Mammals. Cambridge University Press, Cambridge, U.K.

Gunnell, G. F., and J. F. Bloch. 2008. Insectivorous mammals summary; pp. 49–62 in C. M. Janis, G. F. Gunnell, and M. D. Uhen (eds.), Evolution of Tertiary Mammals of North America, Volume 2: Small Mammals, Xenarthrans and Marine Mammals. Cambridge University Press, Cambridge, U.K.

Gunnell, G. F., T. M. Bown, and J. I. Bloch. 2008. Leptictida; pp. 82–88 in C. M. Janis, G. F. Gunnell, and M. D. Uhen (eds.), Evolution of Tertiary Mammals of North America, Volume 2: Small Mammals, Xenarthrans and Marine Mammals. Cambridge University Press, Cambridge, U.K.

Gunnell, G. F., T. M. Bown, J. I. Bloch, and D. M. Boyer. 2008a. Proteutheria; pp. 63–81 in C. M. Janis, G. F. Gunnell, and M. D. Uhen (eds.), Evolution of Tertiary Mammals of North America, Volume 2: Small Mammals,

Xenarthrans and Marine Mammals. Cambridge University Press, Cambridge, U.K.

Gunnell, G. F., T. M. Bown, J. H. Hutchison, and J. I. Bloch. 2008b. Lipotyphla; pp. 89–125 in C. M. Janis, G. F. Gunnell, and M. D. Uhen (eds.), Evolution of Tertiary Mammals of North America, Volume 2: Small Mammals, Xenarthrans and Marine Mammals.

Gustafson, E. P. 1986. Preliminary biostratigraphy of the White River Group (Oligocene, Chadron and Brule Formations) in the vicinity of Chadron, Nebraska. Transactions of the Nebraska Academy of Sciences 14:7–19.

Halffter, G. 1959. Etologia y paleontologia de Scarabaeinae (Coleoptera, Scarabaeidae). Ciencia 19:165–178.

Halffter, G. 1977. Evolution of nidification in the Scarabaeinae (Coleoptera, Scarabaeidae). Quaestiones Entomologicae 13:231–253.

Halffter, G., and E. G. Matthews. 1966. The natural history of dung beetles of the subfamily Scarabaeinae (Coleoptera, Scarabaeidae). Folia Entomologica Mexicana 12–14:1–312.

Hall, J., and F. B Meek. 1855. Descriptions of new species of fossils from the Cretaceous formations of Nebraska, with observations upon Baculites ovatus and B. compressus, and the progressive development of the septa in baculites, ammonites, and scaphites. American Academy of Arts and Sciences Memoirs, n.s. 5:379–411.

Hanley, J. H., 1976. Paleosynecology of nonmarine Mollusca from the Green River and Wasatch formations (Eocene), southwestern Wyoming and northwestern Colorado; pp. 235–261 in R. W. Scott and R. R. West (eds.), Structure and Classification of Paleocommunities. Dowden, Hutchinson, and Ross Inc., Stroudsburg, Pennsylvania.

Hanley, J. H., 1987. Taphonomy and paleoecology of nonmarine Mollusca: Indicators of alluvial plain and lacustrine sedimentation, upper part of the Tongue River Member, Forth Union Formation (Paleocene), northern Powder River Basin, Wyoming and Montana. Palaios, 2:479–496.

Harksen, J. C. 1974. Miocene channels in the Cedar Pass area, Jackson County, South Dakota. South Dakota Geological Survey Report of Investigations 111:1–10.

Harksen, J. C., and J. R. Macdonald. 1969. Guidebook to the major Cenozoic deposits of southwestern South Dakota. South Dakota Geological Survey Guidebook 2, 103 pp.

Harksen, J. C., J. R. Macdonald, and W. D. Sevon. 1961. New Miocene formation in South Dakota. American Association of Petroleum Geologists Bulletin 45:674–678.

Hartley, A. J., G. S. Weissman, G. J. Nichols, and G. L Warwick. 2010. Large distributive fluvial systems: characteristics, distribution, and controls on development. Journal of Sedimentary Research 80:167–183.

Harvey, C. 1960. Stratigraphy, sedimentation and environment of the White River Group of the Oligocene of northern Sioux County, Nebraska. Ph.D. dissertation, University of Nebraska, Lincoln, Nebraska, 151 pp.

Hatcher, H. B. 1893. The Titanotherium beds. American Naturalist 27:204–221.

Hatcher, H. B. 1902. Origin of the Oligocene and Miocene deposits of the Great Plains. American Philosophical Society Proceedings 41:113–131.

Hay, O. P. 1908. The fossil turtles of North America. Carnegie Institution of Washington Publication 75:568 pp.

Hayden, F. V. 1856. A brief sketch of the geological and physical features of the region of the upper Missouri, with some notes on its soil, vegetation, animal life, etc.; in pp. 63–79 in G. K. Warren (ed.), Explorations in the Dacota Country in the Year 1855. United States Senate Executive Document 76, Washington, D.C.

Hayden, F. V. 1858. Notes on the geology of the Mauvais Terres of White River, Nebraska. Proceedings of the Philadelphia Academy of Natural Sciences 9:151–158.

Hayden, F. V. 1869. On the geology of the Tertiary formations of Dakota and Nebraska in J. Leidy, On the extinct mammalian fauna of Dakota and Nebraska. Philadelphia Academy of Natural Sciences Journal, 2nd Series 7:9–21.

Hayden, F. V. 1880. The Great West: Its attractions and resources, Charles R. Brodix, Bloomington, Illinois, 617 pp.

Heaton, T. H. 1993. The Oligocene Rodent Ischyromys of the Great Plains: replacement mistaken for anagenesis. Journal of Paleontology 67:297–308.

Heaton, T. H., and R. J. Emry. 1996. Leptomerycidae; pp. 581–608 in D. R. Prothero and R. J. Emry (eds.), The Terrestrial Eocene–Oligocene Transition in North America. Cambridge University Press, Cambridge, U.K.

Hecht, M. K. 1959. Amphibians and reptiles; pp. 130–146 in P. O. McGrew (ed.), The geology and paleontology of the Elk Mountain and Tabernacle Butte area. Wyoming Bulletin of the American Museum of Natural History 117.

Hermanson, J. W., and B. J. MacFadden. 1992. Evolutionary and functional morphology of the shoulder region and stay-apparatus in fossil and extant horses (Equidae). Journal of Vertebrate Paleontology 12:377–386.

Hersh, A. H. 1934. Evolutionary relative growth in the Titanotheres. American Naturalist 68:537–561.

Hickey, L. J. 1977. Stratigraphy and paleobotany of the Golden Valley Formation (early Tertiary) of Western North Dakota. Geological Society of America Memoir 150, 183 pp.

Hilgen, F., and K. F. Kuiper. 2009. A critical evaluation of the numerical age of the Eocene–Oligocene Boundary; pp. 139–148 in C. Koeberl and A. Montanari (eds.), The Late Eocene Earth: Hothouse, Icehouse, and Impacts. Geological Society of America Special Paper 452.

Hoff, W., J. Kinkade, J. Mertes, S. Schneiderwind, B. Weihs, and H. Maher Jr. 2007. Scale dependence of chalcedony vein strike uniformity within the Chadron Formation of Badlands National Park. Geological Society of America Abstracts with Programs 39:326.

Hoganson, J. W., E. C. Murphy, and N. F. Forsman. 1998. Lithostratigraphy, paleontology, and biochronology of the Chadron, Brule, and Arikaree formations in North Dakota; pp. 185–196 in D. O. Terry Jr., H. E. LaGarry, and R. M. Hunt Jr. (eds.), Depositional Environments, Lithostratigraphy, and Biostratigraphy of the White River and Arikaree Groups (Late Eocene to Early Miocene, North America). Geological Society of America Special Paper 325.

Holman, J. A. 1972. Herpetofauna of the Calf Creek Local Fauna (Lower Oligocene: Cypress Hills Formation) of Saskatchewan. Canadian Journal of Earth Sciences 9:1612–1631.

Holman, J. A. 1979. A review of the North American Tertiary snakes. Publications of the Museum, Michigan State University, Paleontological Series 1:200–260.

Holman, J. A. 1982. *Geringophis* (Serpentes: Boidae) from the middle Oligocene of Nebraska. Herpetologica 38:489–492.

Holman, J. A. 2000. Fossil Snakes of North America: Origin, Evolution, Distribution, Paleoecology. Indiana University Press, Bloomington, Indiana, 357 pp.

Holroyd, P. A. 2002. New record of Anthracotheriidae (Artiodactyla: Mammalia) from the Middle Eocene Yegua Formation (Claiborne Group), Houston County, Texas. Texas Journal of Science 54:301–308.

Honey, J. G., J. A. Harrison, D. R. Prothero, and M. S. Stevens. 1998. Camelidae; pp. 439–462 in C. M. Janis, K. M. Scott, and L. L. Jacobs (eds.), Evolution of Tertiary Mammals of North America, Volume 1: Terrestrial Carnivores, Ungulates, and Ungulatelike Mammals. Cambridge University Press, Cambridge, U.K.

Hooker, J. J. 1989. Character polarities in early perissodactyls and their significance for *Hyracotherium* and infraordinal relationships; pp. 79–101 in D. R. Prothero and R. M. Schoch (eds.), The Evolution of Perissodactyls. Clarendon Press, Oxford, U.K.

Howe, J. A. 1966. The Oligocene rodent *Ischyromys* in Nebraska. Journal of Paleontology 40:1200–1210.

Hren, M. T., N. D. Sheldon, S. T. Grimes, M. E. Collinson, J. J. Hooker, M. Bugler, and K. C. Lohmann. 2013. Terrestrial cooling in Northern Europe during the Eocene–Oligocene transition. Proceedings of the National Academy of Sciences of the United States of America 110:7562–7567.

Hunt, R. M., Jr. 1974. The auditory bulla in Carnivora: an anatomical basis for reappraisal of carnivore evolution. Journal of Morphology 143:21–76.

Hunt, R. M., Jr. 1989. Evolution of the aeluroid Carnivora: significance of the ventral promontorial process of the petrosal, and the origin of basicranial patterns in the living families. American Museum Novitates 2930:1–32.

Hunt, R. M., Jr. 1990. Taphonomy and sedimentology of Arikaree (lower Miocene) fluvial, eolian, and lacustrine paleoenvironments, Nebraska and Wyoming; a paleobiota entombed in fine-grained volcaniclastic rocks; pp. 69–111 in M. G. Lockley and A. Rice (eds.), Volcanism and Fossil Biotas. Geological Society of America Special Paper 244.

Hunt, R. M., Jr. 1996. Amphicyonidae; pp. 476–485 in D. Prothero and R. J. Emry (eds.), The Terrestrial Eocene–Oligocene Transition in North America. Cambridge University Press, Cambridge, U.K.

Hunt, R. M., Jr. 1998a. Amphicyonidae; pp. 196–221 in C. M. Janis, K. M. Scott, and L. L. Jacobs (eds.), Evolution of Tertiary Mammals of North America, Volume 1: Terrestrial Carnivores, Ungulates, and Ungulatelike Mammals. Cambridge University Press, Cambridge, U.K.

Hunt, R. M., Jr. 1998b. Ursidae; pp. 174–189 in C. M. Janis, K. M. Scott, and L. L. Jacobs (eds.), Evolution of Tertiary Mammals of North America, Volume 1: Terrestrial Carnivores, Ungulates, and Ungulatelike Mammals. Cambridge University Press, Cambridge, U.K.

Hunt, R. M., Jr. 2001. Small Oligocene amphicyonids from North America (*Paradaphoenus,* Mammalia, Carnivora). American Museum Novitates 3331: 1–20.

Hutchison, J. H. 1982. Turtle, crocodilian, and champsosaur diversity changes in the Cenozoic of the north-central region of the Western United States. Palaeogeography, Palaeoclimatology, Palaeoecology 37:149–164.

Hutchison, J. H. 1991. Early Kinosterninae (Reptilia: Testudines) and their phylogenetic significance. Journal of Vertebrate Paleontology 11:145–167.

Hutchison, J. H. 1992. Western North American reptile and amphibian record across the Eocene/Oligocene boundary and its climatic implications; pp. 451–463 in D. R. Prothero and W. A. Berggren (eds.), Eocene–Oligocene Climatic and Biotic Evolution. Princeton University Press, Princeton, New Jersey.

Hutchison, J. H. 1996. Testudines; pp. 337–353 in D. R. Prothero and R. J. Emry (eds.), The Terrestrial Eocene–Oligocene Transition in North America. Cambridge University Press, Cambridge, U.K.

Hyland, E., B. Murphy, P. Varela, K. Marks, L. Colwell, F. Tori, S. Monechi, L. Cleaveland, H. Brinkhuis, C. A. van Mourik, R. Coccioni, D. Bice, and A. Montanari. 2009. Integrated stratigraphic and astrochronologic calibration of the Eocene–Oligocene transition in the Monte Cagnero section (northeastern Apennines, Italy): a potential parastratotype for the Massignano global stratotype section and point (GSSP); pp. 303–322 in C. Koeberl and A. Montanari (eds.), The Late Eocene Earth: Hothouse, Icehouse, and Impacts. Geological Society of America Special Paper 452.

Janis, C. 1982. Evolution of horns in ungulates: ecology and paleoecology. Biological Reviews 57:262–318.

Jannett, P. A., and D. O. Terry Jr. 2008. Stratigraphic expression of a regionally extensive impactite within the Latest Cretaceous Fox Hills Formation of southwest South Dakota; pp. 199–213 in K. Evans, J. W. Horton Jr., D. T. King Jr., and J. R. Morrow (eds.), The Sedimentary Record of Meteorite Impacts. Geological Society of America Special Paper 437.

Jenny, H. 1941. Factors of Soil Formation: A System of Quantitative Pedology. McGraw-Hill, New York, New York, 241 pp.

Joeckel, R. M. 1990. A functional interpretation of the masticatory system and paleoecology of entelodonts. Paleobiology 16:459–482.

Johanson, Z. 1996. New marsupial from the Fort Union Formation, Swain Quarry, Wyoming. Journal of Paleontology 70:1023–1031.

Jones, J. N., D. M. Armstrong, R. S. Hoffman, and C. Jones. 1983. Mammals of the Northern Great Plains. University of Nebraska Press, Lincoln, Nebraska, 379 pp.

Jovane, L., R. Coccioni, A. Marsili, and G. Acton. 2009. The late Eocene greenhouse–icehouse transition: observations from the Massignano global stratotype section and point (GSSP); pp. 149–168 in C. Koeberl and A. Montanari (eds.), The Late Eocene Earth: Hothouse, Icehouse, and Impacts. Geological Society of America Special Paper 452.

Kennedy, R. 2011. Local variability in Early Oligocene paleosols as a result of ancient soil catenary processes, Brule Formation, Toadstool Geologic Park, Nebraska. M.S. thesis, Temple University, Philadelphia, Pennsylvania, 60 pp.

Koeberl, C. 2009. Late Eocene impact craters and impactoclastic layers–an overview; pp. 17–26 in C. Koeberl and A. Montanari (eds.), The Late Eocene Earth: Hothouse, Icehouse, and Impacts. Geological Society of America Special Paper 452.

Koeberl, C., and A. Montanari (eds.). 2009. The Late Eocene Earth: Hothouse, Icehouse, and Impacts. Geological Society of America Special Paper 452, 322 pp.

Kohn, M. J., A. Zanazzi, and J. A. Josef. 2010. Stable isotopes of fossil teeth and bones at Gran Barranca as monitors of climate change and tectonics; pp. 341–361 in R. H. Madden, A. A. Carlini, M. G. Vucetich, and R. F. Kay (eds.), The Paleontology of Gran Barranca. Cambridge University Press, Cambridge, U.K.

Kohn, M. J, J. A. Josef, R. Madden, R. F. Kay, G. Vucetich, and A. A. Carlini. 2004. Climate stability across the Eocene–Oligocene transition, southern Argentina. Geology 32:621–624.

Korth, W. W. 1984. Earliest Tertiary evolution and radiation of rodents in North America. Bulletin of the Carnegie Museum of Natural History 24:1–71.

Korth, W. W. 1993. The skull of *Hitonkala* (Florentiamyidae, Rodentia) and relationships within Geomyoidea. Journal of Mammalogy 74:168–174.

Korth, W. W. 1994a. Middle Tertiary marsupials (Mammalia) from North America. Journal of Paleontology 68:376–397.

Korth, W. W. 1994b. The Tertiary Record of Rodents in North America. Plenum Press, New York, New York, 319 pp.

Korth, W. W. 2001. Comments on the systematics and classification of beavers (Rodentia, Castoridae). Journal of Mammalian Evolution 8:279–296.

Korth, W. W. 2008. Marsupialia; pp. 39–47 in C. M. Janis, G. F. Gunnell, and M. D. Uhen (eds.), Evolution of Tertiary Mammals of North America, Volume 2: Small Mammals, Xenarthrans and Marine Mammals. Cambridge University Press, Cambridge, U.K.

Korth, W. W., and R. J. Emry. 1991. The Skull of *Cedromus* and a review of the Cedromurinae (Rodentia, Sciuridae). Journal of Paleontology 65:984–994.

Korth, W. W., J. H. Wahlert, and R. J. Emry. 1991. A new species of *Heliscomys,* and recognition of the family Heliscomyidae (Geomyoidea: Rodentia). Journal of Vertebrate Paleontology 11:247–256.

Krishtalka, L., and R. K. Stucky. 1983. Paleocene and Eocene marsupials of North America. Annals of the Carnegie Museum 52:229–263.

Kron, D. G., and E. Manning. 1998. Anthracotheriidae; pp. 381–388 in C. M. Janis, K. M. Scott, and L. L. Jacobs (eds.), Evolution of Tertiary Mammals of North America, Volume 1: Terrestrial Carnivores, Ungulates, and Ungulatelike Mammals. Cambridge University Press, Cambridge, U.K.

Kurtén, B., and E. Anderson. 1980. Pleistocene Mammals of North America. Columbia University Press, New York, New York.

LaGarry, H. E. 1998. Lithostratigraphic revision and redescription of the Brule Formation, White River Group, western Nebraska; pp. 63–91 in D. O. Terry Jr., H. E. LaGarry, and R. M. Hunt Jr. (eds.), Depositional Environments, Lithostratigraphy, and Biostratigraphy of the White River and Arikaree Groups (Late Eocene to Early Miocene, North America). Geological Society of America Special Paper 325.

LaGarry, H. E. 2004. Taphonomic evidence of bone processing from the Oligocene of northwestern Nebraska. University of Nebraska–Lincoln School of Natural Resources Professional Paper 2:1–35.

LaGarry, H. E., L. A. LaGarry, and D. O. Terry Jr. 1996. New vertebrate fauna from the base of the Chamberlain Pass Fm. (Eocene), Sioux County, Nebraska. Proceedings of the 106th Annual Nebraska Academy of Sciences, Earth Science Section 45.

LaGarry, H. E., W. E. Wells, and D. O. Terry. 1998. The Toadstool Park Trackway Site, Oglala National Grassland, Nebraska; pp. 91–106 in J. E. Martin, J. W. Hoganson, and R. C. Benton (eds.), Partners Preserving Our Past, Planning Our Future. Proceedings of the Fifth Conference on Fossil Resources. Dakoterra 5.

Lander, B. 1998. Oreodontoidea; pp. 402–425 in C. M. Janis, K. M. Scott, and L. L. Jacobs (eds.), Evolution of Tertiary Mammals of North America, Volume 1: Terrestrial Carnivores, Ungulates, and Ungulatelike Mammals. Cambridge University Press, Cambridge, U.K.

Larson, E. E., and E. Evanoff. 1998. Tephrostratigraphy and sources of the tuffs of the White River sequence; pp. 1–14 in D. O. Terry Jr., H. E. LaGarry, and R. M. Hunt Jr. (eds.), Depositional Environments, Lithostratigraphy, and Biostratigraphy of the White River and Arikaree Groups (Late Eocene to Early Miocene, North America). Geological Society of America Special Paper 325.

Leidy, J. 1847. On a new genus and species of fossil Ruminantia, *Poebrotherium wilsoni.* Philadelphia Academy of Natural Sciences Proceedings 3:322–326.

Leidy, J. 1852. Descriptions of the remains of extinct Mammalia and Chelonia from Nebraska Territory, collected during the geological survey under the direction of Dr. D. D. Owen; pp. 534–572 in D. D. Owen (ed.), Report of a Geological Survey of Wisconsin, Iowa and Minnesota and Incidentally a Portion of Nebraska Territory. Lippincott, Grambo and Co., Philadelphia, Pennsylvania.

Leidy, J. 1853. The ancient fauna of Nebraska: a description of remains of extinct Mammalia and Chelonia from the Mauvaises Terres of Nebraska. Smithsonian Contributions to Knowledge 6:1–126, 25 plates.

Leidy, J. 1869. On the extinct mammalian fauna of Dakota and Nebraska. Philadelphia Academy of Natural Sciences Journal, 2nd Series 7:1–502.

Lemley, R. E. 1971. Notice of new finds in the Badlands. Proceedings of the South Dakota Academy of Science 50:70–74.

Licht, A., M. van Cappelle, H. A. Abels, J. B. Ladant, J. Trabucho Alexandre, C. France-Lanord, Y. Donnadieu, J. Vandenberghe, T. Rigaudier, C. Lécuyer, D. Terry Jr., R. Adriaens, A. Boura, Z. Guo, N. S. Aung, J. Quade, G. Dupont-Nivet, and J.-J. Jaeger. 2014. Asian monsoons in a late Eocene greenhouse world. Nature 513:501–506. doi:10.1038/nature13704

Lillegraven, J. A. 1993. Correlation of Paleogene strata across Wyoming–a user's guide. Geology of Wyoming, Geological Survey of Wyoming Memoir 5:414–477.

Lillegraven, J. A., M. C. McKenna, and L. Krishtalka. 1981. Evolutionary relationships of middle Eocene and species *Centetodon* (Mammalia, Insectivora, Geolabidae) with a description of the dentition of *Ankylodon* (Adapisoricidae). University of Wyoming Publications 45:1–115.

Lindsay, E. H. 2008. Cricetidae; pp. 456–479 in C. M. Janis, G. F. Gunnell, and M. D. Uhen (eds.), Evolution of Tertiary Mammals of North America, Volume 2: Small Mammals, Xenarthrans and Marine Mammals. Cambridge University Press, Cambridge, U.K.

Liu, Z., M. Pagani, D. Zinniker, R. DeConto, M. Huber, H. Brinkhuis, S. R. Shah, R. M. Leckie, and A. Pearson. 2009. Global cooling during the Eocene–Oligocene climate transition. Science 323:1187–1189.

Lockley, M. G., and D. D. Gillette. 1991. Dinosaur tracks and traces: an overview; pp. 3–10 in D. D. Gillette and M. G. Lockley (eds.), Dinosaur Tracks and Traces, Cambridge University Press, Cambridge, U.K.

Loomis, L. B. 1904. Two new river reptiles from the Titanothere Beds. American Journal of Science, Series 4, 18:427–432.

Lucas, S. G., R. J. Emry, and S. E. Foss. 1998. Taxonomy and distribution of *Daeodon,* an Oligocene–Miocene enteledont (Mammalia, Artiodactyla) from North America. Proceedings of the Biological Society of Washington 111:425–435.

Lukens, W. E. 2013. Paleopedology and paleogeomorphology of the early Oligocene Orella and Whitney Members, Brule Formation, White River Group, Toadstool Geologic Park, Nebraska. M.S. thesis, Temple University, Philadelphia, Pennsylvania, 173 pp.

Lukens, W. E., and D. O. Terry. 2013. Geomorphologic controls on climatic interpretations in a mixed fluvial–aeolian system: a deep time perspective from the Paleogene White River Group of northwest Nebraska. Geological Society of America Abstracts with Programs 45:145.

Lukens, W. E., D. O. Terry Jr., D. E. Grandstaff, and B. A. Beasley. 2010. Tracking stolen fossils: a study to determine the utility of rare earth elements as a tool for vertebrate fossil protection. Geological Society of America Abstracts with Programs 42:10.

Lull, R. S. 1921. New camels in the Marsh collection. American Journal of Science 1:392–404.

Lundberg, J. G. 1975. The Fossil Catfishes of North America. Claude W. Hibbard Memorial Volume 2, Papers on Paleontology No. 11, Museum of Paleontology. University of Michigan, Ann Arbor, Michigan, 51 pp.

Luterbacher, H. P., J. R. Ali, H. Brinkhuis, F. M. Gradstein, J. J. Hooker, S. Monechi, J. G. Ogg, J. Powell, U. Röhl, A. Sanfilippo, and B. Schmitz. 2004. The Paleogene period; pp. 384–408 in F. Gradstein, J. Ogg, and A. Smith (eds.), A Geologic Time Scale. Cambridge University Press, Cambridge, U.K.

Macdonald, J. R. 1956. The North American anthracotheres. Journal of Paleontology 30:615–645.

Macdonald, J. R. 1963. The Miocene faunas from the Wounded Knee area of western South Dakota. Bulletin of the American Museum of Natural History 125:141–238.

MacFadden, B. J. 1986. Fossil horses from "Eohippus" (*Hyracotherium*) to *Equus:* scaling, Cope's Law, and the evolution of body size. Paleobiology 12:355–369.

MacFadden, B. J. 1998. Equidae; pp. 537–559 in C. M. Janis, K. M. Scott, and L. L. Jacobs (eds.), Evolution of Tertiary Mammals of North America, Volume 1: Terrestrial Carnivores, Ungulates, and Ungulatelike Mammals. Cambridge University Press, Cambridge, U.K.

Maddox, D., and W. P. Wall. 1998. A systematic review of the fossil lizards and snakes (Squamata) from the White River Group of Badlands National Park; pp. 4–7 in V. Santucci and L. McClelland (eds.), National Park Service Paleontological Research, Technical Report NPS/NRGRD/GRDTR-98/01.

Mader, B. J. 1998. Brontotheriidae; pp. 525 – 536 in C. M. Janis, K. M. Scott, and L. L. Jacobs (eds.), Evolution of Tertiary Mammals of North America, Volume 1: Terrestrial Carnivores, Ungulates, and Ungulatelike Mammals. Cambridge University Press, Cambridge, U.K.

Maher, H., Jr., and R. Shuster. 2012. Chalcedony vein horizons and clastic dikes in the White River Group as products of diagenetically driven deformation. Lithosphere 4:167–186.

Martin, J. E. 1987a. The White River Badlands of South Dakota. Geological Society of America, Rocky Mountain Section Centennial Field Guide, pp. 233–236.

Martin, J. E., and D. C. Parris, eds. 2007. The geology and paleontology of the Late Cretaceous marine deposits of the Dakotas. Geological Society of America Special Paper 427, 256 pp.

Martin, L. D. 1987b. Beavers from the Harrison Formation (early Miocene) with a revision of *Euhapsis.* South Dakota School of Mines and Technology. Dakoterra 3:73–91.

Martin, L. D. 1992. A new miniature saber-toothed nimravid from the Oligocene of Nebraska. Annales Zoologici Fennici 28:341–348.

Martin, L. D. 1998. Nimravidae; pp. 228–235 in C. M. Janis, K. M. Scott, and L. L. Jacobs (eds.), Evolution of Tertiary Mammals of North America, Volume 1: Terrestrial Carnivores, Ungulates, and Ungulatelike Mammals. Cambridge University Press, Cambridge, U.K.

Mason, J. A. 2001. Transport direction of Peoria Loess in Nebraska and implications for loess sources on the central Great Plains. Quaternary Research 56:79–86.

Mathis, J. E., and B. J. MacFadden. 2010. Quantifying *Leptomeryx* (Mammalia, Artiodactyla) enamel surface area across the Eocene–Oligocene transition in Nebraska. Palaios 25:682–687.

Matthew, W. D. 1899. A provisional classification of the freshwater Tertiary of the west. Bulletin of the American Museum of Natural History 12:19–75.

Matthew, W. D. 1901. Fossil mammals of the Tertiary of northeastern Colorado. Memoirs, American Museum of Natural History 1:355–447.

Matthew, W. D. 1902. The skull of *Hypisodus,* the smallest artiodactyl with a revision of the Hypertragulidae. Bulletin of the American Museum of Natural History 16:311–316.

Matthew, W. D. 1904. Notice of two new Oligocene camels. Bulletin of the American Museum of Natural History 20:211–215.

Matthew, W. D. 1905. Notice of two new genera of mammals from the Oligocene of South Dakota. Bulletin of the American Museum of Natural History 21:21–26.

Matthew, W. D. 1918. A Tertiary alligator. American Museum Journal 18:503–506.

Matthews, E. G., and G. Halffter. 1968. New data on American *Corpris,* with a discussion of a fossil species. Ciencia 26:147–162.

McCoy, M. J. 2002. Controls on paleosol morphology in a basin wide marker bed in the Scenic Member of the Brule Formation, Badlands National Park, South Dakota. M.S. thesis, Temple University, Philadelphia, Pennsylvania, 88 pp.

McDermott, J. F. (ed.). 1952. Journal of an expedition to the Mauvaises Terres and the Upper Missouri in 1850 by Thaddeus A. Culbertson. Smithsonian Institution Bureau of American Ethnology Bulletin 147:1–164.

McDowell, S. B. 1958. The greater Antillean insectivores. Bulletin of the American Museum of Natural History 115:113–214.

McKenna, M. C. 1960. The Geolabidinae, a new subfamily of Early Cenozoic erinaceoid insectivores. University of California Publications in Geological Sciences 37:131–164.

McKenna, M. C. 1966. Synopsis of Whitneyan and Arikareean camelid phylogeny. American Museum Novitates 2253, 11 pp.

McKenna, M. C., and S. K. Bell. 1997. Classification of Mammals: Above the Species Level. Columbia University Press, New York, New York, 631 pp.

McKinney, M. L., and R. M. Schoch. 1985. Titanothere allometry, heterochrony, and biomechanics: revising an evolutionary classic. Evolution 39:1352–1363.

Mead, A. J. 1994. Jaw mechanics and cranio-dental ecological implications of two Oligocene rhinocerotoids (Perissodactyla), *Hyracodon* and *Subhyracodon.* M.S. thesis, Georgia College, Milledgeville, Georgia, 107 pp.

Mead, A. J. 1998. An embedded tooth in an oreodont cranium: evidence for feeding habits of Oligocene enteledonts. Dakoterra 5:73–75.

Meehan, T., and L. D. Martin. 2012. New large leptictid insectivore from the Late Paleogene of South Dakota, U.S.A. Acta Palaeontologica Polonica 57:509–518.

Meek, F. B. 1876. Invertebrate Cretaceous and Tertiary fossils of the upper Missouri country.

United States Geological Survey of the Territories 9, 629 pp.

Meek, F. B., and F. V. Hayden. 1858. Descriptions of new species and genera of fossils collected by Dr. F. V. Hayden in Nebraska Territory under direction of Lieut. G. K. Warren, U.S. topographical engineer, with some remarks on the Tertiary and Cretaceous formations of the north-west and the parallelism of the latter with those of other portions of the United States and Territories. Academy of Natural Sciences of Philadelphia Proceedings 9:117–148.

Meek, F. B., and F. V. Hayden. 1862. Descriptions of new lower Silurian (Primordial), Jurassic, Cretaceous and Tertiary fossils, collected in Nebraska Territory, with some remarks on the rocks from which they were obtained. Academy of Natural Sciences of Philadelphia Proceedings 13:415–447.

Metcalf, A. L., and R. A. Smartt. 1997. Land snails of New Mexico. New Mexico Museum of Natural History and Science Bulletin 10, 145 pp.

Metzger, C. A. 2004. Use of paleopedology, sedimentology, and rare earth element geochemistry to determine original depositional environment of fossil bone: Badlands National Park, South Dakota. M.S. thesis, Temple University, Philadelphia, Pennsylvania, 209 pp.

Metzger, C. A., D. O. Terry Jr., and D. E. Grandstaff. 2004. The effect of paleosol formation on rare earth element signatures in fossil bone. Geology 32:497–500.

Meyer, H. W. 2003. The Fossils of Florissant. Smithsonian Books, Washington, D.C., 258 pp.

Meylan, P. A. 1987. The phylogenetic relationships of soft-shelled turtles (Family Trionychidae). Bulletin of the American Museum of Natural History 186:1–101.

Michener, C. D. 1974. The Social Behavior of Bees. Harvard University Press, Cambridge, Massachusetts, 404 pp.

Mihlbachler, M. C., S. G. Lucas, and R. J. Emry. 2004. The holotype specimen of *Menodus giganteus,* and the "insoluble" problem of Chadronian brontothere taxonomy; pp. 129–135 in S. G. Lucas, K. E. Zeigler, and P. E. Kondrashov (eds.), Paleogene Mammals. New Mexico Museum of Natural History and Science Bulletin 26.

Miller, K. G., J. D. Wright, M. E. Katz, B. S. Wade, J. V. Browning, B. S. Cramer, and Y. Rosenthal. 2009. Climate threshold at the Eocene–Oligocene transition: Antarctic ice sheet influence on ocean circulation; pp. 169–178 in C. Koeberl and A. Montanari (eds.), The Late Eocene Earth: Hothouse, Icehouse, and Impacts. Geological Society of America Special Paper 452.

Miller, K. G., M. A. Kominz, J. V. Browning, J. D. Wright, G. S. Mountain, M. E. Katz, P. J. Sugarman, B. S. Cramer, N. Christie-Blick, and S. F. Pekar. 2005. The Phanerozoic record of global sea-level change. Science 310:1293–1298.

Miller, M. T. 2010. Sample analysis of *Archaeotherium* (Artiodactyla: Entelodontidae) from the Conata Picnic Ground, Badlands National Park, South Dakota. M.S. thesis, South Dakota School of Mines and Technology, Rapid City, South Dakota, 74 pp.

Mintz, J. S. 2007. The terrestrial response to the post Eocene–Oligocene climatic transition,

Poleslide Member, Brule Formation, Badlands National Park, South Dakota. M.S. thesis, Temple University, Philadelphia, Pennsylvania, 75 pp.

Mintz, J. S., D. O. Terry Jr., and G. Stinchcomb. 2007. The terrestrial response to the post Eocene–Oligocene climatic transition, Poleslide Member, Brule Formation, Badlands National Park, South Dakota. Geological Society of America Abstracts with Programs 39:193.

Mook, C. C. 1934. The evolution and classification of the Crocodilia. Journal of Geology 42:295–304.

Muhs, D. R., E. A. Bettis III, J. N. Aleinikoff, J. P. McGeehin, J. Beann, G. Skipp, B. D. Marshall, H. M. Roberts, W. C. Johnson, and R. Benton. 2008. Origin and paleoclimatic significance of late Quaternary loess in Nebraska: evidence from stratigraphy, chronology, sedimentology, and geochemistry. Geological Society of America Bulletin 120:1378–1407.

Munthe, K. 1998. Canidae; pp. 124–143 in C. M. Janis, K. M. Scott, and L. L. Jacobs (eds.), Evolution of Tertiary Mammals of North America, Volume 1: Terrestrial Carnivores, Ungulates, and Ungulatelike Mammals, Cambridge University Press, Cambridge, U.K.

Myers, J. A. 2003. Terrestrial Eocene–Oligocene vegetation and climate in the Pacific Northwest; pp. 171–185 in D. R. Prothero, L. C. Ivany, and E. A. Nesbitt (eds.), From Greenhouse to Icehouse: The Marine Eocene–Oligocene Transition. Columbia University Press, New York, New York.

Nicknish, J. M., and J. R. Macdonald. 1962. Basal Miocene ash in White River Badlands, South Dakota. American Association of Petroleum Geologists Bulletin 46:685–690.

Nixon, D. A. 1996. Stone tracks. Nebraskaland (June): 26–29.

Novacek, M. J. 1986. The skull of leptictid insectivorans and the higher level of classification of eutherian mammals. Bulletin of the American Museum of Natural History 183:1–112.

O'Harra, C. C. 1910. The Badlands Formations of the Black Hills Region. South Dakota State School of Mines Bulletin 9, 152 pp.

O'Harra, C. C. 1920. The White River Badlands. South Dakota School of Mines Bulletin 13, 181 pp.

Osborn, H. F. 1929. Titanotheres of ancient Wyoming, Dakota, and Nebraska. United States Geological Survey Monograph 55:1–701.

Osborn, H. F., and J. L. Wortman. 1892. Characters of *Protoceras* (Marsh), a new artiodactyl from the Lower Miocene. Bulletin of the American Museum of Natural History 4:351–372.

Pagani, M., J. Zachos, K. H. Freeman, S. Bohaty, and B. Tipple. 2005. Marked change in atmospheric carbon dioxide concentrations during the Oligocene. Science 309:600–603.

Parker, H. W. 1929. Two fossil frogs from the Lower Miocene of Europe. Annals and Magazine of Natural History 4:270–281.

Parris, D. C., and M. Green. 1969. *Dinohyus* (Mammalia: Entelodontidae) in the Sharps Formation, South Dakota. Journal of Paleontology 43:1277–1279.

Parris, D. C., and J. A. Holman. 1978. An Oligocene snake from a coprolite. Herpetologica 34:258–264.

Patton, T. H. 1969. An Oligocene land fauna from Florida. Journal of Paleontology 43:543–546.

Patton, T. H., and B. E. Taylor. 1973. The Protoceratinae (Mammalia, Tylopoda, Protoceratidae). Bulletin of the American Museum of Natural History 150:347–414.

Pearson, H. S. 1923. Some skulls of *Perchoerus* (*Thinohyus*) from the White River and John Day Formations. Bulletin of the American Museum of Natural History 48:61–96.

Pearson, P. N., I. K. McMillan, B. S. Wade, T. D. Jones, H. L. Coxall, P. R. Bown, and C. H. Lear. 2008. Extinction and environmental change across the Eocene–Oligocene boundary in Tanzania. Geology 36:179–182.

Peterson, O. A. 1909. A revision of the Entelodontidae. Memoirs of the Carnegie Museum 9:41–158.

Pettyjohn, W. A. 1966. Eocene paleosol in the northern Great Plains. United States Geological Survey Professional Paper 550C:61–65.

Pilsbry, H. A. 1939. Land mollusca of North America (north of Mexico). Academy of Natural Sciences of Philadelphia, Monograph 3, 1:1–573.

Poag, C. W., E. Mankinen, and R. D. Norris. 2003. Late Eocene impacts: geologic record, correlation, and paleoenvironmental consequences; pp. 495–510 in D. R. Prothero, L. C. Ivany, and E. A. Nesbitt (eds.), From Greenhouse to Icehouse: The Marine Eocene–Oligocene Transition. Columbia University Press, New York, New York.

Premoli, S. I., and D. G. Jenkins. 1993. Decision on the Eocene–Oligocene boundary stratotype. Episodes 16:379–382.

Preston, F. W. 1968.The shapes of birds' eggs: mathematical aspects. Auk 85:454–463.

Preston, F. W. 1969. Shapes of birds' eggs: extant North American families. Auk 86:246–264.

Preston, R. E. 1979. Late Pleistocene cold-blooded vertebrate faunas from the mid-continental United States. University of Michigan Papers on Paleontology 19:1–53.

Prothero, D. R. 1985. Correlation of the White River Group by magnetostratigraphy; pp. 265–276 in J. E. Martin (ed.), Fossiliferous Cenozoic deposits of western South Dakota and northwestern Nebraska. Dakoterra 2.

Prothero, D. R. 1994. The Eocene–Oligocene Transition: Paradise Lost. Columbia University Press, New York, New York, 291 pp.

Prothero, D. R. 1996. Magnetostratigraphy of the White River Group in the High Plains; pp. 262–277 in D. R. Prothero and R. J. Emry (eds.), The Terrestrial Eocene–Oligocene Transition in North America. Cambridge University Press, Cambridge, U.K.

Prothero, D. R. 1998a. Hyracodontidae; pp. 589–594 in C. M. Janis, K. M. Scott, and L. L. Jacobs (eds.), Evolution of Tertiary Mammals of North America, Volume 1: Terrestrial Carnivores, Ungulates, and Ungulatelike Mammals. Cambridge University Press, Cambridge, U.K.

Prothero, D. R. 1998b. Protoceratidae; pp. 431–438 in C. M. Janis, K. M. Scott, and L. L. Jacobs (eds.), Evolution of Tertiary Mammals of North America, Volume 1: Terrestrial Carnivores, Ungulates, and Ungulatelike Mammals. Cambridge University Press, Cambridge, U.K.

Prothero, D. R. 1998c. Rhinocerotidae; pp. 595–605 in C. M. Janis, K. M. Scott, and L. L. Jacobs (eds.), Evolution of Tertiary Mammals of North America, Volume 1: Terrestrial Carnivores, Ungulates, and Ungulatelike Mammals. Cambridge University Press, Cambridge, U.K.

Prothero, D. R. 2005. The Evolution of the North American Rhinoceroses. Cambridge University Press, Cambridge, U.K., 218 pp.

Prothero, D. R., and R. J. Emry (eds.). 1996. The Terrestrial Eocene–Oligocene Transition in North America. Cambridge University Press, Cambridge, U.K., 688 pp.

Prothero, D. R., and R. J. Emry. 2004. The Chadronian, Orellan, and Whitneyan North American Land Mammal Ages; pp. 156–168 in M. O. Woodburne (ed.), Late Cretaceous and Cenozoic mammals of North America. Columbia University Press, New York, New York.

Prothero, D. R., and F. Sanchez. 2008. Systematics of the Leptauchenine Oreodonts (Mammalia: Artiodactyla) from the Oligocene and earliest Miocene of North America. New Mexico Museum of Natural History and Science Bulletin 44:335–355.

Prothero, D. R., and N. Shubin. 1989. The evolution of mid-Oligocene horses; pp. 142–175 in D. R. Prothero and R. M. Schoch (eds.), The Evolution of Perissodactyls. Clarendon Press, Oxford, U.K.

Prothero, D. R., and C. C. Swisher III. 1992. Magnetostratigraphy and geochronology of the terrestrial Eocene–Oligocene transition in North America; pp. 46–73 in D. R. Prothero and W. A. Berggren (eds.), Eocene–Oligocene Climatic and Biotic Evolution. Princeton University Press, Princeton, New Jersey.

Prothero, D. R., and K. E. Whittlesey. 1998. Magnetic stratigraphy and biostratigraphy of the Orellan and Whitneyan land-mammal "ages" in the White River Group; pp. 39–61 in D. O. Terry Jr., H. E. LaGarry, and R. M. Hunt Jr. (eds.), Depositional Environments, Lithostratigraphy, and Biostratigraphy of the White River and Arikaree Groups (Late Eocene to Early Miocene, North America). Geological Society of America Special Paper 325.

Prothero, D. R., L. C. Ivany, and E. A. Nesbitt (eds.). 2003. From Greenhouse to Icehouse: The Marine Eocene–Oligocene Transition. Columbia University Press, New York, New York, 541 pp.

Prout, H. A. 1846. Gigantic *Palaeotherium*. American Journal of Science 2:88–89.

Prout, H. A. 1847. A description of a fossil maxillary bone of a *Palaeotherium* from near White River. American Journal of Science 3:248–250.

Pye, K., and H. Tsoar. 1987. The mechanics and geological implications of dust transport and deposition in deserts with particular reference to loess formation and dune sand diagenesis in the northern Negev, Israel; pp. 139–156 in L. Frostick and I. Reid (eds.), Desert Sediments, Ancient and Modern. Geological Society of London Special Publication 35.

Rasmussen, D. T., T. M. Bown, and E. L. Simons. 1992. The Eocene–Oligocene transition in continental Africa; pp. 548–566 in D. R. Prothero and W. A. Berggren (eds.), Eocene–Oligocene Climatic and Biotic Evolution.

Princeton University Press, Princeton, New Jersey.

Rawling, J. E., III, G. G. Fredlund, and S. Mahan. 2003. Aeolian cliff-top deposits and buried soils in the White River Badlands, South Dakota, U.S.A. Holocene 13:121–129.

Reed, C. A. 1951. Locomotion and appendicular anatomy in three soricoid insectivores. American Midland Naturalist 45:513–671.

Reed, C. A., and W. D. Turnbull. 1965. The mammalian genera *Arctoryctes* and *Cryptoryctes* from the Oligocene and Miocene of North America. Fieldiana Geology 15:95–170.

Reed, K. M. 1961. The Proscalopinae, a new subfamily of talpid insectivores. Bulletin of the Museum of Comparative Zoology 125:471–494.

Reig, O. A., J. A. W. Kirsch, and L. G. Marshall. 1985. New conclusions on the relationships of the opossum-like marsupials, with an annotated classification of the Didelphimorphia. Ameghiniana 21:335–343.

Retallack, G. J. 1983a. A paleopedological approach to the interpretation of terrestrial sedimentary rocks: the mid-Tertiary fossil soils of Badlands National Park, South Dakota. Geological Society of America Bulletin 94:823–840.

Retallack, G. J. 1983b. Late Eocene and Oligocene paleosols from Badlands National Park, South Dakota. Geological Society of America Special Paper 193:1–82.

Retallack, G. J. 1984. Trace fossils of burrowing beetles and bees in an Oligocene paleosol, Badlands National Park, South Dakota. Journal of Paleontology 58:571–592.

Retallack, G. J. 2007. Cenozoic paleoclimate on land in North America. Journal of Geology 115:271–294.

Retallack, G. J., E. A. Bestland, and T. J. Fremd. 2000. Eocene and Oligocene paleosols of central Oregon. Geological Society of America Special Paper 344, 192 pp.

Reynolds, P. S. 2002. How big is a giant? The importance of method in estimating body size of extinct mammals. Journal of Mammalogy 83:321–332.

Rich, T. H. V. 1981. Origin and history of Erinaceinae and Brachyericinae (Mammalia, Insectivora) in North America. Bulletin of the American Museum of Natural History 171:1–116.

Rich, T. H. V., and D. L. Rasmussen. 1973. North American Erinaceine Hedgehogs (Mammalia: Insectivora). Occasional Papers of the Museum of Natural History, University of Kansas 21:1–54.

Ritter, J. R., and R. G. Wolfe. 1958. The channels sandstones of the eastern section of the Big Badlands of South Dakota. South Dakota Academy of Sciences Proceedings 37:184–191.

Rose, K. D. 1999. Postcranial skeleton of Eocene Leptictidae (Mammalia), and its implications for behavior and relationships. Journal of Vertebrate Paleontology 19:355–372.

Rothecker, J., and J. E. Storer. 1996. The marsupials of the Lac Pelletier lower fauna, middle Eocene (Duchesnean) of Saskatchewan. Journal of Vertebrate Paleontology 68:376–397.

Roth-Nebelsick, A., T. Utescher, V. Mosbrugger, L. Diester-Haass, and H. Walther. 2004. Changes in atmospheric CO_2 concentrations and climate from the Late Eocene to Early Miocene: palaeobotanical reconstruction based on fossil floras from Saxony, Germany. Palaeogeography, Palaeoclimatology, Palaeoecology 205:43–67.

Ruddiman, W. F., J. E. Kutzbach, and I. C. Prentice. 1997. Effects of Cenozoic uplift and CO_2 on vegetation; pp. 203–235 in W. F. Ruddiman (ed.), Tectonic Uplift and Climate Change. Plenum, New York, New York.

Sahy, D., A. Fischer, D. O. Terry Jr., D. J. Condon, and K. Kuiper. 2010. New high-precision geochronology from the Late Eocene of North America and Italy. Geological Society of America Abstracts with Programs 42:394.

Sahy, D., A. U. Fischer, D. J. Condon, D. O. Terry Jr., J. Hiess, H. Abels, S. K. Hüsing, and K. F. Kuiper. 2012. High precision radioisotopic age constraints on the Late Eocene–Early Oligocene geomagnetic polarity time scale; pp. 64–65 in Abstracts and Programs of High Fidelity: The Quest for Precision in Stratigraphy and Its Applications, May 16–17, 2012, London, U.K.

Sakagami, S. F., and C. D. Michener. 1962. The nest architecture of the sweat bees (Halictinae). A comparative study of behavior. University of Kansas Press, Lawrence, Kansas, 135 pp.

Santucci, V. L. 2009. National Park Service Geologic Resources Division: Paleontology Program Action Plan. Geologic Resources Division Internal Report, 21 pp.

Schlaikjer, E. M. 1933. Contributions to the stratigraphy and paleontology of the Goshen Hole area, Wyoming, 1. A detailed study of the structure and relationships of a new Zalambdodont insectivore from the middle Oligocene. Bulletin of the Museum of Comparative Zoology 76:1–27.

Schuchert, C., and C. M. LeVene. 1940. O. C. Marsh, Pioneer in Paleontology. Yale University Press, New Haven, Connecticut, 349 pp.

Schultz, C. B., and C. H. Falkenbach. 1956. Miniochoerinae and Oreonetinae, two new subfamilies of oreodonts. Bulletin of the American Museum of Natural History 109:373–482.

Schultz, C. B., and T. M. Stout. 1955. Classification of Oligocene sediments of Nebraska. Bulletin of the University of Nebraska State Museum 4:17–52.

Scott, W. B. 1890. The dogs of the American Miocene. Princeton College Bulletin 2:37–39.

Scott, W. B. 1891. On the osteology of Mesohippus and Leptomeryx with observations on the modes and factors of evolution in mammalia. Journal of Morphology 5:301–402.

Scott, W. B. 1962. A History of the Land Mammals of the Western Hemisphere. Hafner Publishing, New York, New York, 786 pp.

Scott, W. B., and G. L. Jepsen. 1936. The mammalian fauna of the White River Oligocene, part 1: Insectivora and Carnivora. Transactions of the American Philosophical Society 28:1–152. Plates 1–22.

Scott, W. B., and G. L. Jepsen. 1940. The mammalian fauna of the White River Oligocene, part 4: Artiodactyla. Transactions of the American Philosophical Society 28:363–746. Plates 36–78.

Seefeldt, D. R., and M. O. Glerup. 1958. Stream channels of the Brule Formation, western Big Badlands, South Dakota. South Dakota Academy of Sciences Proceedings 37:195–202.

Sheldon, N. D., and G. J. Retallack. 2004. Regional paleoprecipitation records from the Late Eocene and Oligocene of North America. Journal of Geology 112:487–494.

Sheldon, N. D., R. L. Mitchell, M. E. Collinson, and J. J. Hooker. 2009. Eocene–Oligocene transition paleoclimate record from paleosols, Isle of Wight (U.K.); pp 241–248 in C. Koeberl and A. Montanari (eds.), The Late Eocene Earth: Hothouse, Icehouse, and Impacts. Geological Society of America Special Paper 452.

Shelton, S. Y., C. Branciforte, J. Darbyshire, S. Johnson, J. Mead, M. Miller, M. Pinsdorf, and M. Sauter. 2009. Pig Dig final report. On file at Badlands National Park and the South Dakota School of Mines and Technology, Rapid City, South Dakota, 30 pp.

Shipman, P. 1981. Life History of a Fossil: An Introduction to Taphonomy and Paleoecology. Harvard University Press, Cambridge, Massachusetts, 222 pp.

Shuler, J. 1989. A revelation called the Badlands: building a national park, 1909–1939. Badlands Natural History Association, Interior, South Dakota, 48 pp.

Simpson, G. G. 1945. The principles of classification and a classification of mammals. Bulletin of the American Museum of Natural History 85, 1–350.

Sinclair, W. J. 1921a. Entelodonts from the Big Badlands of South Dakota in the Geological Museum of Princeton University. Proceedings of the American Philosophical Society 60:467–495.

Sinclair, W. J. 1921b. The "Turtle-Oreodon Layer" or "Red Layer," a contribution to the stratigraphy of the White River Oligocene; results of the Princeton University 1920 expedition to South Dakota. American Philosophical Society Proceedings 60:456–466.

Sinclair, W. J. 1923. The faunas of the concretionary zones of the Oreodon Beds, White River Oligocene. American Philosophical Society Proceedings 63:94–133.

Slaughter, B. H. 1978. Occurrence of didelphine marsupials from the Eocene and Miocene of the Texas Gulf Plain. Journal of Paleontology 52:744–746.

Smith, G. R., and W. P. Patterson. 1994. Mio-Pliocene seasonality on the Snake River plain: comparison of faunal and oxygen isotopic evidence. Palaeogeography, Palaeoclimatology, Palaeoecology 107:291–302.

Smith, K. G. 1952. Structure plan of clastic dikes. American Geophysical Union Transactions 33:869–892.

Soil Survey Staff. 1999. Soil taxonomy: a basic system of soil classification for making and interpreting soil surveys. 2nd edition. Natural Resources Conservation Service. United States Department of Agriculture Handbook 436, 871 pp.

Stamm, J. F., R. R. Hendricks, J. F. Sawyer, S. A. Mahan, and B. J. Zaprowski. 2013. Late Quaternary stream piracy and strath terrace formation along the Belle Fourche and lower Cheyenne Rivers, South Dakota and Wyoming. Geomorphology 197:10–20.

Stevens, K. K. 1996a. Taphonomy of an early Oligocene (Orellan) fossil assemblage in the Scenic Member, Brule Formation, White River Group, Badlands National Park, South Dakota.

M.S. thesis, South Dakota School of Mines and Technology, Rapid City, South Dakota, 103 pp.

Stevens, K. K. 1996b. Taphonomy of an Orellan (early Oligocene) fossil assemblage in the Scenic Member, Brule Formation, White River Group, Badlands National Park, South Dakota. Geological Society of America Abstracts with Programs 28:39.

Stevens, M. S., and J. B. Stevens. 1996. Merycoidodontinae and Miniochoerinae; pp. 499–573 in D. R. Prothero and R. J. Emry (eds.), The Terrestrial Eocene–Oligocene Transition in North America. Cambridge University Press, Cambridge, U.K.

Stevens, M. S., and J. B. Stevens. 2007. Family Merycoidodontidae; pp. 157–168 in D. R. Prothero and S. E. Foss (eds.), The Evolution of Artiodactyls. Johns Hopkins University Press, Baltimore, Maryland.

Stinchcomb, G. E. 2007. Paleosols and stratigraphy of the Scenic–Poleslide member boundary, early Oligocene Brule Formation, White River Group, Badlands National Park, South Dakota, U.S.A. M.S. thesis, Temple University, Philadelphia, Pennsylvania, 94 pp.

Stinchcomb, G., D. O. Terry Jr., and J. Mintz. 2007. Paleosols and stratigraphy of the Scenic–Poleslide Member boundary: implications for pedofacies analysis and regional correlation of the early Oligocene Brule Formation, South Dakota, U.S.A. Geological Society of America Abstracts with Programs 39:305.

Stirton, R. A. 1940. Phylogeny of the North American Equidae. University of California Publications, Bulletin of the Department of Geological Sciences 25:165–198.

Stirton, R. A., and J. M. Rensberger. 1964. Occurrence of the insectivore genus Micropternodus in the John Day Formation of central Oregon. Bulletin of the Southern California Academy of Sciences 63:57–80.

Stock, C. 1935. Artiodactyla from the Sespe of the Las Posas Hills, California. Publications of the Carnegie Institute of Washington 453:119–125.

Stoffer, P. W., P. Messina, J. A. Chamberlain Jr., and D. O. Terry. 2001. The Cretaceous–Tertiary Boundary Interval in Badlands National Park, South Dakota. U.S. Geological Survey Open-File Report 01–56, 49 pp.

Storch, G., B. Engresser, and M. Wuttke. 1996. Oldest fossil record of gliding in rodents. Nature 379:439–441.

Stovall, J. W., and W. S. Strain. 1936. A hitherto undescribed coprolite from the White River Badlands of South Dakota. Journal of Mammalogy 17:27–28.

Sullivan, R. M. 1979. Revision of the Paleocene genus Glyptosaurus (Reptilia, Anguidae). Bulletin of the American Museum of Natural History 163:1–72.

Sullivan, R. M., and J. A. Holman. 1996. Squamata; pp. 354–372 in D. R. Prothero and R. J. Emry (eds.), The Terrestrial Eocene–Oligocene Transition in North America. Cambridge University Press, Cambridge, U.K.

Sundell, C. 1997. Orellan burrows and associated fauna from Converse County, Wyoming. Journal of Vertebrate Paleontology, Abstract Supplement 17:80A.

Swinehart, J. B., V. L. Souders, H. M. DeGraw, and R. F. Diffendal. 1985. Cenozoic paleogeography of western Nebraska; pp. 209–229 in

R. M. Flores and S. Kaplan (eds.), Cenozoic Paleogeography of the West-Central United States. Society of Economic Paleontologists and Mineralogists, Rocky Mountain Section, Special Publication 3.

Symonds, M. R. E. 2005. Phylogeny and life histories of the "Insectivora." Controversies and consequences. Biological Reviews 80:93–128.

Tanner, L. G. 1973. Notes regarding skull characteristics of *Oxetocyon cuspidatus* Green, (Mammalia, Canidae). Transactions of the Nebraska Academy of Sciences 2:66–69.

Tedford, R. H., X. Wang, and B. E. Taylor. 2009. Phylogenetic systematics of the North American fossil Caninae (Carnivora: Canidae). Bulletin of the American Museum of Natural History 325, 218 pp.

Tedford, R. H., J. Swinehart, D. R. Prothero, C. C. Swisher III, S. A. King, and T. E. Tierney. 1996. The Whitneyan–Arikareean transition in the High Plains; pp 312–334 in D. R. Prothero and R. J. Emry (eds.), The Terrestrial Eocene–Oligocene Transition in North America. Cambridge University Press, Cambridge, U.K.

Tedford. R. H., L. B. Albright III, A. D. Barnosky, I. Ferrusquia-Villafranca, R. M. Hunt Jr., J. E. Storer, C. C. Swisher III, M. R. Voorhies, S. D. Webb, and D. P. Whistler. 2004. Mammalian biochronology of the Arikareean through Hemphillian interval (late Oligocene through early Pliocene epochs); pp. 169–231 in M. O. Woodburne (ed.), Late Cretaceous and Cenozoic Mammals of North America Biostratigraphy and Geochronology. Columbia University Press, New York, New York.

Terry, D. O., Jr. 1996a. Stratigraphic and paleopedologic analysis of depositional sequences within the Pig Wallow Site, Badlands National Park. National Park Service/University of Nebraska–Lincoln Cooperative Agreement CA-1300-5-9001, Badlands National Park Archives, South Dakota, 164 pp.

Terry, D. O., Jr. 1996b. Stratigraphy, paleopedology, and depositional environment of the Conata Picnic Ground Bone Bed (Orellan), Brule Formation, Badlands National Park, South Dakota. Geological Society of America Abstracts with Programs 28:A-40.

Terry, D. O., Jr. 1998. Lithostratigraphic revision and correlation of the lower part of the White River Group: South Dakota to Nebraska; pp. 15–37 in D. O. Terry Jr., H. E. LaGarry, and R. M. Hunt Jr. (eds.), Depositional Environments, Lithostratigraphy, and Biostratigraphy of the White River and Arikaree Groups (Late Eocene to Early Miocene, North America). Geological Society of America Special Paper 325.

Terry, D. O., Jr. 2001. Paleopedology of the Chadron Formation of Northwestern Nebraska: implications for paleoclimatic change in the North American Midcontinent across the Eocene–Oligocene boundary. Palaeogeography, Palaeoclimatology, Palaeoecology 168:1–38.

Terry, D. O., Jr. 2010. Untangling regional records of Eocene–Oligocene climate change across Wyoming, Nebraska, and South Dakota; pp. 32–33 in Abstract Volume for SEPM-NSF Workshop Paleosols and Soil Surface Analog Systems.

Terry, D. O., Jr., and J. E. Evans. 1994. Pedogenesis and paleoclimatic implications of the Chamberlain Pass Formation, Basal White River Group, Badlands of South Dakota. Palaeogeography, Palaeoclimatology, Palaeoecology 110:197–215.

Terry, D. O., Jr., and H. E. LaGarry. 1998. The Big Cottonwood Creek Member: a new member of the Chadron Formation in northwestern Nebraska; pp 117–141 in D. O. Terry Jr., H. E. LaGarry, and R. M. Hunt Jr. (eds.), Depositional Environments, Lithostratigraphy, and Biostratigraphy of the White River and Arikaree Groups (Late Eocene to Early Miocene, North America). Geological Society of America Special Paper 325.

Terry, D. O., Jr., and J. I. Spence. 1997. Documenting the extent and depositional environment of the Chadron Formation in the South Unit of Badlands National Park. National Park Service/University of Nebraska–Lincoln Cooperative Agreement CA-1300-5-9001, Badlands National Park Archives, South Dakota, 94 pp.

Terry, D. O., Jr., D. E. Grandstaff, W. E. Lukens, and A. E. Drewicz. 2010. Recent advances in geochemical fingerprinting of fossil vertebrates from the Eocene–Oligocene White River sequence. Geological Society of America Abstracts with Programs 42:459.

Thorpe, M. R. 1937. The Merycoidodontidae an extinct group of ruminant mammals. Memoirs of the Peabody Museum of Natural History 3, 428 pp.

Tihen, J. A. 1964. Tertiary changes in the herpetofaunas of temperate North America. Senckenbergiana Biologica 45:265–279.

Tordoff, H. B., and J. R. Macdonald. 1957. A new bird (Family Cracidae) from the early Oligocene of South Dakota. Auk 74:174–184.

Troxell, E. L. 1920. Entelodonts in the Marsh collection. American Journal of Science 50:243–245, 361–386, 431–455.

Troxell, E. L. 1921. The American bothriodonts. American Journal of Science, ser. 5 1:325–339.

Troxell, E. L. 1925. Fossil logs and nuts of hickory. Scientific Monthly 21:570–572.

Van Valkenburgh, B. 1985. Locomotor diversity in past and present guilds of large predatory mammals. Paleobiology 11:406–428.

Vondra, C. F. 1956. Remarks on the lower Chadron in northwestern Nebraska; p. 12 in Abstracts with Programs of the Nebraska Academy of Sciences.

Wahlert, J. H. 1974. The cranial foramina of protrogomorphous rodents; an anatomical and phylogenetic study. Bulletin of the Museum of Comparative Zoology 146:363–410.

Wahlert, J. H. 1983. Relationships of the Florentiamyidae (Rodentia, Geomyoidea) based on cranial and dental morphology. American Museum Novitates 2769:1–23.

Wall, W. P., and W. Hickerson. 1995. A biomechanical analysis of locomotion in the Oligocene rhinocerotoid, *Hyracodon;* pp. 19–26 in V. L. Santucci and L. McClelland (eds.), National Park Service Paleontological Research. Technical Report NPS/NRPO/NRTR-95/16.

Wang, X. 1994. Phylogenetic systematics of the Hesperocyoninae (Carnivora: Canidae). Bulletin of the American Museum of Natural History 221:1–207.

Wang, X., and R. H. Tedford. 2008. Dogs: Their Fossil Relatives and Evolutionary History. Columbia University Press, New York, New York, 219 pp.

Wang, X., T. E. Cerling, and B. J. MacFadden. 1994. Fossil horses and carbon isotopes: new evidence for Cenozoic dietary, habitat, and ecosystem changes in North America. Palaeogeography, Palaeoclimatology, Palaeoecology 107:269–179.

Wang, X., R. H. Tedford, and B. E. Taylor. 1999. Phylogenetic systematics of the Borophaginae (Carnivora: Canidae). Bulletin of the American Museum of Natural History 243:1–391.

Wanless, H. R. 1921. Lithology of the White River sediments. American Philosophical Society Proceedings 61:184–203.

Wanless, H. R. 1923. Stratigraphy of the White River beds of South Dakota. American Philosophical Society Proceedings 62:190–269.

Ward, F. 1922. The geology of a portion of the Badlands. South Dakota Geological and Natural History Survey Bulletin 11:1–80.

Warren, G. K. 1856. Reconnaissances in the Dacota Country, 1855. Scale 1:600,000, 1 sheet in G. K. Warren's Explorations in the Dacota Country in the year 1855. Senate Executive Document 76, 34th Congress, 1st Session. P. S. Duval & Co. Lithographers, Philadelphia, Pennsylvania.

Webb, S. D. 1998. Hornless ruminants; pp. 463–476 in C. M. Janis, K. M. Scott, and L. L. Jacobs (eds.), Evolution of Tertiary Mammals of North America, Volume 1: Terrestrial Carnivores, Ungulates, and Ungulatelike Mammals. Cambridge University Press, Cambridge, U.K.

Weissmann, G. S, A. J. Hartley, G. J. Nichols, L. A. Scuderi, M. Olson, H. Buehler, and R. Banteah. 2010. Fluvial form in modern continental sedimentary basins: distributive fluvial systems. Geology 38:39–42.

Welty, J. C., and L. Baptista. 1988. The Life of Birds. Saunders College Publishing, New York, New York, 581 pp.

Western, D. 1975. Water availability and its influence on the structure and dynamics of a savannah large mammal community. East African Wildlife Journal 13:265–286.

Wetmore, A. 1927. Fossil birds from the Oligocene of Colorado. Proceedings of the Colorado Museum of Natural History 7:3–13.

Wetmore, A. 1940. Fossil bird remains from Tertiary deposits in the United States. Journal of Morphology 66:25–37.

Wetmore, A. 1942. Two new fossil birds from the Oligocene of South Dakota. Smithsonian Miscellaneous Collections 101:1–6.

Wetmore, A. 1956. A fossil guan from the Oligocene of South Dakota. Condor 58:234–235.

Wetmore, A., and E. C. Case. 1934. A new fossil hawk from the Oligocene beds of South Dakota. Contributions from the Museum of Paleontology, University of Michigan 4:129–132.

Whelan, M. P., J. S. Hamre, and A. T. Hardy. 1996. Preliminary field and laboratory observations relating to the origin of clastic dikes, Badlands National Park, South Dakota. Geological Society of America Abstracts with Programs 19:42.

White, C. A. 1883. A review of the non-marine fossil Mollusca of North America. U.S. Geological Survey Annual Report 3:403–550.

Wieland, G. R. 1935. The Cerro Cuadrado Petrified Forest. Carnegie Institute of Washington Publication 449, 180 pp.

Wilhelm, P. B. 1993. Morphometric analysis of the limb skeleton of generalized mammals in relation to locomotor behavior with applications to fossil mammals. Ph.D. dissertation, Brown University, Providence, Rhode Island, 210 pp.

Williams, E. E. 1952. A new fossil tortoise from Mona Island, West Indies, a tentative arrangement of the tortoises of the world. Bulletin of the American Museum of Natural History 99:541–560.

Wilson, M. H. V. 1977. New records of insect families from the freshwater middle Eocene of British Columbia. Canadian Journal of Earth Sciences 14:1139–1155.

Wilson, R. W. 1949. On some White River rodents. Carnegie Institution of Washington Publication 584:27–50.

Wilson, R. W. 1975. The National Geographic Society–South Dakota School of Mines and Technology expedition into the Big Badlands of South Dakota 1940. National Geographic Research Reports 1890–1954 Projects, 79–85.

Wolfe, J. A. 1992. Climatic, floristic, and vegetational changes near the Eocene/Oligocene boundary in North America; pp. 421–436 in D. R. Prothero and W. A. Berggren (eds.), Eocene–Oligocene Climatic and Biotic Evolution. Princeton University Press, Princeton, New Jersey.

Wolfe, J. A. 1994. Tertiary climatic changes at middle latitudes of western North America.

Palaeogeography, Palaeoclimatology, Palaeoecology 108:195–205.

Wood, A. E. 1937. Rodentia; part 2 in W. B. Scott and G. L. Jepsen (eds.), The mammalian fauna of the White River Oligocene. Transactions of the American Philosophical Society 28:155–269. Plates 23–33.

Wood, A. E. 1940. Lagomorpha; part 3 in W. B. Scott and G. L. Jepsen (eds.), The mammalian fauna of the White River Oligocene. Transactions of the American Philosophicai Society 28:271–362.

Wood, A. E. 1980. The Oligocene rodents of North America. Transactions of the American Philosophical Society 70:1–68.

Wood, H. E., II, R. W. Chaney Jr., J. Clark, E. H. Colbert, G. L. Jepsen, J. B. Reeside Jr., and C. Stock. 1941. Nomenclature and correlation of the North American continental Tertiary. Geological Society of America Bulletin 52:1–48.

Wood, R. C. 1977. Evolution of the emydine turtles *Graptemys* and *Malaclemys* (Reptilia, Testudines, Emydidae). Journal of Herpetology 11:415–421.

Woodburne, M. O. (ed.). 2004. Late Cretaceous and Cenozoic Mammals of North America. Columbia University Press, New York, 391 pp.

Woodburne, M. O., and C. C. Swisher III. 1995. Land mammal high resolution geochronology, intercontinental overland dispersals, sea-level, climate, and vicariance; pp. 335–364 in W. A. Berggren, D. V. Kent, M.-P. Aubry, and J. Hardenbol (eds.), Geochronology, Time-Scales and Stratigraphic Correlation; Framework for

an Historical Geology. Society of Stratigraphic Geology Special Publication 54.

Wortman, J. L. 1893. On the divisions of the White River or lower Miocene of Dakota. American Museum of Natural History Bulletin 5:95–105.

Wortman, J. L., and C. Earle. 1893. Ancestors of the tapir from the lower Miocene of Dakota. Bulletin of the American Museum of Natural History 5:159–180.

Xiao, G. Q., H. A. Abels, Z. Q. Yao, G. Dupont-Nivet, and F. J. Hilgen. 2010. Asian aridification linked to the first step of the Eocene–Oligocene climate transition (EOT) in obliquity-dominated terrestrial records (Xining Basin, China). Climate of the Past 6:501–513.

Zanazzi, A., M. J. Kohn, and D. O. Terry Jr. 2009. Biostratigraphy and paleoclimatology of the Eocene–Oligocene boundary section at Toadstool Park, northwestern Nebraska, U.S.A.; pp. 197–214 in C. Koeberl and A. Montanari (eds.), The Late Eocene Earth: Hothouse, Icehouse, and Impacts. Geological Society of America Special Paper 452.

Zanazzi, A., M. J. Kohn, B. MacFadden, and D. O. Terry. 2007. Abrupt temperature drop across the Eocene–Oligocene Transition, central North America. Nature 445:639–642.

Zweifel, R. G. 1956. Two pelobatid frogs from the Tertiary of North America and their relationship to fossil and recent forms. American Museum Novitates 1762:1–45.

Index

RACHEL C. BENTON is Park Paleontologist at Badlands National Park. She started her work with the National Park Service as a volunteer at Delaware Water Gap National Recreation Area in 1984. She has since worked at Big Bend National Park, Wind Cave National Park, and Fossil Butte National Monument. In 1994 she started her position as the first park paleontologist at Badlands National Park, where she has built an extensive program involving paleontological research and resource management. Her doctorate is from the University of Iowa, and her research interests focus on paleoecology and taphonomy. She is a member of the Society of Vertebrate Paleontology and the Society for the Preservation of Natural History Collections. Her primary contributions to the book include the chapters on systematics, paleoecology and taphonomy, and paleontology resource management.

DENNIS O. TERRY JR. is Associate Professor in the Department of Earth and Environmental Science at Temple University in Philadelphia, where he teaches geology of the national parks, physical geology, facies models, and a graduate class in soils and paleosols. He became fascinated with Badlands National Park in the summer of 1986 while attending a geology field camp with Ball State University. The following summer he was hired as an interpretive ranger by Badlands National Park through the Student Conservation Association. It was during this first summer as a park ranger that his research into the geologic history of the Badlands began, thanks primarily to his interactions with the chief interpretive ranger for the park at that time, Jay Shuler. Through these interactions a series of questions regarding the geologic history of some of the oldest strata in the park would become the focus of his master's thesis research at Bowling Green State University, Bowling Green, Ohio. Over the next two summers he was employed by Badlands National Park as the F. V. Hayden Intern, working as an interpretive ranger and conducting research. After completing his master's degree, he began his doctoral research at the University of Nebraska–Lincoln and expanded upon his observations from South Dakota into correlative deposits of northwest Nebraska. He continues his research to this day on the Badlands of South Dakota and correlative deposits in Nebraska, Wyoming, and

North Dakota. His specialties include depositional environments, stratigraphy, and the analysis of ancient soils in the rock record (paleosols) and their application to interpreting the formation of bone beds, paleoenvironmental conditions, and paleoclimatic change across the Eocene–Oligocene boundary. During his career he has generated over 100 publications (abstracts, field guides, journal articles, and books) on the geology and paleontology of the Badlands in collaboration with numerous graduate and undergraduate students, colleagues, and federal agencies, including the National Park Service, the U.S. Forest Service, and the Bureau of Land Management. He is a member of the American Association of Petroleum Geologists, the Geological Society of America, and the Society for Sedimentary Geology. His primary contributions to the book include the chapters on paleopedology, stratigraphy, deposition, postdepositional history, and the Badlands in space and time, with additional contributions to the sections on resource management and vertebrate taphonomy.

EMMETT EVANOFF is Associate Professor in the Department of Earth and Atmospheric Sciences at the University of Northern Colorado in Greeley, Colorado, where he teaches historical geology, paleontology, sedimentary geology, and regional geology. He started his career in paleontology studying fossil freshwater and land snails but gradually shifted his interests into the stratigraphy and origin of early Cenozoic distal volcaniclastic sequences, or fine-grained volcanic-rich rocks that were deposited by rivers, in lakes, and by wind during the first half of the age of mammals. His doctoral dissertation at the University of Colorado, Boulder, was on the stratigraphy, sedimentology, and fossil land snails of the White River Formation near Douglas, Wyoming. After graduation he worked as a geology instructor at the University of Colorado and the Denver Museum of Nature and Science, and also worked as a consulting paleontologist on paleontologic resource evaluation projects and as a sedimentary geologist on fluvial architecture studies. He has worked and continues to work on the middle Eocene Bridger Formation of southwest Wyoming, and the White River rocks of Wyoming, Colorado, Nebraska, and South Dakota. Badlands

National Park is an area of special interest to him, for he has now worked on the stratigraphy of the White River Group continuously for more than 15 summers. A recent topic of interest for him is the history of geologic studies in the Bridger Formation and the White River Group in the South Dakota Badlands. In the current book, he has written the chapter on the history of geological and paleontological studies of the Badlands and contributed information to the stratigraphy, sedimentology (fluvial and eolian), general geology, and systematics of mollusks. He is a member of the Geological Society of America, SEPM, the Society for Sedimentary Geology, the Paleontological Society, the Society of Vertebrate Paleontology, and the Colorado Scientific Society.

H. GREGORY MCDONALD is Senior Curator of Natural History in the National Park Service Museum Management Program. He started with the park service as the paleontologist at Hagerman Fossil Beds National Monument and was the paleontology program coordinator in the park service's Geologic Resources Program. Before joining the park service, he was curator of vertebrate paleontology at the Cincinnati Museum of Natural History. He currently serves as Associate Editor for Paleontology of the *Journal of Cave and Karst Studies* and is a past associate editor of the *Journal of Vertebrate Paleontology*. He is a member of the Society of Vertebrate Paleontology, American Society of Mammalogists, and the Society for the Preservation of Natural History Collections. His doctorate is from the University of Toronto, and his research interests have focused on Plio-Pleistocene mammals in North and South America. His primary contributions to this book include the chapters on systematics and paleontology resource management.